THE CHROMSOC SEPARATION SCIENCE SERIES

This series of high-level reference works provides a comprehensive look at key subjects in the field of separation science. The aim is to describe cutting-edge topics covering all aspects and applications of this important discipline. Each book is a vital technical resource for scientists and researchers in academia and industry.

Published titles

Monolithic Chromatography and its Modern Applications
Edited by Perry G. Wang

Analytical Characterisation and Separation of Oligonucleotides and their Impurities
Edited by George Okafo, David Elder and Mike Webb

Endorsed by

The Chromatographic Society

Analysis of Oligonucleotides and their Related Substances

Edited by George Okafo, David Elder and Mike Webb

ILM Publications

Published in 2013 by ILM Publications
Oak Court Business Centre, Sandridge Park,
Porters Wood, St Albans, Hertfordshire,
AL3 6PH, UK

www.labmate-online.com/books

ILM Publications is a trading division of International Labmate Limited

British Library Cataloguing in Publication Data

A catalogue record for this book is available from the British Library

ISBN 978-1-906799-14-4

Typeset by Keytec Typesetting Ltd, Dorset, UK

Printed and bound by CPI Group (UK) Ltd, Croydon, CR0 4YY

Table of Contents

List of Acronyms

1X PBS	phosphate buffered saline with the same molarity as standard saline
2-AP	2-aminopurine
2D	2-dimensional
3-HPA	3-hydroxypicolinic acid
A	adenosine
AA	atomic absorption spectroscopy
Ac	acetyl
ACN	acetonitrile
ADTT	1,2-dithiazole-5-thione (xanthane hydride)
AEX	anion exchange chromatography
API	active pharmaceutical ingredient
ATT	2-amino thiothymine
BAP	bacterial alkaline phosphatase
BDMAB	butyldimethylammonium bicarbonate
BEH	ethylene bridged hybrid
BNA	bicyclic nucleic acid
BSE	bovine spongiform encephalopathy
Bz	benzoyl
C	cytidine
C&D	cleavage and deprotection
CD	circular dichromism
cDNA	complementary DNA
CDTA	*trans*-1,2-diaminocyclohexane-*N,N,N′,N′*-tetraacetic acid
CE	cyanoethyl
CGE	capillary gel electrophoresis
CID	collision-induced dissociation
CNET	N^3-cyanoethyl-thymidine
COSY	correlation spectroscopy
CPG	controlled pore glass
CQAs	critical quality attributes

CSP	calf spleen phosphodiesterase
dA	deoxy adenosine
Da	Dalton, unit of atomic mass
DABP	3,4-diaminobenzophenone
dC	deoxy cytidine
DCA	dichloroacetic acid
DCI	4,5-dicyanoimidazole
DCM	dichloromethane
DDTT	3-((dimethylamino-methylidene)amino)-3*H*-1,2,4-dithiazole-3-thione
DE	delayed extraction
DEA	diethylamine
dG	deoxy guanosine
DLS	dynamic light scattering
DMAP	4-N,N-dimethylamino pyridine
DMT	dimethoxytrityl
DNA	deoxyribonucleic acid
DOSY	diffusion-ordered spectroscopy
DQF-COSY	double quantum filtered-COSY
DSC	differential scanning calorimetry
dsRNA	double-stranded RNA
DTD	dimethylthiuram disulfide
EDITH	3-ethoxy-1,2,4-dithiazoline-5-one
EDTA	ethylenediaminetetraacetic acid
ESI-MS	electrospray ionisation mass spectrometry
FDA	US Food and Drug Administration
FID	free induction decay
FLP	full-length product
FLP(PO)	phosphorothioate oligonucleotide having a single phosphodiester linkage
FTICR	Fourier transform ion cyclotron resonance
FT-NMR	Fourier transform NMR spectroscopy technique
FWHM	full width at half maximum
G	guanosine
Gapmer	class of oligonucleotide having a stretch of deoxynucleoside bases flanked by LNA, 2′OMe 2′-F or MOE at either end
GC	gas chromatography
HA	hexylamine
HAA	hexylammonium acetate
HFIP	1,1,1,3,3,3,-hexafluoro-2-propanol

HFIP/TEA	hexafluoroisopropanol with triethylamine
HMBC	heteronuclear multiple-bond correlation spectroscopy
HPLC	high-performance liquid chromatography
HPLC-MS	HPLC mass spectrometry
HSQC	heteronuclear single-quantum correlation spectroscopy
iBu	isobutyryl
IC	ion chromatography
ICH	International Conference on Harmonization
ICP-MS	inductively coupled plasma mass spectrometry
ICP-OES	ICP optical emission spectroscopy
IDMS	isotope dilution mass spectrometry
IEX	ion exchange chromatography
IM-MS	ion mobility-mass spectrometry
IP	ion-pairing
IP-RP	ion-pairing reversed phase
ITC	isothermal titration calorimetry
LAL	*Limulus* amoebocyte lysate
LC	liquid chromatography
LCAA	long chain alkyl amine
LC-MS	combination of chromatographic techniques with mass spectrometric detection
LIT	linear ion trap
LNA	locked nucleic acid
L-TOF	linear time-of-flight
MALDI-TOF	matrix-assisted laser desorption/ionisation– time of flight
MD	molecular dynamics
MMT	monomethoxytrityl
MOE	oligonucleotide having O-methoxyethyl substitution at the 2′-position
mRNA	messenger RNA
MS	mass spectrometry
MS^n	tandem mass spectrometry
$n - 1$	impurity characterised by FLP minus a single base
$n - 1$(PS)	impurity characterised as FLP minus the 3′-terminal base having a 3′-terminal phosphorothioate monoester
ND	not defined/not determined
Nd:YAG	neodymium-doped yttrium aluminium garnet
NMI	*N*-methylimidazole
NMR	nuclear magnetic resonance (spectroscopy)
NN	nearest neighbour
NOE	nuclear Overhauser effect

NOESY	nuclear Overhauser effect spectroscopy
PADS	phenylacetyl disulfide
PAGE	polyacrylamide gel electrophoresis
PBS	phosphate buffered saline
PCR	polymerase chain reaction
PEG	polyethylene glycol
POS	polyorg sulfa
PS	polystyrene
PTFE	polytetrafluorethylene
QbD	quality by design
QIT	quadrupole ion trap
QqQ	triple quadrupole
rMD	restrained molecular dynamics
RNA	ribonucleic acid
ROESY	^{1}H-^{1}H rotational frame NOESY
RP	reversed phase
RTIL	room-temperature ionic liquid
S/N	signal-to-noise ratio
SAX	strong anion exchange
SEC	size exclusion chromatography
siRNA	short interfering ribonucleic acid
SNP	single nucleotide polymorphism
SPE	solid-phase extraction
SRM	selected reaction monitoring
SVP	snake venom phosphodiesterase
T	thymidine
TBDMS	*tert*-butyldimethylsilyl
TCA	trichloroacetic acid
TEA	triethylammonium
TEA: 3HF	triethylamine-trihydrogen fluoride
TEAA	triethylammonium acetate
TEAB	triethylammonium bicarbonate
THAP	2,4,6-trihydroxyacetophenone
THF	tetrahydrofuran
T_m	melting temperature
TMP	trimethyl phosphate
TOCSY	total coherance transfer spectroscopy
TOF	time-of-flight
TSP	3-trimethylsilyl-propionate

U	uridine
UPLC	ultra-high-pressure liquid chromatography
UV	ultra-violet
UVRR	ultra-violet/resonance Raman spectroscopy
VT	variable temperature

The Editors

George Okafo studied chemistry and biochemistry at Imperial College of Science, Technology and Medicine, London and continued at Imperial College to undertake research for a PhD in chemical carcinogenesis. Dr Okafo then continued his research as a postdoctoral research fellow at the University of Toronto, Canada in the Institute of Medical Biophysics, where he studied the mechanism for nitrosamine-induced chemical carcinogenesis. Dr Okafo has more than 22 years' experience in the pharmaceutical industry in legacy companies of GSK (SK&F and SB). His current role is a science director in the externalisation group (SCINOVO) in GSK R&D that provides expert drug development consultancy to GSK external drug discovery collaborators. Dr Okafo has published 45 papers, authored three book chapters focused on analytical and separation sciences (HPLC, GC, CZE, MEKC, LC-MS) and detection modes (UV, fluorescence), and is the co-owner of two patents on fluorescence detection. Dr Okafo's recent publications have focused on analytical strategies for characterising synthetic oligonucleotides and, in 2010, he co-organised an international analytical symposium on characterising therapeutic oligonucleotides. George is a member of the Royal Society of Chemistry and an Associate of the Royal School of Chemistry, London.

David Elder studied chemistry (BSc) and analytical chemistry (MSc) at Newcastle upon Tyne, before moving to Edinburgh University to undertake research for a PhD in crystallography. Dr Elder has 34 years' experience in the pharmaceutical industry at a variety of different companies (Sterling, Syntex and GSK). For the last 19 years he has been employed by GSK. He is currently a director in the externalisation group (SCINOVO) in GSK R&D. Dr Elder is a member of the British Pharmacopoeia (Expert Advisory Group PCY: Pharmacy), a member of the Analytical Division Council (Royal Society of Chemistry, UK) and a council member of the Joint Pharmaceutical Analysis Group, UK. He is also a member of the PhRMA and EfPIA sub-groups on genotoxic impurities and was part of the PQRI group that assessed the control strategies for alkyl mesylates. He has published over 40 papers and given over 60 presentations at international fora on a variety of pharmaceutical topics. He has authored six book chapters focused on degradation, impurity identification/control (both genotoxic and standard impurities) and *in vitro* approaches to assess genotoxicity. David is a fellow of the Royal Society of Chemistry (FRAC), a chartered scientist (CSci) and chartered chemist (CChem). He is also a member of the Chartered Quality Institute (MCQI) and Chartered Quality Professional (CQP).

Mike Webb was awarded his BSc in chemistry from the University of Essex. He went on to a masters degree in molecular spectroscopy at Kingston University and a PhD in molecular recognition at Imperial College of Science, Technology and Medicine in London. After a short career in academia, Dr Webb joined the pharmaceutical industry, where he has remained for 32 years. During the bulk of this time Dr Webb has worked in positions of increasing responsibility in analytical chemistry, specialising in spectroscopy and later separation sciences. Dr Webb has published a number of papers, presented at international meetings and has edited two books on the analysis of pharmaceuticals. In January 2010 Dr Webb took a short secondment into a small group looking at oligonucleotide delivery. In October 2010 he took up his current role as Vice-President of API Chemistry & Analysis UK. He is responsible for the synthetic design, scale-up and analytical control of active drug substances in development in the UK.

The Contributors

Hagen Cramer, PhD

Hagen Cramer, PhD, has worked in the field of oligonucleotides since 1989. He has published 22 peer-reviewed papers including a book chapter, holds four patents and has authored one book. Dr Cramer received his PhD from the University of Konstanz, Germany in the laboratory of Professor Dr Wolfgang Pfleiderer in 1995. After completing his PhD, he joined Dr Paul Torrence's group at the National Institutes of Health for post doctorate work on chemically modified 2-5A analogues and 2-5A antisense. Before joining Girindus in 2005, Dr Cramer served as Scientific Director for Gemini Technologies and as Director of Chemistry for Ridgeway Biosystems. At NITTO DENKO Avecia he currently is Director of Operations where he directs the company's operations through a team of functional managers and group leaders, with a focus of timely delivery of Active Pharmaceutical Ingredients (APIs), adherence to cGMP procedures, safety, customer satisfaction, employee relations as well as setting and achieving plant performance metrics.

Kevin Finn, PhD

Kevin Finn completed his undergraduate education in chemistry at Ohio University where he carried out undergraduate research with Professor Mark McMills in the area of Rh-catalysed tandem reactions. Kevin received his PhD from Brock University (Ontario, Canada) in 2006 under the supervision of Professor Tomáš Hudlický. His thesis work involved the development of chemo-enzymatic methods toward the synthesis of synthesis of morphine alkaloids. Kevin accepted a post-doctoral position in Freiburg, Germany with Professor Reinhard Brückner, where he was engaged in total synthesis of light-harvesting carotenoids. Since 2008, Kevin has been employed at NITTO DENKO Avecia in the Oligonucleotides Process Development Group. His title is Scientist, and his focus is scale-up, technology transfer and process validation of therapeutic oligonucleotides.

Nanda Sinha, PhD

Dr Nanda D. Sinha received his doctoral degree in organic chemistry from Patna University (Patna) India. His postdoctoral works at the university of Massachusetts and Yale University were to develop newer methods for synthesis of natural products and their analogues. Later, he joined the laboratory of Professor Hubert Koester at the University of Hamburg to pursue research work in the field of nucleic acids chemistry and oligonucleotides. There he co-invented ß-cyanoethyl phosphoramidite chemistry for the synthesis of oligonucleotides in 1983 and also developed a method for the solid phase synthesis of methylphosphono-oligonucleosides. In 1986, Dr Sinha joined Biosearch Inc. as a Research Scientist in California. Subsequently at Millipore MA he was in charge of nucleic acids chemistry. There he developed many newer reagents and chemistries for the refinement of oligonucleotides synthesis. Dr Sinha, in 1996, co-founded Boston Biosystems Inc. for large-scale synthesis of oligonucleotides under cGMP to support therapeutic and diagnostic applications. The company was acquired by Avecia Inc. in 1999. At Avecia, he served as a Vice-President of Research & Development. He was instrumental in the scale-up siRNA synthesis, purification and developed a very efficient activator for synthesis based on Saccharin. Dr Sinha has contributed four chapters in different books in the field of oligonucleotides and analogues. He has co-authored more 45 scientific publications and is co-inventor and inventor of more than 10 US Patents.

Bernhard Noll, PhD

Dr Bernhard Noll is a Consultant for Analytics & CMC (Chemistry, Manufacturing & Controls) (www.NBChem.de) in the Boston area (MA, USA), providing strategic and technical expertise on analytics, validation, process development, scale-up and technology transfer. Prior to his current role, he served as CMC Project Manager for ThromboGenics NV (Leuven, Belgium), contributing to the company's successful first BLA submission. His responsibilities included planning and coordinating of CMC activities at external contractor sites, and writing and review of technical and regulatory documentation.

Prior to joining Thrombogenics, he served for several years as Associate Director Analytics CMC, at Roche Kulmbach GmbH (Kulmbach, Germany), playing a leading role in the IND preparation of Roche's first siRNA therapeutic. Dr Noll started his professional work in 1998 as a Scientist at Coley Pharmaceutical GmbH, where he helped the company to evolve from a small start-up to a major player in the field of nucleic acid therapeutics. In his last position at Coley, he served as Associate Director in Chemistry and Analytics in Düsseldorf, Germany. Dr Noll obtained his diploma in Chemistry at the Technische Universität Darmstadt in 1993 and received his PhD in Biochemistry at the University of Mainz in 1997.

Ingo Röhl, PhD

Dr Röhl studied chemistry at the University of Oldenburg (Germany) from 1992 to 1997, after which he gained his Diplom thesis work in combinatorial organic chemistry. From 1998 to 2000, he completed his PhD work, focussing on the isolation and identification of new sex pheromones from marine invertebrates. From 2001 to 2003, Dr Röhl became head of the analytical chemistry group at the NOXXON Pharma AG, Berlin. The company is focused on the development of the so-called Spiegelmers – RNA aptamers based on the non-natural L-ribose configuration in the sugar backbone. From 2003 to 2007, he moved to Alnylam Europe AG, where he was head of analytical chemistry with a focus on siRNA analytics. From 2007 to 2011, he joined Roche Kulmbach GmbH as associate director of analytical chemistry at the Roche RNA Center of Excellence. His main responsibilities include leading the analytical QC group to support the in-house synthesis department, CMC development for clinical RNA candidates and DMPK and tissue distribution of siRNA. Currently, Dr Röhl is Director of Analytics and DMPK at the Axolabs GmbH in Kulmbach, Germany.

Patrick Limbach, PhD

Patrick A. Limbach was born in 1966 and obtained his BS degree in 1988 from Centre College (Kentucky, USA). In 1992 he obtained his PhD degree in Analytical Chemistry from The Ohio State University, under the supervision of Alan G. Marshall. He spent two years as a post-doctoral researcher in the laboratory of James A. McCloskey at the University of Utah. In 1995 he accepted his first faculty position in the Department of Chemistry at Louisiana State University in Baton Rouge, Louisiana. In 2001 he moved to his current location in the Department of Chemistry at the University of Cincinnati. He is a Professor of Chemistry and is an Ohio Eminent Scholar. His research interests include the development of new mass spectrometry methods for analysing RNAs, the identification of unknown modified nucleosides, the characterisation of RNA-protein complexes and investigating the role of modified nucleosides in biological systems.

Elena Bichenkova, PhD

Dr Bichenkova graduated in Chemistry (BSc – 1983) with a PhD in NMR structural studies of nucleic acids (1993, Russia), Dr Elena Bichenkova continued her research in the USA at Purdue University (1992, 1994) and then at the University of Texas (1996) collaborating with the first-rank NMR laboratory of Professor David G. Gorenstein. After being awarded a Royal Society/NATO Postdoctoral Fellowship in 1996, she joined the University of Manchester as a Research Fellow. In January 2004 Elena was appointed by School of Pharmacy & Pharmaceutical Sciences as a Lecturer in Medicinal Chemistry followed by promotion to the Senior Lecturer level in 2009.

Daren Levin, PhD

Dr Levin received his MSc from Northeastern University in Chemistry and Chemical Biology and his Ph.D. from Northeastern University in Analytical Chemistry with a focus on the development of differential mobility spectrometry – mass spectrometry instrumentation and its use for the analysis of proteins peptides and oligosaccharides. Prior to joining GlaxoSmithKline he worked at Alkermes Inc. on the analytical development and control strategy of their novel inhaled and injectable microsphere products encompassing small molecule, protein and peptide therapeutics. Since joining GlaxoSmithKline in 2006 he has taken a lead role in the CMCs of oligonucleotide therapeutics. He has co-authored an industry and regulatory agency led white paper on oligonucleotide therapeutic drug substance specifications and has also authored several peer-reviewed articles and presented at several conferences on the analytical method development/validation, control strategy and stability of oligonucleotide therapeutics.

Foreword

Working in the field of synthetic oligonucleotides is extremely fascinating. Synthetic oligonucleotides are an exciting class of therapeutic products that are under development for a variety of indications. Many of them have been developed to address significant medical needs that are so far unmet.

Oligonucleotides are a unique class of molecules with diverse mechanisms of action different from small molecules but also different from biopharmaceuticals. The specific chemical properties and the large sizes compared to small molecules make the analysis of synthetic oligonucleotides and the quantification of their related substances and degradation products one of the most challenging tasks for analytical chemists.

Characterisation and quality control testing are required throughout the clinical development of oligonucleotides intended for therapeutic use. Detailed information on structural characterisation studies that supports the designation of these properties or characteristics should be provided in submissions to regulatory agencies worldwide.

What to control is one of the key questions in connection with regulatory submissions during clinical development and marketing authorisation applications.

Identifying critical quality attributes (CQAs) is vital during pharmaceutical development. A critical quality attribute is a physical, chemical, biological or microbiological property or characteristic that should be within an appropriate limit, range or distribution to ensure the desired product quality. Impurities are an important class of potential drug substance CQAs because of their potential impact on drug product safety. This is described in detail in the International Conference on Harmonisation Quality Guidelines on pharmaceutical development (ICH Q8) and the development and manufacture of drug substances (ICH Q11). Consequently, a major focus is knowledge and control of impurities. How to apply these principles to the analysis of oligonucleotides is a core feature of the present book.

Synthetic oligonucleotides are not included in the scope of the ICH specification guidelines – neither for synthetic drug substances (ICH Q6A) nor for products of biotechnology (ICH Q6B). They are also excluded from the ICH Guidelines on impurities (ICH Q3A and Q3B) and therefore consequently from the scope of ICH Q11. Nevertheless, in ICH Q11 it is clearly stated that this guideline might also be appropriate for other types of products, such as oligonucleotides, following consultation with the appropriate regulatory authorities.

Critical quality attributes are part of the overall target product profile that is based on the desired clinical performance. The extent of characterisation is linked to the level of risk associated with each phase of drug development. It is expected that the oligonucleotide molecule will have been well-characterised before a marketing authorisation application is submitted to the regulatory agencies.

New and improved state-of-the-art analytical technologies and techniques are becoming available on an on-going basis and should be applied during development for both characterisation and routine control. Analytical methods for characterisation during development and control for release and/or stability testing are covered excellently and comprehensively in the present book. The authors of each chapter are practitioners of the art and leaders in the field of oligonucleotide analysis.

Clinical qualification is considered the most important aspect when setting acceptance criteria in specifications. For critical attributes the acceptance criteria should not be wider than what has been qualified to yield a safe and efficacious product. Therefore, identifying differences in impurity profiles using orthogonal analytical methods for the detection and quantification of impurities in early stages of drug development and the application of good science will be extremely helpful in the course of development.

For successful research and development, it is necessary to acquire new skills and knowledge in the field of oligonucleotide analysis. Therefore, there was a dire need for such a book as this which provides various analytical technologies that are in frequent use in modern research.

The book is subdivided into seven chapters. An introduction to oligonucleotides is followed by a comprehensive chapter on oligonucleotide impurities and their origin. Chapter 3 provides an overview on the separation of oligonucleotides and related substances. Chapter 4 deals with the analytical characterisation of oligonucleotides by mass spectrometry while Chapter 5 is concerned with the analytical characterisation of oligonucleotides by NMR spectroscopy. Chapter 6 focuses on the analytical characterisation of oligonucleotides using thermal melting curves. The last chapter of the book is devoted to very important topics in connection with oligonucleotide stability and degradation. Each chapter is composed as an independent unit enabling the reader to pick the topic of immediate interest.

The technological achievements over the last decades are tremendous; this book provides the reader with broad perspectives and a wealth of current knowledge related to oligonucleotide research.

The reader of this book is also familiarised with the actual status of instrumentation, i.e. the current state of the art and the different latest techniques. All techniques are described in a clear manner and by means of examples and case studies including explanations of the theoretical background.

The authors must be congratulated for producing what is considered a truly outstanding and unique work, thereby rendering a most valuable service to scientists working in the field of oligonucleotides in support of therapeutic development.

Bonn, December 2012

Dr René Thürmer, Pharmaceutical Assessor
BfArM – Federal Institute for Drugs and Medical Devices
Kurt-Georg-Kiesinger-Allee 3, 53177 Bonn, Germany

Preface

Over the last two decades, research into the development of synthetic oligonucleotides as therapeutic medicines has grown significantly. These biopolymers have the potential to be used as highly specific, targeted medicines to treat a wide range of diseases, particularly where there is a genetic origin or basis, such as Duchene muscular dystrophy, thrombotic thrombocytopenic purpura and cystic fibrosis.

This growth in oligonucleotide research has been catalysed by a number of key historical events. First was the 2006 Nobel Prize for Fire and Mello's ground-breaking discovery of RNA interference, which signalled the possibility of silencing genes. Second was the sequencing of the human genome, which has helped science to create a map of key gene groups that is implicated in many genetic disorders. Third was the regulatory approval of oligonucleotide-based drugs such as fomivirisen sodium (Vitravene) in 1997, an antisense oligonucleotide used to treat cytomegalovirus retinitis in acquired immune deficiency syndrome patients, and, in 2004, pegaptanib octasodium (Macugen), an angiogenic pegylated aptamer approved for the treatment of neovascular age-related macular degenerative disease. A wide range of oligonucleotide therapeutic classes now exist; these include short interference RNA (siRNA), antisense oligonucleotides (ASOs), oligonucleotide ligands (aptamers and Speigelmers), immunomodulatory oligonucleotides (IMOs), micro interference RNA-blocking oligonucleotides, RNA decoys, ribozymes and DNA enzymes. By far the most cited molecules are the siRNAs, ASOs, aptamers and IMOs.

Synthetic oligonucleotides have a relatively simple linear structure consisting of repeating nucleotides of between 10 and 21 units (or mers; i.e. 10mer) in length. Their molecular weights range from 3.5 to 7 KDa. Uniquely, the biological properties of oligonucleotides are principally a function of their primary and secondary structure. Each nucleotide unit is linked together by a phosphate backbone and consists of a sugar residue (ribose for RNA or deoxyribose for DNA) and an organic purine (guanine and adenine) or pyrimidine (uracil, thymine and thymine) base. In many of the current oligonucleotide therapeutics, the nucleotides have been chemically modified at either the 2′ position of the sugar, the phosphate linkage or on the organic base. These changes serve to enhance their biological stability and resistance to endogenous nuclease attack and also offer improved pharmacokinetic profiles. Oligonucleotides are manufactured using well-established solid-phase synthesis methods; the inherent variability and inefficiency of this process can give rise to numerous impurities and often lowers the synthetic yield. Hence, an ability to monitor, characterise and control the quality of these products during their manufacture is vital to establish consistency and assure safety to the patient. The judicious selection of analytical methodology either as a single tool or combinations of these tools in a synergistic fashion can help to achieve this important goal.

Much of the recent scientific literature relating to synthetic oligonucleotides has been focused on understanding their metabolism, pharmacokinetics and safety profiles. However, in the case of analytical characterisation, many of the reported papers either cover specific analytical areas or spotlight individual techniques. There are relatively few articles or reviews that provide a comprehensive overview of all the analytical techniques (particularly, when used synergistically) and strategies that are tailored for characterising oligonucleotides. This book helps to address this deficiency. Moreover, this book provides a critical review of current developments and future aspects relating to analytical technology when applied to oligonucleotides.

An international panel of experts in the field of oligonucleotide analysis has been assembled to review the current scientific literature and to bring examples from their own work to describe the most up-to-date analytical strategies for synthetic oligonucleotides. In each chapter, the authors highlight how analytical techniques have been applied (and adapted in some cases) for these biopolymers, offer method development approaches and provide recommendations and applications to therapeutic and diagnostic oligonucleotides. The structure of the book is designed to follow a logical order, starting with an introductory chapter that describes the importance of oligonucleotides as potential therapeutic medicines from a pharmaceutical industry perspective. The next chapter provides a detailed account of different types of oligonucleotide-related impurities and their origins. The subsequent chapters provide a comprehensive overview of analytical strategies for separation methods, mass spectrometric approaches, nuclear magnetic resonance and thermal methods. The final chapter concludes with a discussion on the degradation characteristics of oligonucleotides, including mechanistic pathways for degradation, the types of degradants and recommendations for analysis. Other analytical techniques such as Karl Fisher, residue on ignition, gas chromatography, infrared, ultraviolet absorbance and microbiological analysis have been deliberately excluded from the book because they can be applied generically to small molecules as well as oligonucleotides, and require minimal method development.

This book serves to provide a comprehensive, current and complementary scientific review of the scientific literature that exists in the area of oligonucleotide analytical characterisation.

Introduction to Oligonucleotides

1

George Okafo, David P. Elder and Mike Webb

1.1 What Are Oligonucleotides?

1.1.1 Structure and Origin of Oligonucleotides

Oligonucleotide means 'many' nucleotides and the single unit from which oligonucleotides are made is the nucleotide (see Figure 1.1). Oligonucleotides, as well as being synthetically derived medicines, are naturally occurring molecules. Genes are deoxyribonucleic acid (DNA) containing the genetic instructions that control the development and functioning of all known organisms, and are central to the mechanism of protein synthesis. DNA is a double helix structure formed from two strands of oligonucleotides (see Figure 1.2).

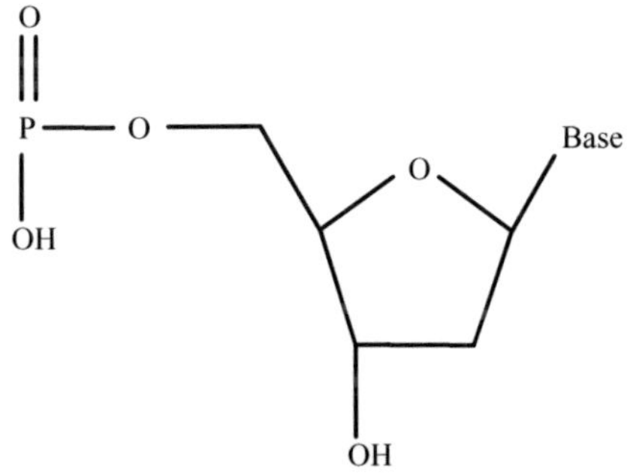

Figure 1.1 A single nucleotide. The nucleotide is composed of a phosphate or thiophosphate linker group, a ribose sugar and a Watson–Crick base pair. The base can be adenine (A), cytosine (C), guanine (G), thymine (T) or uracil (U).

Analysis of Oligonucleotides and their Related Substances, edited by George Okafo, David Elder and Mike

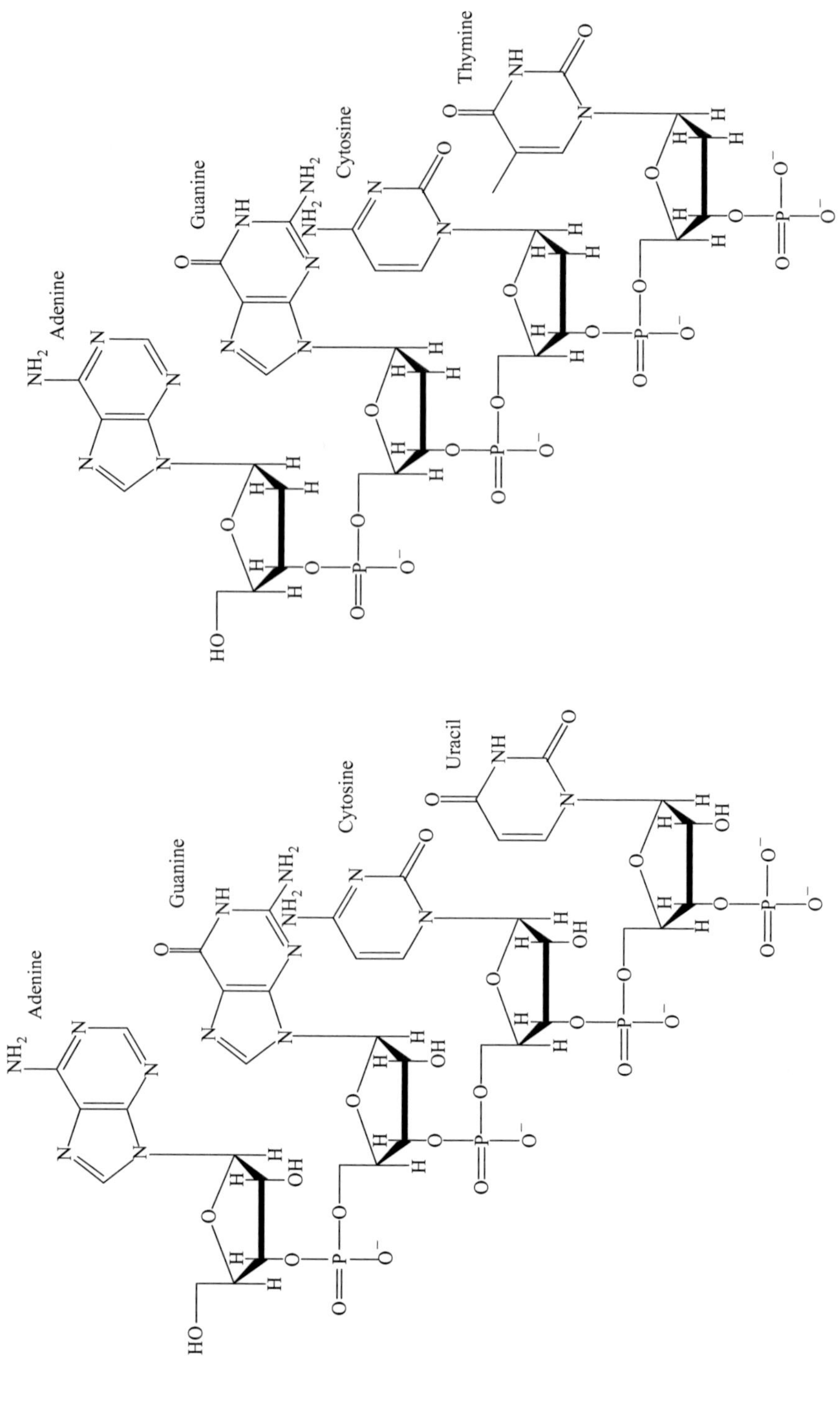

Figure 1.2 Oligonucleotides: left is portion of RNA (ribonucleic acid) and right is a portion of DNA.

1.1.2 Role of Oligonucleotides in Organisms

DNA contains all of the genetic information needed to build an organism. The overall structure of DNA was determined by Watson and Crick in 1953 [1]. However, it took another half a century before the human genome project completed the decoding of human DNA. Over 13 years, three billion base pairs of DNA and approximately 30 000 genes were decoded and geneticists began to understand which genes were involved in which biological processes [2].

DNA exists within the nucleus of a cell and the genetic code is transcribed into ribonucleic acid (RNA), which can leave the nucleus into the cell cytoplasm. Enzymes translate this code into a functional protein. The link between the genetic code of DNA, the structure of RNA, the protein it forms and the protein's function provides a platform for molecular biologists to understand the origin and nature of many disease states. This unidirectional, sequential flow of biological information from DNA to RNA to protein forms the basis of the so-called 'Central dogma of molecular biology', first enunciated by Crick in 1970 [3]. Recent advances in gene cloning, polymerase chain reaction and our growing understanding of viral nucleic acids have led to discoveries that challenge the validity of the central dogma (Figure 1.3). Key to this was the discovery of reverse transcription of the originating nucleic acid to form complementary DNA *via* the retroviral enzyme reverse transcriptase [4]. Also, the existence of self-replicating viral RNA [5] heralded the fact that RNA can sometimes be transcribed back to DNA. A further finding, in 2011, was the discovery by Li *et al.* [6] that in cellular RNA, some of the organic bases were not the ones expected from the DNA sequence used to make the RNA read-out. This was thought to be due to cellular editing of the RNA after transcription. More recently, Lee *et al.* [7] identified small bits of RNA, termed microRNA, which can bind to messenger RNA, essentially switching off protein production. MicroRNA has been implicated in many disease states, for example heart disease, diabetes, some cancers and neurodegenerative diseases.

This is summarised schematically in Figure 1.3.

1.2 Oligonucleotides as Drugs

Traditional small molecule drugs frequently, but not exclusively, work by interacting with active sites on a protein molecule outside the cell in order to modify the way the

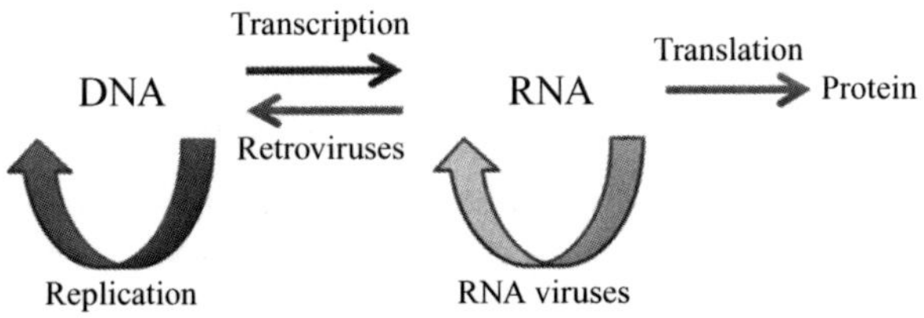

Figure 1.3 The transcription of DNA to RNA and translation of RNA to form a protein.

protein works. However, with the advent of the insight gained from the human genome project, oligonucleotide drugs can now be developed that interact with RNA either to block or activate their translation to form a protein (i.e. blocking the formation of a harmful protein or promoting the formation of a beneficial protein). One important application of oligonucleotide drugs uses Watson–Crick base pairing [1]. Oligonucleotide drugs can bind strongly to RNA to inhibit its translation processes. These kinds of oligonucleotide drugs are classed as antisense oligonucleotide drugs [8]. These drugs are typically short strands of DNA or RNA or mixtures of two up to 20 bases long (20-mers).

Oligonucleotides chemically differ from traditional small molecule drugs in a number of significant ways. These differences are summarised in Table 1.1. Significantly, the medical indications for oligonucleotide drugs may often be different to small molecules. Potentially, oligonucleotides can treat undruggable illnesses such as those arising from genetic defects, for example muscular dystrophy [9], cystic fibrosis [10] or certain cancers [11]. It is also true that oligonucleotides may present better options to treat illnesses that currently have small molecule or biopharmaceutical medicines to treat them (e.g. psoriasis [12], lupus [13] and cancer [11]).

In addition to antisense oligonucleotide drugs, there are 'short interfering RNA' oligonucleotide drugs (siRNAs), which utilise the RNA interference mechanism [14] (RNAi, see later), microRNA oligonucleotide drugs [6] and exon skipping oligonucleotide drugs [9]. These categories all work by mediating the RNA translation process and need to gain entry to the cell cytoplasm to work. Another category of oligonucleotide drug are those that act as chemical antibodies known as aptamers [15] and spiegelmers [16] (these utilise Watson–Crick binding to themselves to build up a structural motif, which can resemble the active motif of a biopharmaceutical).

There are oligonucleotide drugs which act on the immunostimulatory system [17]; these drugs can mediate inflammatory disease and can be used as adjuvants for vaccines. Oligonucleotides can also alter cell regulation by binding with, and hence blocking, transcription factor proteins, thereby acting as decoy oligonucleotides [18].

Table 1.1 Comparison of different attributes of oligonucleotides and small molecules.

Attribute	Small molecule drug	Oligonucleotide drug
Molecular weight	Typically 300–400 Da	Typically 7000–20 000 Da
Charge	Neutral molecules or simple mono- or divalent salts	Usually highly charged 20–50 negatively charged phosphate salts
Solubility	Typically low (< 50 μg/ml) as a consequence of combinatorial chemistry and high-throughput screening	Typically high (> 50 mg/ml) as a consequence of multiple charged state
Synthesis	Batch manufacture *via* solution chemistry	Solid-phase synthesis followed by ion exchange purification
Cost of goods	£1–20/g	£1000–2000/g
Dose	10 μg (mcg)–2 g/day	50–200 mg/week

Finally, immunosuppressant oligonucleotides suppress the immune response [19]. The categories of oligonucleotide medicines are summarised in Table 1.2.

1.3 The Discovery of the Cell Mechanism to Make Use of Double-Stranded Oligonucleotides

In addition to the human genome project, a more recent breakthrough was gene-silencing [20], and the pioneers (Craig Mello and Andrew Fire) received the Nobel Prize four years later. Mello and Fire identified a natural gene silencing mechanism that used very small amounts of short, double-stranded oligonucleotides (siRNAs) that 'knock down' gene expression by a significant amount. This is referred to as RNA interference or RNAi and is summarised in Figure 1.4.

This breakthrough encouraged a number of small biotechnology companies to start up and focus on developing RNAi drugs, such as Alnylam and SiRNA. Large pharmaceutical companies such as Merck, Novatis, Pfizer and GSK began collaborations with some of these companies. In 2007, Merck bought SiRNA, a company specialising in RNAi, for $1.1 billion [21].

1.4. The Development of Oligonucleotides as Medicines

In theory any dysfunctional gene can be targeted by oligonucleotide medicines. However, in the two or three decades since the search for these drugs started, only two

Table 1.2 Types of oligonucleotide drug and their therapeutic activity.

Type of oligonucleotides	Single or double stranded	Action	Reference
Antisense	Single	Blocks messenger RNA to prevent protein formation	[8]
siRNA	Double	Uses a natural process in the cell to block protein formation	[14]
Micro-RNA	Single	Works on the cell's complex gene regulation and trafficking	[6]
Exon skipping	Single	Blocks part of pre-messenger RNA to promote protein formation	[9]
Aptamers and spieglemers	Single	Mimics biopharmaceutical drugs (e.g. antibodies)	[15, 16]
Decoy	Double	Blocks the function of transcription factor proteins	[18]
Immunostimulatory	Single (typically)	Adjuvents for vaccines	[17]
Immunosuppressant	Single (typically)	Suppresses the immune system	[19]

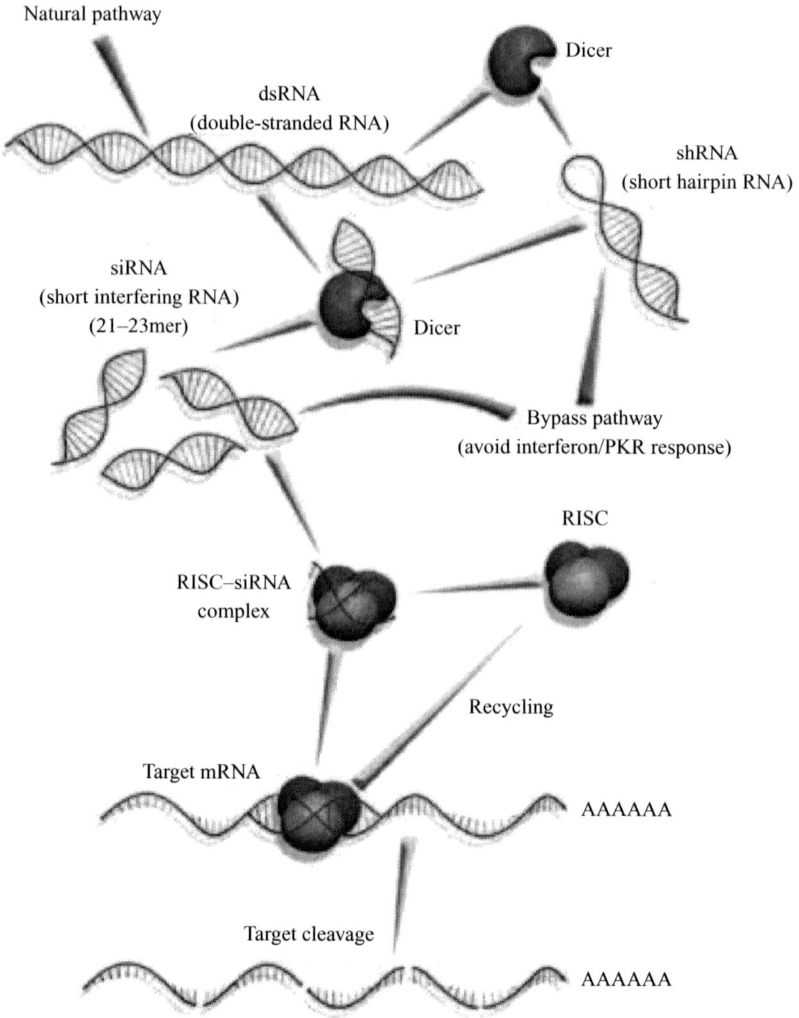

Figure 1.4 Natural targeting mechanism of RNAi. (where RISC is RNA-Induced Silencing Complex and PKR is Protein Kinase R)

drugs have reached the market [22]. An overview of the different types of oligonucleotide drugs and when development was initiated is shown in Figure 1.5.

1.4.1 The First Oligonucleotides on the Market (Antisense and Aptamer)

Despite this intense focus on oligonucleotide drug development over 35 years, the first drug to be approved by the US Food and Drug Administration (FDA) was a decade ago. This was an antisense oligonucleotide, Vitravene, from Isis Pharmaceuticals for the treatment of cytomegalovirus retinitis in acquired immune deficiency syndrome

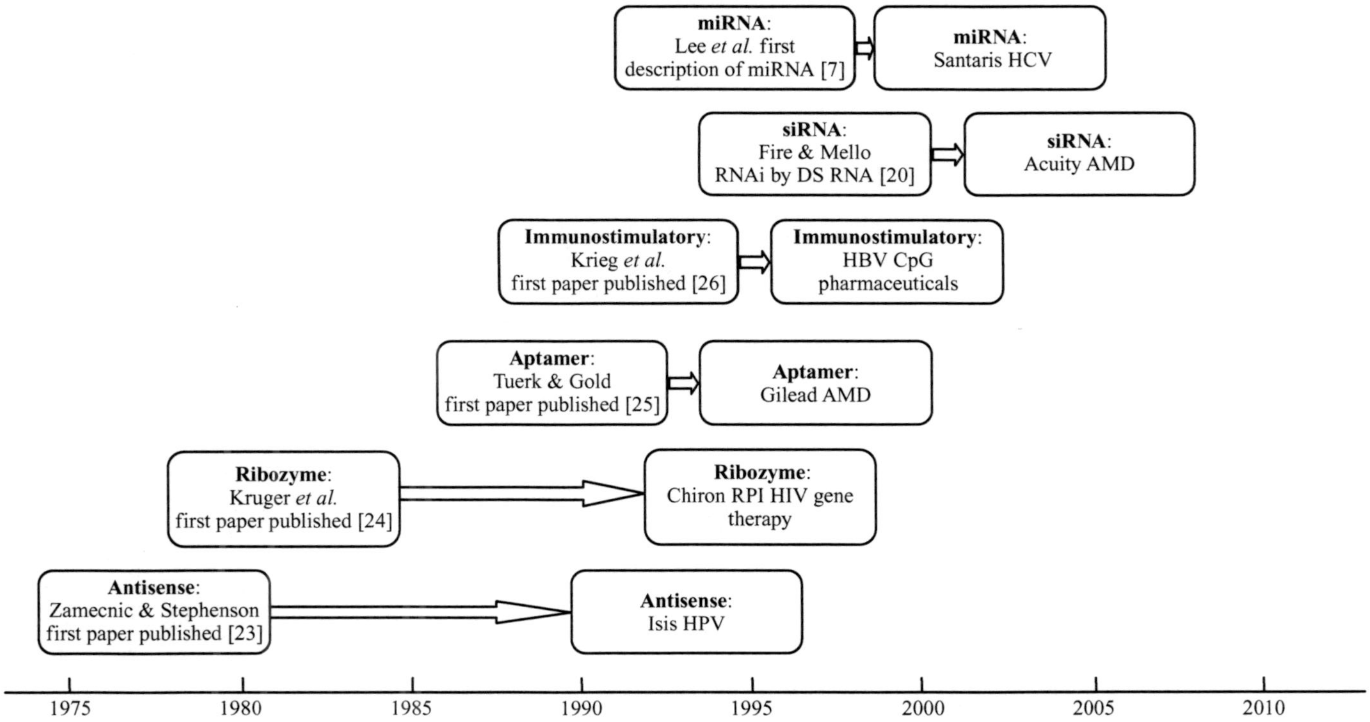

Figure 1.5 A timeline of the development of oligonucleotide medicines. (Source: Agilent Technologies, Inc 2010. Reproduced with permission, courtesy of Agilent Technologies, Inc.)

patients (subsequently withdrawn due to poor sales) [27]. The only other successful filing was Macugen, an oligonucleotide aptamer (oligonucleotide with antibody properties), indicated for wet, age-related macular degeneration [28].

1.4.2 The Challenges of Delivery to the Cell

The major limitation to RNAi and antisense oligonucleotide drugs is that they only work within the cell cytoplasm. Therefore, the drugs have to cross a number of biological barriers within the body to be effective. Most oligonucleotide drugs are injected into the blood stream *via* intravenous/subcutaneous injection. A few are administered directly to the site of infection *via* topical application to the eye, the skin or the lungs (the latter by way of a nebulised solution for inhalation). Oligonucleotides are generally not expected to be absorbed *via* the oral route (arising from poor permeability due to large molecular size and charge) and hence solid dosage forms, at the time of writing, are rare. Taking the example of oligonucleotides injected into the systemic system, the multiple biological barriers that these drugs need to cross are summarised in Figure 1.6.

Oligonucleotides are large molecules and therefore show poor permeability; hence, they have difficulty in migrating from the systemic system into cellular targets, in the same way that a small molecule can. For this reason, oligonucleotides tend to concentrate in the clearance organs, such as the liver and kidney, or in tumours where some blood vessels are 'leaky' (fenestrations). Therefore, many of the most advanced systemic oligonucleotide medicines are indicated for liver conditions or carcinomas. In addition to the problem of migrating from the systemic system, oligonucleotides must concentrate in the tissues of interest and be presented to the cell surface. Once at the cell surface, most oligonucleotides (unlike many small molecule drugs) need to enter the cell cytoplasm in order to work. This is a major hurdle, as the oligonucleotide is a large multiply charged molecule, with poor permeability, which will not readily cross the cell surface membrane. This barrier is particularly difficult for the siRNAs.

1.4.3 Potential Solutions to the Delivery Problem (Especially siRNAs)

Classically, most oligonucleotides have not undergone targeted delivery to the cell. Oligonucleotides have been delivered typically at large doses with the expectation that this will saturate the body's clearance mechanisms, resulting in large concentrations within the systemic system that will translate into therapeutic concentrations at the site of action, for example cellular systems. This approach is predicated on the good systemic safety of oligonucleotides, but typically injection site reactions have resulted from these large doses. More recently, targeted delivery has been initiated. However, an in-depth overview of these strategies is beyond the scope of this book. The interested reader should access the many overviews that are available in the specialist literature. A particularly good review was published in *Advanced Drug Delivery* in 2009 [29].

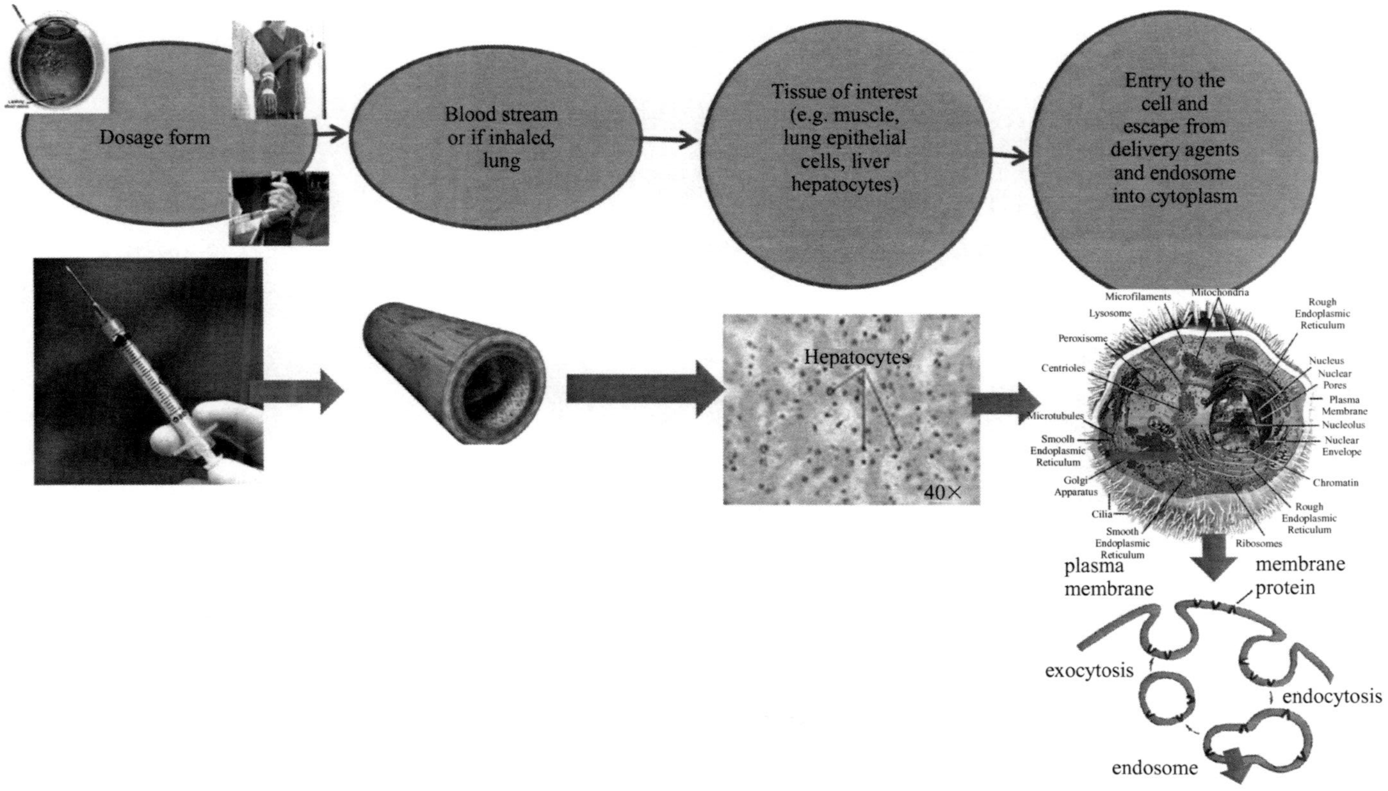

Figure 1.6 Barriers to effective delivery of oligonucleotides to the cell.

1.4.4 The Growth of Oligonucleotides in Recent Years

During the last decade the pharmaceutical industry has seen a steady growth of oligonucleotide research programmes, as shown in Figure 1.7. This growth has flattened in the last three years, which could be attributed to the current economic downturn [30].

Of the 245 programmes publicly disclosed, 145 programmes are in clinical development, with eight programmes in phase III clinical trials. These data show that oligonucleotide research is maturing towards commercialisation world-wide. The data in Figures 1.8 and 1.9 (see colour insert for Figure 1.9) show that siRNA programmes make up a significant amount of both research and clinical phases of development [30].

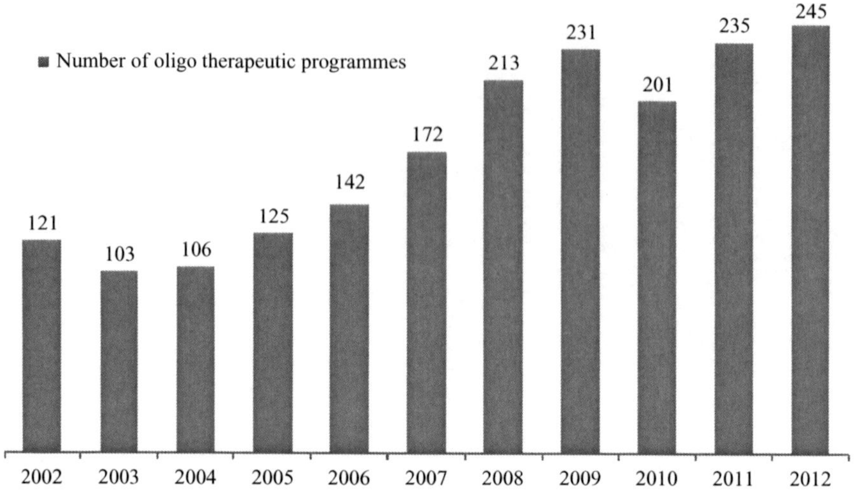

Figure 1.7 Growth of oligonucleotide programmes worldwide between 2002 and 2012. (Source: ©Agilent Technologies, Inc 1213. Reproduced with permission, courtesy of Agilent Technologies, Inc.)

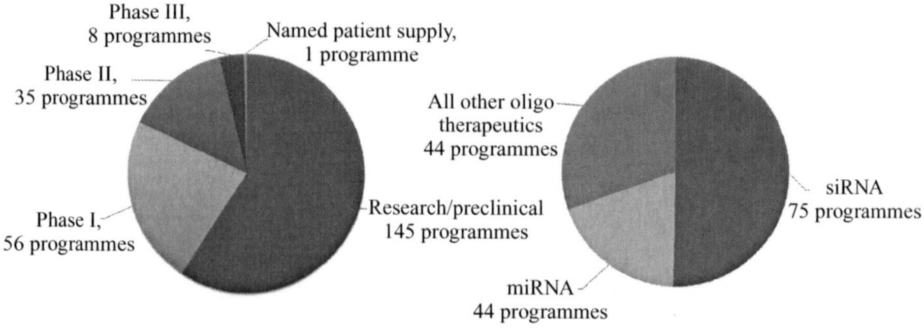

Figure 1.8 The state of oligonucleotide research and development world-wide in 2009. (Source: ©Agilent Technologies, Inc 2010. Reproduced with permission, courtesy of Agilent Technologies, Inc.)

However, siRNA oligonucleotides are the most difficult to deliver to the cell. There is only one siRNA in phase III.

1.5 Oligonucleotide Suppliers

1.5.1 Primary Process Overview

All known oligonucleotide suppliers use the same basic synthetic cycle (see Figures 1.10 and 1.11) to make oligonucleotides, using automated synthesisers that are commercially available at a number of manufacturing scales. The oligonucleotide is built up on a solid support (glass or polystyrene beads) using a cycle of four steps per base unit and is highly automated. Therefore, there could be 80+ individual chemical transformations for a typical 20-base (or 20-mer) oligonucleotide. Despite the high chemical efficiency of each step, it is still possible to make shorter (shortmers) or

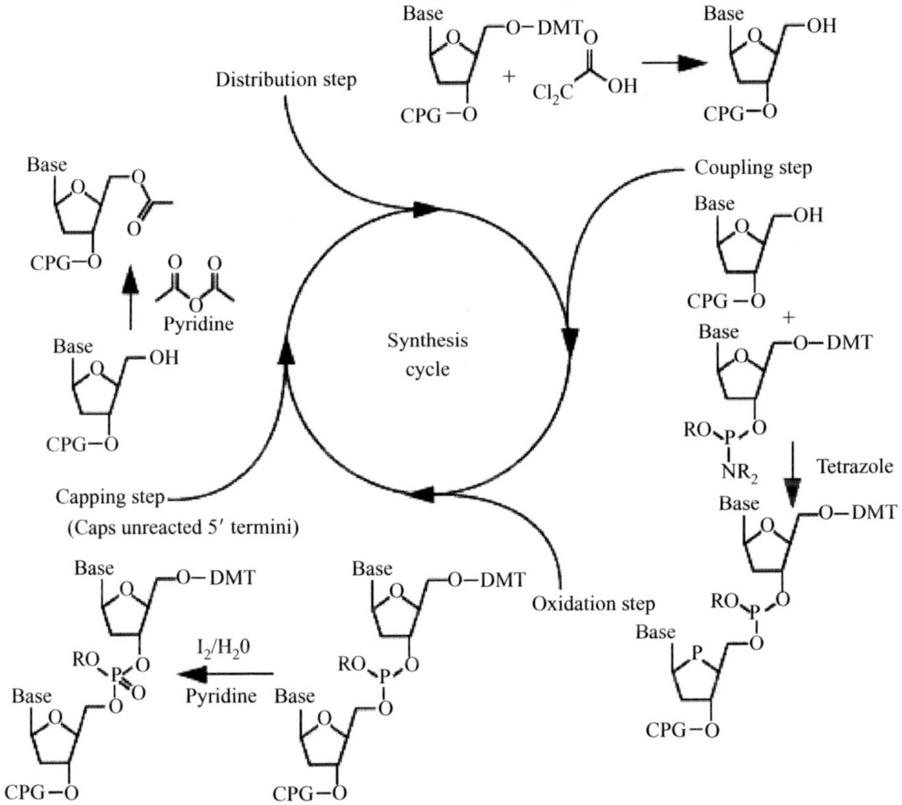

• Each cycle consists of 4 chemical steps → 20mer synthesis: 4 × 20 = 80 chemical steps
• Each chemical step potentially causes impurity formation

Figure 1.10 Automatic solid-state synthesis of oligonucleotides.

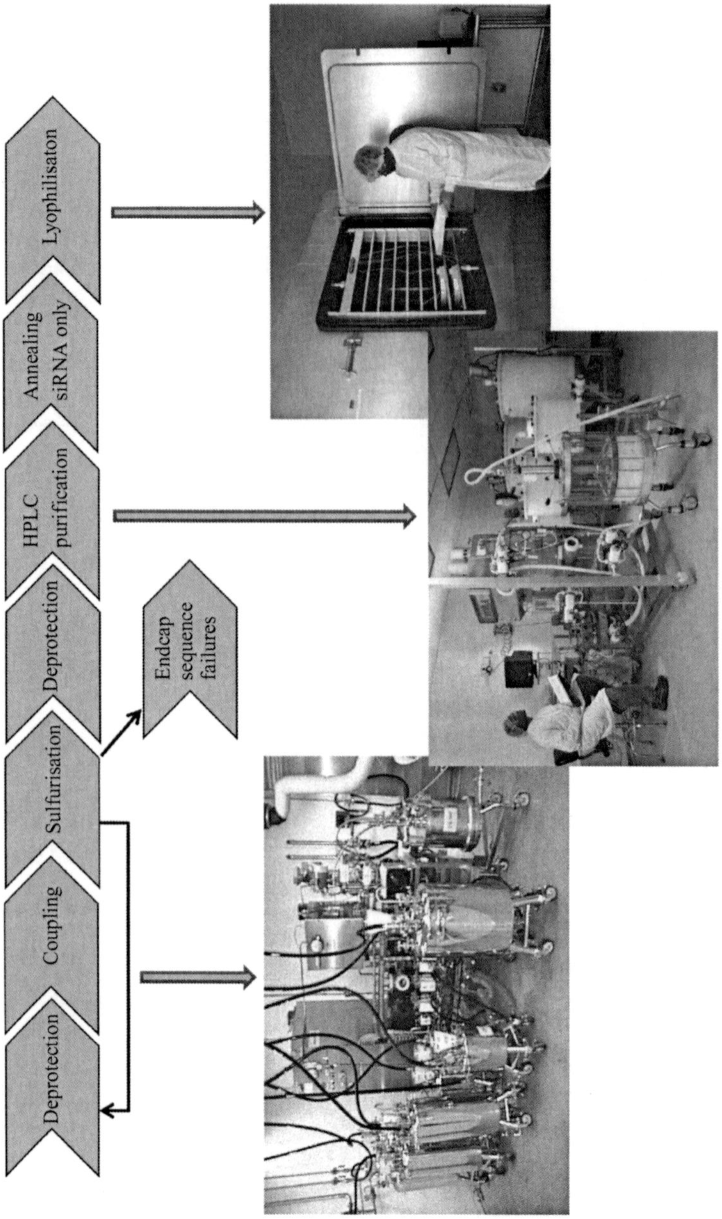

Figure 1.11 The basic process steps to form a pure oligonucleotide API.

longer (longmers) by-products after the product is cleaved from the column and the protecting groups are taken off; the required purity is only reached using an anion exchange preparative chromatography purification step. A more detailed description of typical oligonucleotide impurities is provided in Chapter 2.

After the purification, the product may be concentrated and then de-salted and microbiologically purified by ultra-filtration. The product in solution is then lyophilised to remove water and to provide a white to off-white hygroscopic powder, which is stored and shipped at low temperature for secondary manufacture.

1.5.2 Purification

The crude oligonucleotide will be contaminated with shortmers ($n - 1$, $n - 2$, etc.) or longmers ($n + 1$, $n + 2$, $2n + 1$, etc.), as well as other impurities (see Chapter 2 for more details). Purification is usually carried out using a chromatography column. The crude oligonucleotide, once cleaved off the solid support of the synthesis column using ammonia in methanol (which also deprotects the exocyclic amines on the bases), will still retain the final dimethyltrityl (DMT) protecting group. This crude product can be loaded onto a reverse phase chromatography column and separated using a reverse phase mode, or the last DMT can be deprotected at the head of an anion exchange column before ion exchange chromatography is performed. The fractions from the chromatography steps are analysed individually and this analysis is combined mathematically to make 'mock pools' which simulate the quality of the combined fractions. Once a satisfactory combination is ascertained the pools are combined before concentration, ultra-filtration and lyophilisation.

1.6 Quality by Design Applied to Oligonucleotide Manufacture

Quality by design (QbD) is an approach developed by the regulators and the pharmaceutical industry and is described in the *International Conference on Harmonisation Quality Guidelines* (ICH Q8) [31]. QbD is an approach which designs quality into manufacturing processes rather than testing it in at the end of the process. It usually involves a systematic approach, with predefined quality objectives, which leads to a detailed understanding of the processing parameters which affect the quality of the product as defined by the critical quality attributes (CQAs) of that product. Risk management principles are employed which lead directly to an approach to quality management. Quality in turn is linked to the impact of the product on the patient in terms of safety, quality and efficacy. Figure 1.12 summarises the key elements and connectivities of QbD.

The result of the QbD work should result in an understanding of critical process parameters (CCP) and how they affect the critical quality attributes (CQAs) and a control strategy to ensure the quality of the product. Focusing on the oligonucleotide drug substance itself, an example of a CQA would be the shortmer ($n - 1$) impurities. This CQA could be controlled by a number of parameters, as described in Figure 1.13.

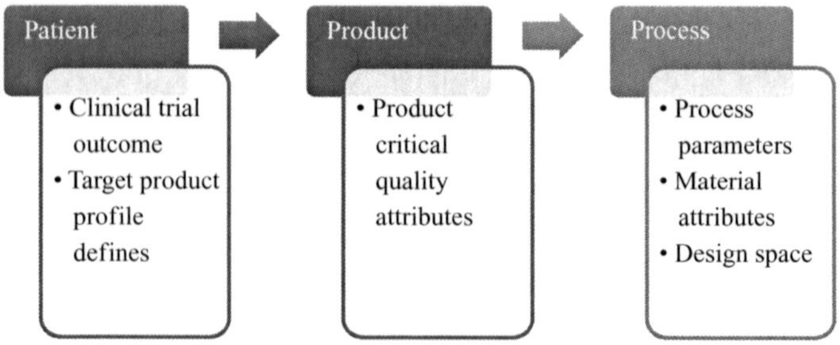

Figure 1.12 The basic principles of quality by design.

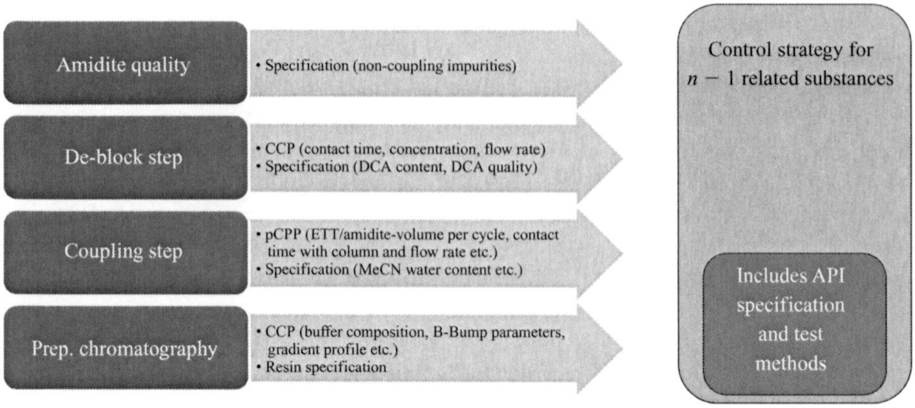

Figure 1.13 A potential control strategy for the CQA $n-1$ impurities (where DCA is Dichloroacetic Acid and ETT is 5-Ethyl thio-1H-tetrazole).

The QbD approach is a very good way of ensuring control of the quality of synthetically synthesised oligonucleotides. It should ensure consistent quality. However, the quality will only be determined by accurate and reproducible measurement of the CQAs as determined by the analytical methods and approaches described in the chapters of this book.

1.7 Regulatory Guidance

There is no specific regulatory guidance covering oligonucleotide quality. Although there is ICH guidance for small molecules, for example ICH Q3A [32], ICH Q3B [33], and large bio-molecules, ICH Q6B [34]; oligonucleotides are specifically excluded from the former and are not covered by the latter. This is because this class of compounds shows characteristics typical of both small and large molecules, that is they are chemically synthesised, with impurities derived from the synthesis, but they show higher order structure (secondary, tertiary and higher order) and the impurities often have biological activity. An industry expert group is currently compiling a white paper on quality aspects of oligonucleotides [35] and is working with regulatory agencies to develop appropriate guidance in this area. Thus, oligonucleotides are not covered by existing regulatory guidance for specifications for active substances, ICH Q6A [36] or Q6B [34]. The industry expert group have provided some guidance on the 'typical' tests (with the caveat that other tests may be applicable on a case-by-case basis) that should be included on an oligonucleotide specification. The industry expert group deliberately has not sought to provide acceptance criteria for specific tests, the main hurdle being that, like classical bio-molecules, (i) oligonucleotide impurities are intrinsically biologically active (often with similar activity to the parent) and (ii) specificity of individual impurities is impossible. Owing to the solid-phase synthesis groups or families of related impurities, for example shortmers ($n-1$, $n-2$, etc.) or longmers ($n+1$, $n+2$, etc.), are formed.

These tests are further classified into single-stranded oligonucleotides (Table 1.3) and siRNAs (Table 1.4). When a test expected of the double-stranded API is performed at the single-strand intermediate stage, the results will still constitute a part of API release and are, therefore, subject to the same scrutiny as any result generated on the API. This is aligned with ICH Q6A, Section 2.3, 'In-process tests' [36] (this

Table 1.3 Tests and example analytical procedures for single-stranded oligonucleotide API (reproduced with permission of Industry Expert Group [35]).

Test	Example analytical procedures
Appearance	Visual inspection
Identification: Mass determination	Mass spectrometry (MS)
Identification: Sequence confirmation	Tandem mass spectrometry (MS^n) Melting temperature analysis (T_m) Failure sequence analysis Real-time monitoring of synthesiser output
Counterion testing	Inductively coupled plasma mass spectrometry (ICP-MS) ICP optical emission spectroscopy (ICP-OES) Atomic absorption spectroscopy (AA) Ion chromatography (IC)
Assay	Ultraviolet spectrometry (UV) High-performance liquid chromatography (HPLC) HPLC mass spectrometry (HPLC-MS) Capillary gel electrophoresis (CGE) Anion exchange chromatography (AEX)
Purity and impurity profile	HPLC HPLC-MS CGE AEX

(*continued*)

Table 1.3 (*continued*)

Test	**Example analytical procedures**
Residual solvents	Gas chromatography (GC)
Heavy metals	ICP-MS ICP-OES AA
Bacterial endotoxins	*Limulus* amoebocyte lysate (LAL) as described in USP<85>
Microbial limits testing	As described in USP<61>

Table 1.4 Tests and typical analytical procedures for double-stranded oligonucleotide API (reproduced with permission of Industry Expert Group [35]).

Test	**Example analytical procedures**
Single-strand tests	
Identification: Mass determination	MS
Identification: Sequence confirmation	MS^n T_m Failure sequence analysis Real-time monitoring of synthesiser output
Purity and impurity profile	HPLC HPLC-MS CGE
Double strand tests	
Appearance	Visual inspection
Identification: Identification of single strands	MS Retention time by denaturing HPLC
Identification: Identification of duplex	T_m Circular dichroism spectroscopy Microcalorimetry
Counterion testing	Inductively coupled plasma MS (ICP-MS) ICP-OES) AA IC

(*continued*)

Table 1.4 (*continued*)

Test	Example analytical procedures
Assay	UV Non-denaturing AEX Non-denaturing size exclusion chromatography (SEC) Non-denaturing HPLC Non-denaturing CGE
Purity and impurity profile	Non-denaturing ion exchange chromatography (IEX) Non-denaturing SEC Non-denaturing HPLC Non-denaturing CGE Denaturing HPLC Denaturing CGE Denaturing AEX Denaturing HPLC-MS
Residual solvents	GC
Heavy metals	ICP-MS ICP-OES AA
Bacterial endotoxins	LAL as described in USP<85>
Microbial limits testing	As described in USP<61>

allows for some tests that are listed on the API specification to be performed earlier in the synthesis process).

1.8 Advances in Analytical Methodology

As well as advances in the synthesis of oligonucleotides and strategies for drug development, significant improvements have also occurred in the analytical methodologies used to analyse and characterise these large biopolymers. In the early 1980s, gel electrophoresis coupled with enzymatic analysis (also known as 'Southern Blot') was considered to be the method of choice for analysing impurities in large nucleic acids [37]. Similarly, mass spectrometers and nuclear magnetic resonance (NMR) spectroscopy were limited owing to the technology available at that time, for example small magnets, embryonic acquisition methodology and limiting computing power. Nowadays, analytical methodology has improved significantly and the following chapters will serve to highlight the major advances in high-performance liquid chromatography (HPLC) in the form of ion pair reverse phase HPLC (IP-RP-HPLC), anion exchange HPLC (AEX) and size exclusion HPLC (SEC), as well as hydrophilic interaction chromatography (see Chapter 3) to resolve oligonucleotide impurities. Mass spectrometric methods have also improved with the advent of softer ionisation modes (electrospray, matrix-assisted laser desorption/ionisation–time of flight mass spectrometry (MS) (often referred to as MALDI-TOF)) and hyphenation to HPLC (HPLC-MS, HPLC-MS-MS), which has provided improved accuracy to molecular weight determinations, more comprehensive fragmentation chemistries to help in oligonucleotide sequencing and impurity determination (Chapter 4). Recent developments in NMR have allowed scientists not only to provide structural information on the nucleic acid, but to quantify impurities, and understand and monitor structural and conformation changes under different conditions (Chapter 5). Understanding the stability of an oligonucleotide under degradative conditions (Chapters 6 and 7) and the nature and origin of the impurities present is vital in drug development (Chapter 2) is also possible.

These recent development in analytical methodology will now allow the nucleic acid scientists to analyse and characterise oligonucleotides more accurately and with greater consistency and reproducibility.

References

1. Watson, J.D., Crick, F.H.C., A structure for deoxyribose nucleic acid, *Nature*, 1953, 171, 737–738.
2. International Human Genome Sequencing Consortium, Initial sequencing and analysis of the human genome, *Nature*, 2001, 409, 860–921.
3. Crick, F., Central dogma of molecular biology, *Nature*, 1970, 227(5258), 561–563.
4. Temin, H.M., Mizutani, S., Viral RNA-dependent DNA polymerase: RNA dependent DNA polymerase in virions of rous sarcoma virus, *Nature*, 1970, 226, 1211–1213.
5. Zhou, X., Berglund, P., Rhodes, G., Parker, S.E., Jondal, M., Liljeström, P., Self replicating Semliki Forest virus RNA as recombinant vaccine, *Vaccine*, 1994, 12(16), 1510–1514.
6. Li, M., Wang, I.X., Li, Y., Bruzel, A., Richards, A.L., Toung, J.M., Cheung, V.G., Widespread RNA and DNA sequence differences in the human transcriptome, *Science*, 2011, 333, 53–58.
7. Lee, R.C., Feinbaum, R.L., Ambros, V., The *C. Elegans* heterochronic gene lin-4 encodes small RNAs with antisense complementarity to lin-14, *Cell*, 1993, 75(5), 843–854.
8. Sahu, N.K., Shilakari, G., Nayak, A., Kohli, D.V., Antisense technology: a selective tool for gene expression regulation and gene targeting, *Curr. Pharm. Biotechnol.*, 2007, 8(5), 291–304.
9. Aartmus-rus, A., Kaman, W.E., den Dunnen, J.T., van Ommen, G.J., van Deutekim, J.C., Exploring the frontiers of therapeutic exon skipping for Duchenne muscular dystrophy by double targeting within one or multiple exons, *Mol. Ther.*, 2006, 14(3), 401–407.
10. Sobczak, K., Bangel-Ruland, N., Semmler, J., Lindemann, H., Heermann, R., Weber, W.-M., Antisense oligonucleotides for therapy of cystic fibrosis inhibition of sodium absorption mediated by ENaC in nasal epithelial cells, *HNO*, 2009, 57(11), 1106–1112.
11. Monia, B.P., Johnston, F.J., Geiger, T., Mueller, M., Fabbro, D., Antitumor activity of a phosphorothioate antisense oligodeoxynucleotide targeted against C-*raf* kinase, *Nature Medicine*, 1996, 2, 668–675.
12. Whitley, P.J., Atley L.M., Wraight, C.J., Antisense oligonucleotide treatments for psoriasis, *Expert Opin. Biol. Ther.*, 2004, 4(1), 75–81.
13. Ansari, M.A., Dhar, M., Muthukrishnan, V., Morton, T.L., Bakht, N., Jacobson, J.D., Administra-

tion of antisense oligonucleotides to $G\alpha_{Q/11}$ reduces the severity of murine lupus, *Biochimie*, 2003, 85(6), 627–632.
14. Castanotto, D., Rossi, J.J., The promise and pitfalls of RNA-interference-based therapeutics, *Nature*, 2009, 457(7228), 426–433.
15. Gold, L., Polisky, B,, Uhlenbeck, O.C., Yarus, M., Diversity of oligonucleotide tunctions, *Ann. Rev. Biochem.*, 1995, 64, 763–797.
16. Purschke, G., Radtke, F., Kleinjung, F., Klussmann, S., A DNA Speilgelmer for Staphylococcal enterotoxin B, *Nucleic Acids Res.*, 2003, 31, 3027–3032.
17. Kreig, A.M., Therapeutic potential of toll-like receptor 9 activation, *Nat. Rev. Drug Discov.*, 2006, 5(6), 471–484.
18. Sel, S., Henke, W., Dietrich, A., Herz, U., Renz, H., Treatment of allergic asthma by targeting transcription factors using nucleic-acid based technologies, *Curr. Pharm. Des.*, 2006, 12(25), 3293–3304.
19. Kahan, B.D., Kirken, R.A., Stepkowski, S.M., New approaches to transplant immunosuppression, *Transplantation Proc.*, 2003, 35, 1621–1623.
20. Fire, A., Xu, S., Montgomery, M., Kostas, S., Driver, S., Mello, C., Potent and specific genetic interference by double-stranded RNA in *Caenorhabditis elegans*, *Nature*, 1998, 391, 806–811.
21. USA Today, *Merck to Buy Biotech Firm for $1.1B*, 2006, http://www.usatoday.com/money/industries/health/2006-10-31-merck-sima_x.htm (accessed May 2012).
22. Informa Healthcare, Executive briefing – Oligonucleotide therapeutics: The next big thing. In *Scrip Yearbook 2008*, Wilkinson, A., (Ed.), 2008, Informa Healthcare, USA.
23. Zamecnik, P.C., Stephenson, M.L., Inhibition of Rous sarcoma virus replication and cell transformation by a specific oligodeoxynucleotide, *Proc. Natl Acad. Sci.*, 1978, 75, 280–284.
24. Kruger, K., Grabowski, P.J., Zaug, A.J., Sands, J., Gottschling, D.E., Cech, T.R., Self-splicing RNA: autoexcision and autocyclization of the ribosomal RNA intervening sequence of tetrahymena, *Cell*, 1982, 31(1), 147–157.
25. Tuerk, C., Gold, L., Systematic evolution of ligands by exponential enrichment: RNA ligands to bacteriophage T4 DNA polymerase, *Science*, 1990, 249, 505–510.
26. Weiner, G.J., Liu, H.M., Wooldridge, J.E., Dahle, C.E., Krieg, A.M., Immunostimulatory oligodeoxynucleotides containing the CpG motif are effective as immune adjuvants in tumor antigen immunization, *Proceedings of the National Academy of Sciences of the United States of America*, 1997, 94(20), 10833–10837.
27. US FDA, *Drug Approval Package: Vitravene (Fomivirisen Sodium Intraveal Injectable) Injection*, http://www.fda.gov/ForConsumers/ConsumerUpdates/ucm081477.htm.
28. US FDA, *FDA Approves New Drug Treatment for Age-Related Macular Degeneration*, 2004, http://www.fda.gov/NewsEvents/Newsroom/PressAnnouncements/2004/ucm108385.htm.
29. Takakura, Y., Towards therapeutic application of RNA-mediated gene regulation, *Adv. Drug Delivery*, 2009, 61, 667–776.
30. Carter, G., *2002–2012 Surveys*, 2010 and 2012, Agilent Technologies Inc., Boulder, CO, USA.
31. ICH Expert Working Group, ICH Q8, *ICH Harmonised Tripartite Guideline: Pharmaceutical Development Q8 (R2)*, 2009. http://www.ich.org/fileadmin/Public_Web_Site/ICH_Products/Guidelines/Quality/Q8_R1/Step4/Q8_R2_Guideline.pdf (accessed May 2012).
32. ICH Q3A (R2), *Impurities in Drug Substance*, 26 October 2006.
33. ICH Q3B (R2), *Impurities in New Drug Products (Revised Guideline)*, 2006.
34. ICH Q6B, *Specifications: Test Procedures and Acceptance Criteria for Biotechnological/Biological Products*, 1999.
35. Capaldi, D., Ackley, K., Brooks, D., Carmody, J., Draper, K., Kambhampati, R., Kretschmer, M., Levin, D., McArdle, J., Noll, B., Raghavachari, R., Roymoulik, I., Sharma, B.P., Thuermer, R., Wincott, F., Quality aspects of oligonucleotide drug development: specifications for active pharmaceutical ingredients (API), *Drug Inf. J.*, 14 May 2012 (online), doi: 10.1177/0092861512445311.
36. ICH Q6A, *Specifications:Test Procedures and Acceptance Criteria for New Drug Substances and New Drug Products: Chemical Substances*, 1999.

Oligonucleotide Impurities and their Origin

2

Hagen Cramer, Kevin J. Finn and Nanda D. Sinha

2.1 Introduction

The impurities present in oligonucleotides depend on the synthesis methods utilised during their preparation. Figure 2.1 represents the synthesis cycle based on the phosphoramidite approach used predominantly today. Figure 2.2 depicts the cleavage and deprotection (C&D) reaction where the oligonucleotide is released from the solid support and the protecting groups are removed.

The chain elongation step in the phosphate diester approach used during the early days of oligonucleotide synthesis is very slow and results in branching of the chain as the resulting phosphate diester is exposed to activation and condensation. Subsequently the phosphate triester approach was developed, where the chain elongation step is much faster and the coupling reaction proceeds without the need for activating agents. The phosphate triester formed is not exposed to further activation, thereby mitigating the formation of branched impurities. However, branching or modification can still occur at the heterocyclic bases, especially at the unprotected O-6 position of guanine and the O-4 position of thymine/uracil (see Figure 2.3 for nomenclature discussion). Impurities can further result from strand cleavage during the removal of the phosphate protecting groups.

Two synthetic approaches based on phosphoramidite and H-phosphonate chemistries were predominantly adapted for automated synthesis during the 1980s. During these early stages of automation, instrument related impurities such as deletion of monomers or double additions were observed. H-phosphonate related impurities were based on unwanted capping taking place with the activating reagent pivaloyl chloride or adamantanoyl chloride present during chain elongation resulting in failure sequences.

This chapter will mainly cover the different types of impurities formed, their

Analysis of Oligonucleotides and their Related Substances, edited by George Okafo, David Elder and Mike Webb.

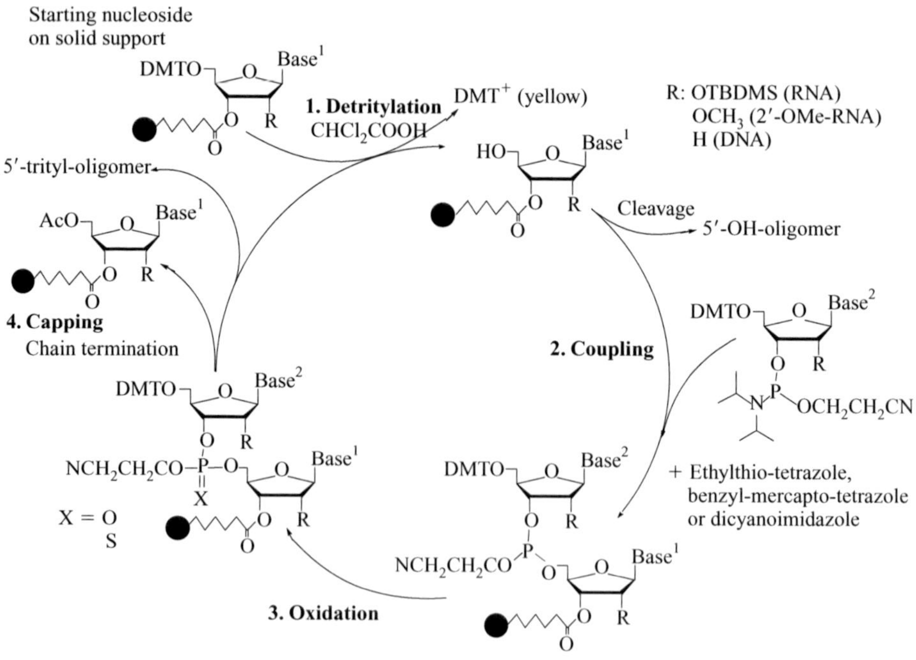

Figure 2.1 Solid-phase synthesis cycle for making oligonucleotides employing the phosphoramidite strategy predominantly in use today. Note that synthesis occurs from 3′ to 5′.

origins and potential mechanisms of formation. Also, the chapter will recommend practical approaches to mitigate or reduce impurity formation. Most of the chemistry discussion will be from the phosphoramidite process, which is most widely used nowadays for oligonucleotide synthesis. As a starting point, the most common impurities, their origins, typical levels and control strategies have been summarised in Table 2.1. Impurities in the oligonucleotide API originate from the raw materials, predominantly from impurities in phosphoramidites, the type of solid support used, ancillary reagents and from the synthesis process. Moisture control is of utmost importance in phosphoramidite-based synthesis as moisture also reacts with the activated phosphoramidite. Any moisture present will hydrolyse the activated phosphoramidite, thereby making the building block inactive and resulting in lower coupling efficiencies.

2.2 Brief Historical Perspective of Oligonucleotide Synthesis

2.2.1 Chemical Challenges

The first synthesis of a dinucleotide with proper 3′-5′ orientation was reported by Michelson and Todd in 1955 [1]. The challenges in the synthesis of DNA were

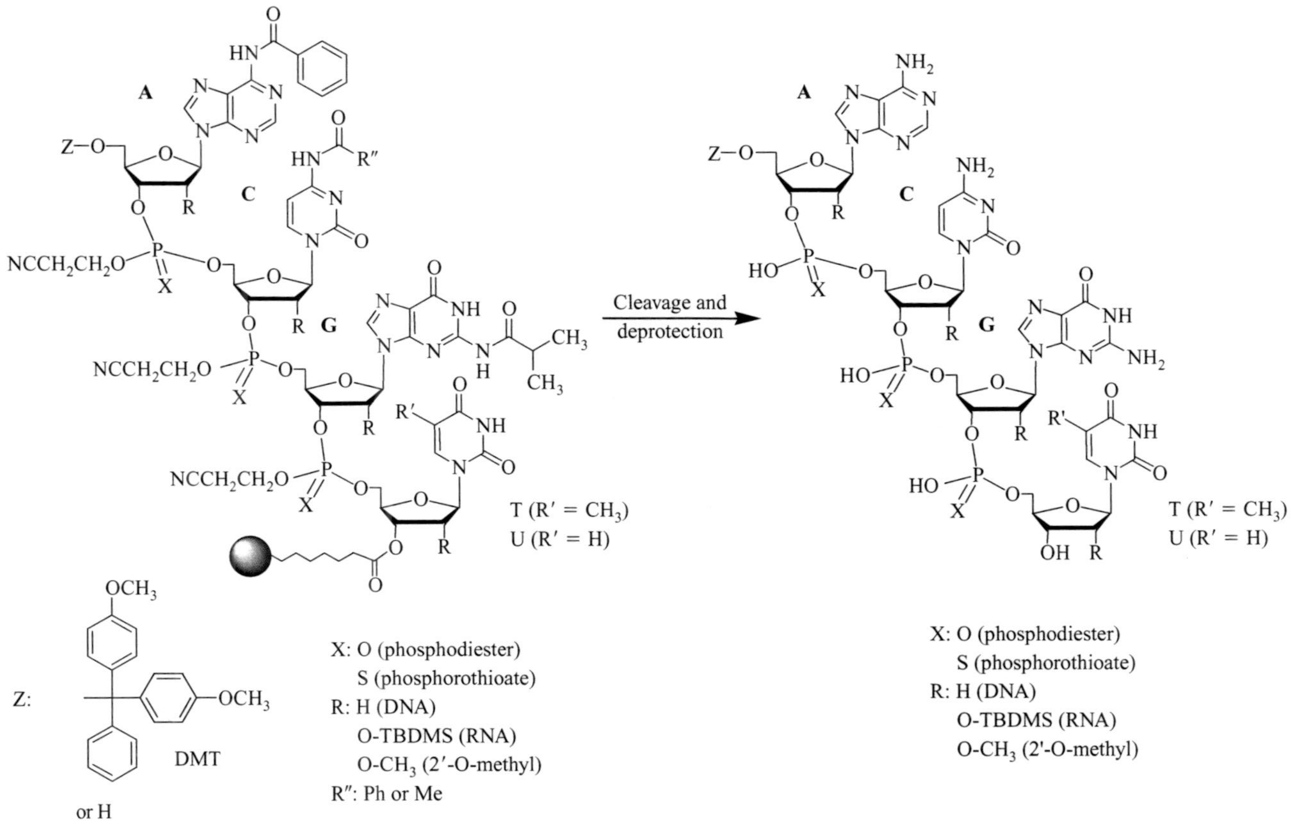

Figure 2.2 Release of oligonucleotide from the solid support and removal of protecting groups.

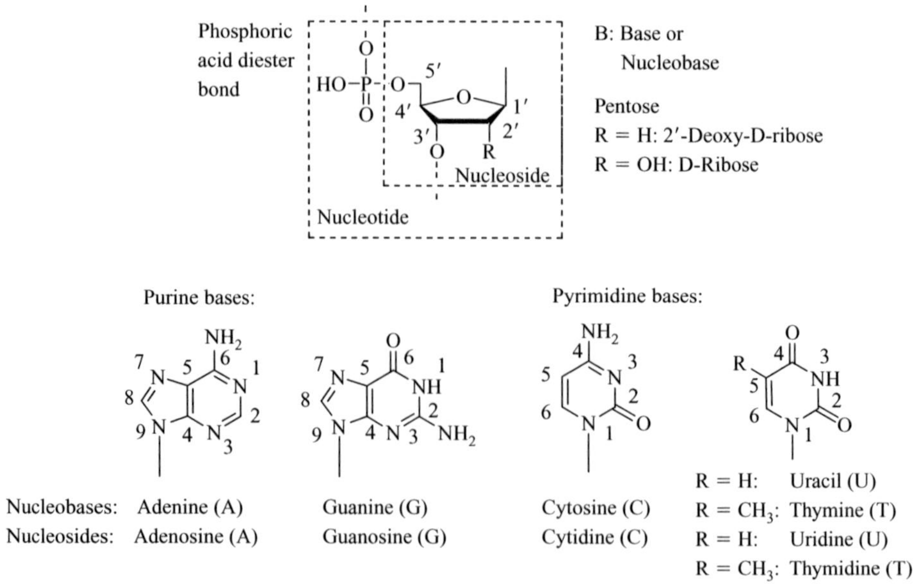

Figure 2.3 Nomenclature and numbering system used to describe nucleosides, nucleotides and nucleic acids.

selective protection and deprotection of various reactive moieties present in deoxyribonucleosides. These challenges were successfully resolved [2] via the introduction of various protecting groups (trityl groups for primary hydroxyl and benzoyl, isobutyryl for exocyclic amines) and this culminated in the first synthesis of sequence defined oligodeoxynucleotide [3, 4]. Interestingly, most of these protecting groups are still in use today. This initial synthetic strategy employed was termed phosphate diester because a phosphate diester group was formed after condensation. In spite of some of the early shortcomings, this strategy was successfully utilised for assembly of a synthetic gene [5].

Further improvements to the synthesis of oligonucleotides include the introduction of phosphate triester methodology, which addressed some of the shortcomings of the phosphate diester approach [6]. This phosphate triester approach was further improved by several groups [7–11] who introduced better coupling agents. While progress was made in solution phase synthesis, the introduction of polymer supported solid-phase synthesis in 1965 [6] simplified oligonucleotide synthesis to the point of having them readily available in the laboratory. The combination of these two strategies, solid support synthesis and phosphate triester chemistry allowed utilisation of synthetic oligonucleotides in molecular biology research, recombinant technology and genetic engineering.

More recent improvements spurred on by newer applications of synthetic oligonucleotides for therapeutic and diagnostic purposes [12] include development of novel synthesis strategies based on P (III) chemistry [13], introduction of methyl-phosphoramidite [14, 15] and β-cyanoethyl-phosphoramidite chemistries [16, 17].

Table 2.1 Control of major oligonucleotide impurities.

Origin	Impurity	Type of impurity (mass change)	Typical level in final API	Control strategy
Amidite starting material	α-Anomeric phosphoramidite ((i) in Figure 2.9)	Incorrect orientation of nucleo-base (no mass difference)	Trace	Tight amidite spec (HPLC, ^{1}H-NMR)
Amidite starting material	3′-*O*-DMT, 5′-*O*-phosphoramidite ((ii) in Figure 2.9)	Incorrect linkage (5′,5′) (no mass difference)	Trace	Tight amidite spec (HPLC, ^{31}P-NMR)
Amidite starting material	*N*,*O*-5′-(*O*)-bis-acetylated nucleoside phosphoramidite ((iii) in Figure 2.9)	Formation of 3′,5′-internucleotidic linkage leading to chain termination (no mass difference)	Trace	Tight amidite spec (HPLC, ^{31}P-NMR)
Amidite starting material	Cyanoethyl-amidito-phosphite ((vii) in Figure 2.9)	Internucleotidic di-phosphate (+80 Da) or di-thioate linkage (+96 Da)	Trace	Tight amidite spec (HPLC, ^{31}P-NMR)
Amidite starting material	3′-*O*-t-BDMSi/OMe-2′-*O*-phosphoramidite ((xi) in Figure 2.9)	Incorrect linkage (2′,5′) (no mass difference)	Trace	Tight amidite spec (HPLC, ^{31}P-NMR)
Amidite starting material	5′-*O*,N-Di-DMT adenosine ((xiii) in Figure 2.9)	Branchmers	Small	Tight amidite spec (RP-HPLC)
Solid support	n − 1 (3′) (typically with terminal phosphate/phosphothioate)	3′-Nucleoside missing	Variable	Proper capping of support
Solid support	n − 1 (3′)	3′-Nucleotide missing	Variable	Detritylation time of Unylinker support

(*continued*)

Table 2.1 (*continued*)

Origin	Impurity	Type of impurity (mass change)	Typical level in final API	Control strategy
Solid support	n + (n − 1 (3′))	Branchmer	Variable, depending on age of support	Use fresh support to minimise loss of base protecting group during storage of nucleoside loaded support
Reagent	Chloral	Internucleotide adduct (+147 Da)	Trace	Chloral spec in DCA
Reagent	β-*N*-methylamino acetamide (MAM)	+85 Da adduct	Trace	NMI impurity in capping reagent
Process	Depurination	Abasic site (−135 Da (G) or −117 Da (A))	Main	Acid contact time/ acid strength
Process	DMT-C phosphate ester	Capture of DMT cation by P(III) centre	Trace	Oxidation/sulfurisation
Process	n − 1	Deletion sequence	Main	Control of moisture; amidite equivalency; contact time; activator
Process	n + 1	Over condensation	Main	Acidity of activator; coupling contact time
Process	PO	Backbone modification (−16 Da)	Main impurity in PS oligos	Sulfurisation/C&D
Process	2,6-Diaminopurine	O-6 modification of guanine (−1 Da)	Trace	Capping

(*continued*)

Table 2.1 (*continued*)

Origin	Impurity	Type of impurity (mass change)	Typical level in final API	Control strategy
Process	G-Ac	Protecting group exchange (+42 Da)	Trace	Capping
Process	CNET	Capture of acrylonitrile (+53 Da)	Main	Amine wash
Process	Transamidation	Transfer of protecting groups to terminal amine	Main	Choice of 5′-amino protecting group; elimination of final capping step
Process	Transamination	Replacement of C-4 amine with methylamine (+14 Da)	Variable	C&D
Process	2′-3′-Isomerisation (RNA)	Incorrect linkage (2′,5′) (no mass difference)	Main	C&D
Process	Depyrimidation (RNA)	Degradation of uridine (−54 Da, −92 Da) and cytidine	Variable	C&D
Process	Fluoride loss	Elimination of 2′-fluoro (−20 Da) or net substitution of F with OH (−2 Da)	Main	C&D
Process	Strand breakage resulting into terminal phosphates or 2′,3′-cyclic phosphates (RNA)	Elimination of 3′ fragment after depurination (low pH) or RNA strand breakage (high pH)	Variable	C&D

During the same time period, a synthetic approach based on H-phosphonate chemistry originally adapted to diribonucleoside synthesis was re-introduced for oligodeoxynucleotide synthesis [18–20].

Along with oligodeoxynucleotide chemistry advances, strategies for synthesis of oligoribonucleotides also progressed, specifically in the area of 2′ hydroxyl protecting groups, namely *t*-butyldimethylsilyl protection [21] and tetrahydropyran and variants [7]. Today, 2′-*O*-*t*-butyldimethylsilyl-protected ribonucleoside phosphoramidites are still the preferred protecting group for oligoribonucleotide synthesis. Several review articles are available on the different approaches used for making oligonucleotides [22–24]. At present, the β-cyanoethyl-phosphoramidite chemistry is predominantly used in solid support-based oligonucleotides synthesis (see Figure 2.1).

The success of any drug development not only depends on satisfactory clinical trials but also on the quality and proper characterisation of the drug substance. However, in order to properly characterise and minimise the impurities in the drug substance it is imperative to understand their origin. This chapter primarily aims to define the sources of impurities which are introduced by impurities in the building blocks (phosphoramidite related), solid supports and ancillary reagents (reagent related) as well as during the manufacturing process (process and chemistry related impurities).

2.2.2 Analytical Challenges

Initially, characterisation of synthetic oligonucleotides was extremely challenging. Only limited methods and techniques were available, for example polyacrylamide gel electrophoresis (PAGE) and, to some extent, chromatographic techniques were used to indicate the presence of 'shortmers' ($n-1$, $n-2$, etc.) or 'failure sequences'. Ultraviolet (UV) spectroscopic methods were used for base composition analysis based on different absorption ratios of the bases at 280 and 260 nm. However, modification of nucleosides during oligonucleotide synthesis could not be easily determined using this approach. Other analytical approaches were time consuming, such as enzymatic digestions (using snake venom phosphodiesterase (SVP) and bacterial alkaline phosphatase) of oligonucleotides followed by reversed-phase high-performance liquid chromatography (RP-HPLC) analysis of the digest [25–27]. Early determinations of the sequence of an oligonucleotide were even more challenging. One of the early sequencing methods was called 'wandering spot analysis', which was based on the characteristic mobility shifts of partial degradation products on two-dimensional chromatograms obtained from high-voltage electrophoresis and homo-chromatography [28, 29].

The rapid development of modern analytical methods such as strong anion exchange HPLC (SAX-HPLC), RP-HPLC, capillary electrophoresis (CE), matrix assisted laser desorption/ionisation–time of flight mass spectroscopy (MALDI-TOF MS) and HPLC-MS for separation and identification have overcome these challenges, allowing the development of manufacturing processes of high-quality oligonucleotides for therapeutic applications [30]. The challenges involved in oligonucleotide analysis are described in detail in Chapter 3.

2.3 Raw Material Related Impurities

2.3.1 Amidite Related Impurities

Initially nucleosides used for the synthesis of phosphoramidites were procured from natural sources, such as salmon fish milt [31]. However, health concerns linked to the outbreak of mad cow disease (bovine spongiform encephalopathy (BSE)) in the UK in 1986 and its link to Creutzfeld Jakob disease in humans (first cases appeared in 1994) [32] rendered natural nucleosides unsuitable, as bovine enzymes (nucleases and phosphatases) were used for the isolation of nucleosides from salmon fish milt. The current supply of nucleosides used in the synthesis of therapeutic oligonucleotides is sourced exclusively from man-made synthetic sources and must be certified to be BSE- and TSE- (transmissible spongiform encephalopathy) free.

Nucleosides possess several reactive sites and need to be selectively protected before these can be converted to phosphoramidites. The protection of 5′-hydroxyl (and 2′-OH in RNA) as well as exocyclic amino groups are prerequisites for the synthesis of oligonucleotides.

Phosphoramidite impurities typically result from the introduction of nucleoside protecting groups, the quality of synthetic nucleosides (α-anomeric, 3′-substituted impurities) and the quality of the phosphitylating reagent. These impurities may influence the overall quality of oligonucleotides as these impurities can be reactive and critical.

In the past, the quality of nucleoside phosphoramidites was controlled by RP-HPLC as well as ^{1}H and ^{31}P-NMR (nuclear magnetic resonance spectroscopy). Utilisation of HPLC-MS-based analytical methods for the characterisation of impurities present in phosphoramidites has made a significant impact on characterising and understanding the nature of impurities. These impurities can be divided into three categories:

- non-reactive and non-critical
- reactive but non-critical
- reactive and critical [33, 34].

2.3.1.1 Non-reactive and Non-critical Impurities

This class of impurities is made up of non-phosphorylated nucleosides and hydrolysed or oxidised phosphite species derived from improper protection of nucleosides, or from synthesis and work up of phosphoramidites. These impurities do not interfere with the synthesis of oligonucleotides. Nucleoside-based impurities are depicted in Figure 2.4, while phosphitylation reagent-based impurities are shown in Figure 2.5.

During 5′-hydroxyl group protection, the 4, 4′-dimethoxytrityl (DMT) group might also react with a free 3′-hydroxyl creating non-reactive impurity (ii) (Figure 2.4). Acrylonitrile is generated during hydrolysis of phosphoramidites resulting into H-phosphonate (iv). Acrylonitrile can then induce the formation of impurity (vi) by an Arbuzov type rearrangement as described previously [35, 36]. The degradation occurs via the same pathway as described for the degradation of the alkoxy-bis-dialkylamino-phosphine (phosphitylating reagent) used for the preparation of phosphoramidites [37]

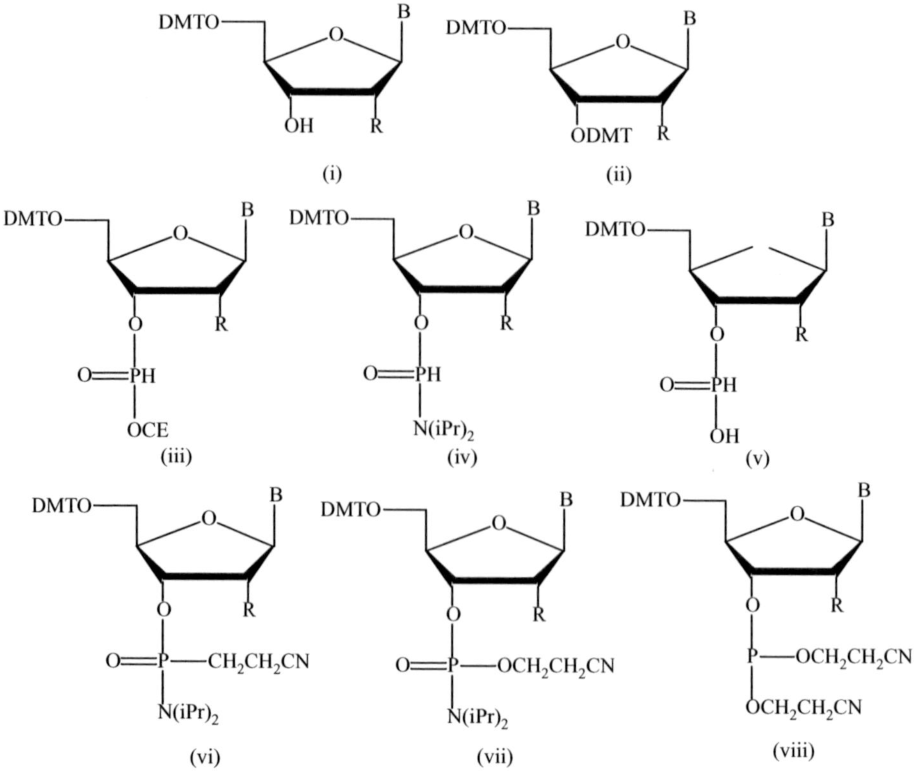

Where B = N-protected A,C,G and T/U; R= H, OTBDMSi, OMe, OMOE, or F

Figure 2.4 Non-reactive and non-critical nucleoside-based impurities in nucleoside phosphoramidites.

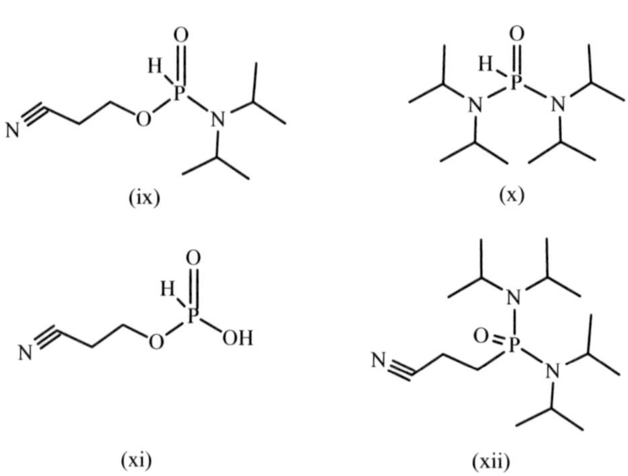

Figure 2.5 Non-reactive and non-critical phosphitylation reagent-based impurities in nucleoside phosphoramidites.

resulting in impurity (xii) (Figure 2.5). Impurity (viii) is the result from the bis-alkoxy-dialkylamino-phosphine impurity in the phosphitylating reagent.

In addition to nucleoside-based impurities, phosphitylation reagent-based impurities are also found in phosphoramidites. Most of these are non-reactive and non-critical and are represented in Figure 2.5.

2.3.1.2 Reactive but Non-critical Impurities

Reactive but non-critical phosphoramidite impurities are generated from nucleoside impurities with partially modified *N*, 5′-*O* or 2′-*O* (RNA) protecting groups. These phosphoramidite impurities will also incorporate into the oligonucleotide chain; these protecting groups display reactivity similar to the correct protecting group and, as such, are removed together with standard protecting groups without impacting the quality of the synthesised oligonucleotide (see Figure 2.6). Examples of such modified protecting groups which are rendered redundant through removal are highlighted in Figure 2.7 and Figure 2.8.

The protecting group of choice for the 5′-hydroxyl group is the DMT group, which is introduced by reacting the free 5′-hydroxyl nucleoside with DMT chloride. Generally DMT chloride is obtained in high purity greater than 96%. The remaining 4% of impurities present are reactive modified trityl chlorides (comprising methoxy, methyl trityl or monochlorodimethoxytrityl) and non-reactive dimethoxytritanol. Apart from dimethoxytritanol, the modified trityl group impurities are also reactive and are introduced into the nucleoside at the 5′-end. Phosphoramidites prepared from these impurities also couple effectively but do not impact the quality of the resulting oligonucleotide. Some examples of this class of impurities are represented in Figure 2.7. Figure 2.7(i) shows a variety of protecting group variants on the 5′-*O*-DMT position. All acid labile variants of the trityl group will be subjected to deprotection along with the DMT protecting group.

Figure 2.7(ii) shows non-critical variants in the alkyl chain of the alkylamine group of the phosphoramidite. These reactive impurities will typically compete with the phosphoramidite starting material, being either more or less reactive than the main component. However, during activation the alkylamine group is being released and is then removed during the coupling reaction and it therefore will not be present in the final oligonucleotide and does not impact its quality.

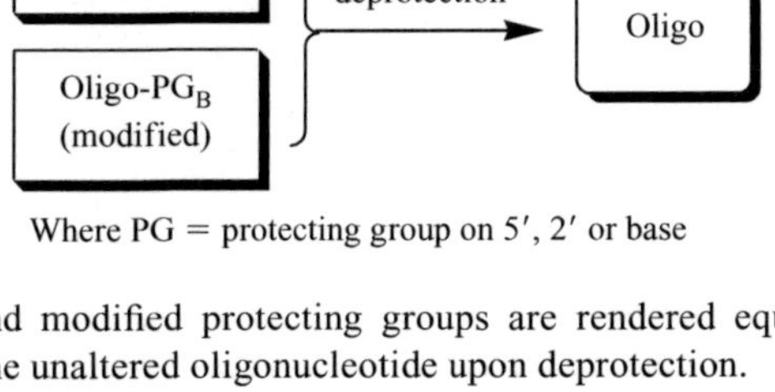

Figure 2.6 Correct and modified protecting groups are rendered equivalent when equally reactive, resulting in the unaltered oligonucleotide upon deprotection.

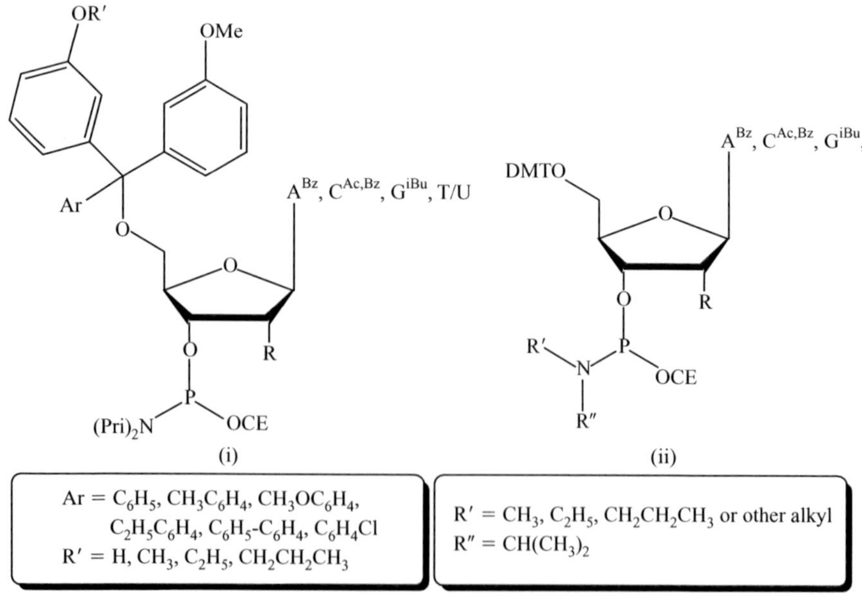

Figure 2.7 Reactive but non-critical impurities in nucleoside phosphoramidites – part I.

Reactive protecting group impurities of another type are those associated with exocyclic amine protection of the nucleoside base. In place of standard protecting groups (Bz on C and A, and iBu on G), chlorobenzoyl protecting groups on cytosine, adenine and n-butyryl on guanosine are the most common impurities (Figure 2.8).

2.3.1.3 Reactive and Critical Impurities

These reactive impurities are generated during synthesis of nucleosides, during protection of the 2′, 3′ and 5′-hydroxyl groups of the sugar and the exocyclic amino group of the base or originate from reactive impurities in the phosphitylating reagent. Incorporation of these reactive impurities has a significant impact on the quality or nature of the oligonucleotide synthesised. Hence, control of the impurities is of utmost importance. The structures of potential impurities in nucleoside phosphoramidites are shown in Figures 2.9 and 2.10.

Some of the reactive and critical impurities in the nucleoside phosphoramidites are generated from impurities in the phosphitylating reagent and are illustrated by structures (iv)–(x) in Figure 2.9. In general, the quality of the phosphitylation reagent has greatly improved over the years and today only negligible levels of these impurities (if present at all) are detected in amidites used in oligonucleotide manufacturing for therapeutic applications. Impurities derived from the protected nucleoside are more commonly observed at detectable levels.

During 5′-hydroxyl group protection of the nucleoside, the DMT group might react with a free 3′-hydroxyl instead of the 5′-hydroxyl group or it reacts also with the

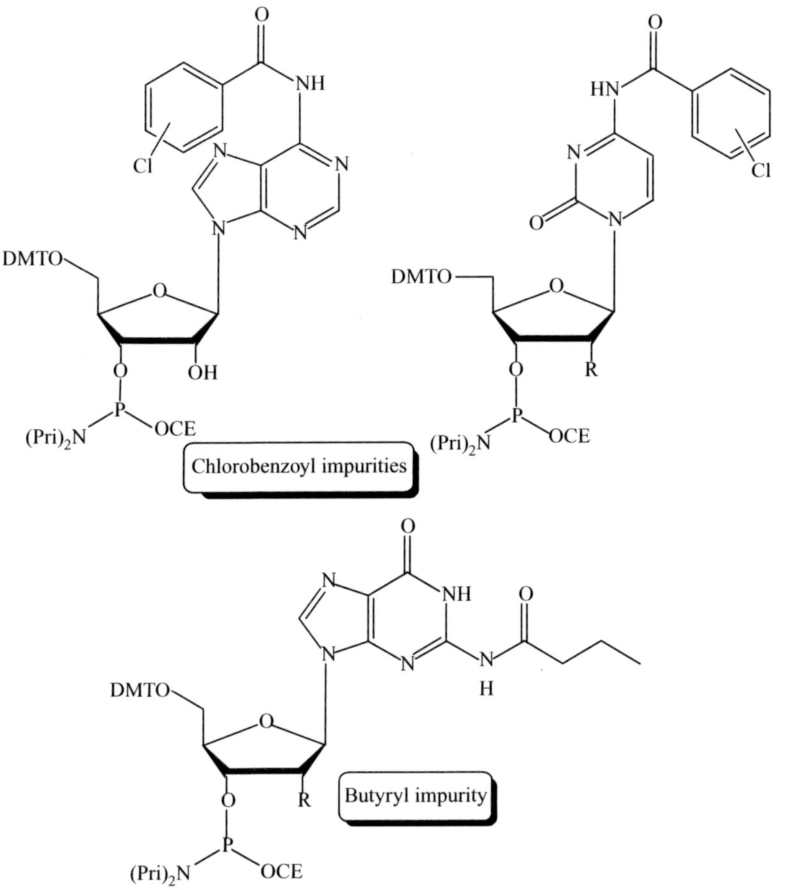

Figure 2.8 Reactive but non-critical impurities in nucleoside phosphoramidites – part II.

free (non-acylated) exocyclic amino group. These nucleoside impurities will be subsequently converted to the phosphoramidite, resulting into critical impurities (ii) and (xiii), respectively. Incorporation of impurity (ii) will create a different orientation of the oligonucleotide chain, 5′, 5′ instead of 3′, 5′, whereas (xiii) would cause branching. When an α-anomer (i) is introduced in the oligonucleotide chain, it may lose the ability to hybridise (bind properly) to the target.

In the case of ribonucleosides, the 2′-hydroxyl group is commonly protected by the *tert.*-butyldimethylsilyl (TBDMSi) group. During its introduction, typically a mixture of 2′- and 3′-isomer is obtained, which needs to be separated chromatographically. In addition, this protecting group can migrate from the 2′- to the 3′-hydroxyl position. RNA phosphoramidite contaminated with 3′-*O*-TBDMSi-phosphoramidite will result in a 2′,5′-linkage instead of the desired 3′,5′-linked oligoribonucleotides and therefore will impact the quality of RNA synthesis. This type of impurity is depicted as structure (xi) in Figure 2.9 (R = TBDMS).

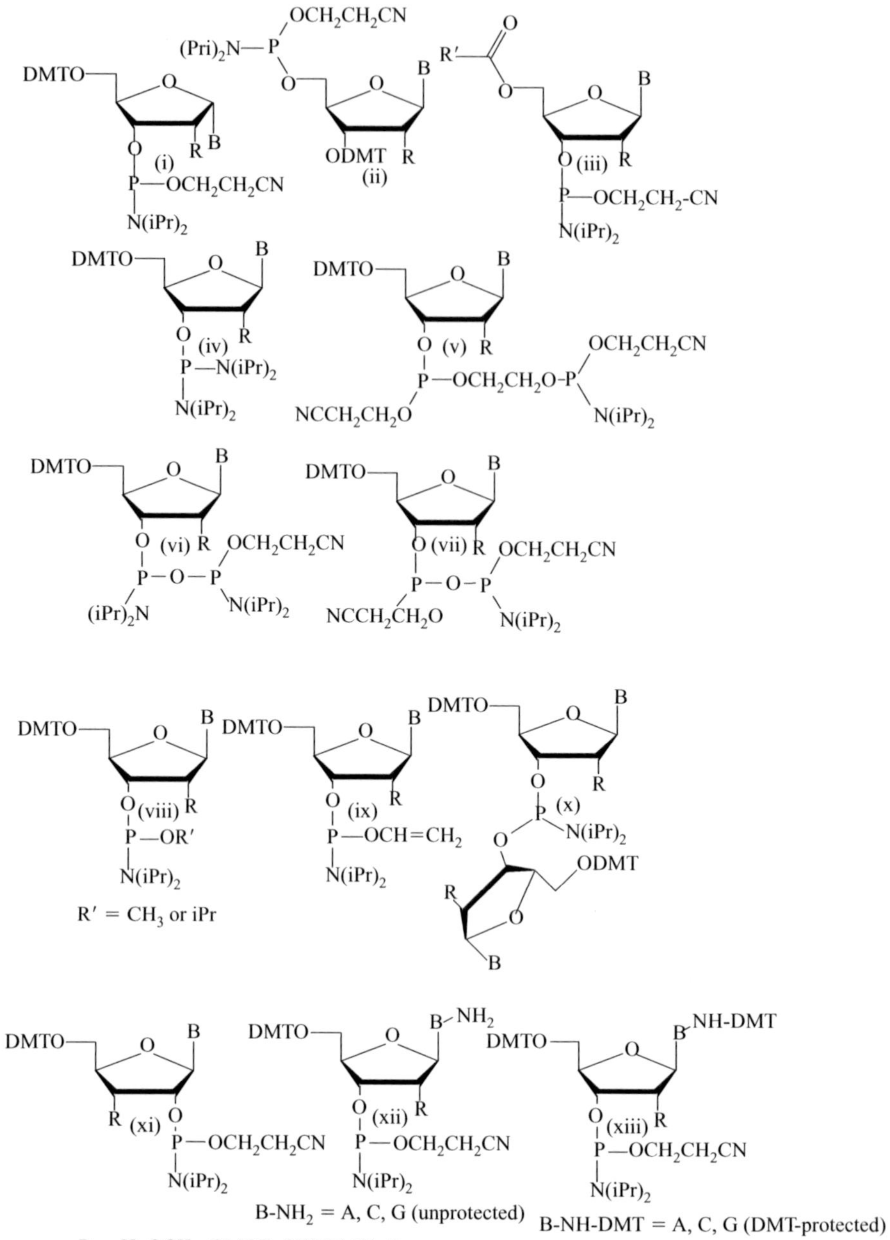

R = H, OCH_3, OMOE, OTBDMSi, F
B = A^{Bz}, $C^{Bz/Ac}$, G^{iBu}, T/U

Figure 2.9 Reactive and critical amidite-based impurities in nucleoside phosphoramidites.

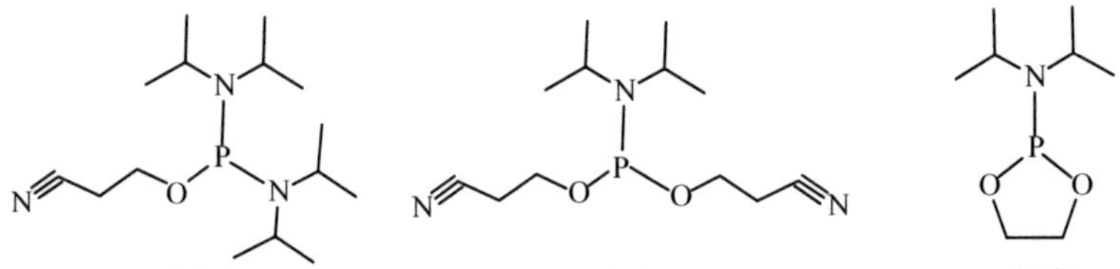

Figure 2.10 Reactive and critical phosphitylation reagent-based impurities in nucleoside phosphoramidites.

Reactive exocyclic amine groups are protected using various acyl groups. If the 5′-hydroxyl group is unprotected during acyl protection of the exocyclic amine group, it will result in a 5′-*O*, *N* bis-acylated nucleoside impurity, as represented by critical impurity (iii) in Figure 2.9. Similarly, phosphoramidites prepared from incompletely protected exocyclic amine nucleosides will result in critical impurity (xii) in Figure 2.9 and will cause branching during oligonucleotide synthesis.

In the past, many impurities were introduced by the phosphitylation reagent used for making the nucleoside phosphoramidite. Phosphitylation dependent impurites are represented by structures (iv)–(x) in Figure 2.9. Impurities (viii) and (ix), when introduced into the growing oligonucleotide chain, may lead to degradation of the oligonucleotide or remain uncleaved during deprotection. Hydroxide ions in the ammonia solution could partially cleave the pentavalent phosphate triester, resulting in strand breakage. Other methyl, isopropyl or ethylenyl groups remain on the final oligonucleotide since standard deprotection procedure cannot remove them completely from the phosphorus atom. When introduced, impurity (iii) would terminate chain elongation.

Impurities (v), (vi) and (vii) in Figure 2.9, added onto the chain, will extend the length by adding a phosphorous based spacer, which may prevent proper hybridisation or binding specificity. Impurities (v), (viii) and (ix) are the result of impurities present in the cyanoethanol used when making the phosphitylation reagent. Nowadays, these impurities are present at much lower abundance, and they are not typically detected in pharmaceutical grade amidites. Impurities (iv) and (x) would cause branching of the oligonucleotides chain. The 5′-hydroxyl groups of two growing chains would react with impurity (iv) to continue as only one chain, while impurity (x) would cause the classical form of branching, where one chain continues growing as two.

Impurity (xi) is a common impurity found in 2′-modified sugars and can form a 2′, 5′- instead of a 3′, 5′-linkage. Since these impurities are phosphoramidite manufacturing process related, their complete elimination is unlikely, but they can be monitored and controlled using proper analytical methods and appropriate specifications for the phosphoramidites. HPLC-MS analysis is an important tool for the accurate identification of impurities affecting the quality of oligonucleotides (see Chapter 4 for more details). ^{1}H-NMR and RP-HPLC have been used for identification of nucleoside phosphoramidites. ^{31}P-NMR and RP-HPLC can be effectively used for controlling reactive impurities in phosphoramidites.

Any phosphitylation reagent (see (xiv) in Figure 2.10), which might have been carried over into the phosphoramidite would also be reactive and critical, resulting in a terminal phosphate/thiophosphate. It could further react with a 5′-hydroxyl of a second growing chain to form a 3′-5′, 5′-3′ symmetric oligonucleotide impurity. If impurity (xv) in the phosphitylation reagent is carried over into the amidite it will result in a terminal phosphate or thiophosphate. Impurity (xvi) is created from ethylene glycol impurites in the cyanoethanol used to make the phosphitylation reagent. However, the quality of the cyanoethanol has improved greatly over the years and this impurity typically is no longer detected.

2.3.1.4 Phosphoramidite Impurity Control

From the above discussion, it can be concluded that phosphoramidite purity control is of utmost importance for ensuring consistently high-quality oligonucleotides. However, it also becomes apparent that not all phosphoramidite impurities contribute to impurity formation in the final oligonucleotide. Of all the impurities described above, only reactive and critical impurities need to be tightly controlled. How should specifications be set?

Traditionally the most widely used analytical tools are ^{31}P-NMR and HPLC. For the detection of small-level impurities ^{1}H-NMR is less suited because of its complexity. Every signal is split into two because of the diastereoisomeric isomers, due to a chiral centre at the phosphorus. Since all critical impurities have a phosphorus atom, ^{31}P-NMR turns out to be a highly valuable tool for the analysis of phosphoramidites. Impurities in the ^{31}P-NMR can be easily categorised into P(III) (100–180 ppm) and P(V) region (−10–50 ppm). Thereby the classification is based on the oxidation state. A hydrolysed phosphoramidite has an oxidation state of V, since its most stable form is the H-phosphonate (see (iii) in Figure 2.4) and hydrogen has a slightly higher electronegativity than phosphorus [38, 39].

By using orthogonal methods such as HPLC and ^{31}P-NMR most impurities can be detected and quantified. In ^{31}P-NMR isomeric impurities (such as (i) and (ii) in Figure 2.7 and (xi) in Figure 2.9) are often poorly separated from the phosphoramidite double signal (both diastereoisemers typically do not have the same chemical shift) and therefore cannot be easily quantified. To reduce the complexity of the spectra, they are usually recorded proton decoupled. However, satellite peaks around the main signal, originating from coupling of P(31) with N(15) and C(13), make the interpretation of the spectra around the main signals difficult. Any nucleoside starting material not having a phosphorus centre cannot be detected with ^{31}P-NMR. In HPLC the phosphoramidite also elutes as two peaks owing to the diasteroisomeric isomers, because of a chiral centre at the phosphorpus atom. They are well separated from the starting material and typically also from isomers (such as (i), (ii) and (xi) in Figure 2.9). However, standard HPLC detectors are not capable of detecting impurities related to the phosphitylation reagent as those do not possess UV active chromophores. Along with phosphitylating reagent impurities, other P(III) and P(V) impurities can easily be distinguished with ^{31}P-NMR, as these appear in different areas of the ^{31}P-NMR spectrum.

While ^{31}P-NMR and HPLC allow for the quantification of impurities, both methods lack the ability to easily characterise the detected impurities. Recently, HPLC-MS methods were developed, which allow for mass assignments of impurities, making impurity characterisation easier [40–42]. An example for an amidite analysed by HPLC and ^{31}P-NMR is illustrated in Figure 2.11(a) and (b). As described above, all impurities are not detectable by both HPLC and ^{31}P-NMR methods. The non-critical, non-reactive nucleoside impurity at a retention time of 11.6 min is 'invisible' in the ^{31}P-NMR spectra (see also impurity (ii) in Figure 2.4). The reactive and critical 5′-*O*-*N*(4)-Di DMT amidite impurity (see also (xii) in Figure 2.9) can be easily quantified since it has a largely different retention time in the HPLC chromatogram. While this impurity can also be detected in the ^{31}P-NMR spectra, quantification of this impurity using ^{31}P-NMR is not possible because it is overlapping with the desired amidite signal. The unidentified impurity at 118 ppm cannot be detected by HPLC. It appears that it either is not stable in the aqueous environment used during HPLC analysis or it does not have an UV active chromophore.

2.3.2 Solid Supports

The solid supports used in oligonucleotide synthesis also present a potential source of impurities and a good understanding of the different types of supports and processing conditions is important to reduce the impurity burden in the final product. Traditionally, silica-based controlled pore glass (CPG) was almost exclusively used in phosphoramidite-based solid-phase oligonucleotide synthesis. CPG supports could be manufactured with predefined pore sizes and space required for a growing oligonucleotide chain. They are non-swelling and inert to the solvents and reagents used during oligonucleotide synthesis. Typically, in oligonucleotides synthesis, a CPG support with a pore size of 500 Å is used for oligonucleotides up to a length of 30 bases and a pore diameter of 1000 Å for up to 50 bases [43, 44]. Therapeutic oligonucleotides are generally under 50 bases long, hence CPG support with 500 Å or 1000 Å pore size satisfies this need. Disadvantages of glass supports, however, include the instability towards basic conditions used during C&D and low loading capacity, which can lead to formation of impurities.

To mitigate CPG's disadvantages, highly cross-linked polystyrene [45, 46] and methacrylate [47] supports were developed. These initially developed supports have low loadings and are still in use for small-scale oligonucleotide synthesis [48, 49]. Later, high-loaded macroporous polystyrene supports with loadings of 200 μmol nucleoside per gramme of support were developed by Pharmacia Biotech (now part of GE Healthcare). Another high-loading solid support, NittoPhase, originally developed in collaboration with ISIS Pharmaceuticals is being sold by Nitto Denko. Latest developments have led to a support called NittoPhase HL, with loadings of 400 μmol nucleoside per gramme. Other high-loaded supports have been developed or are in development. ISIS Pharmaceuticals has developed a new, polystyrene-based, high-loaded support (350 μmol/g), which was recently made available commercially by GE Healthcare under the name Primer Support 5G. The different chemistries used for the

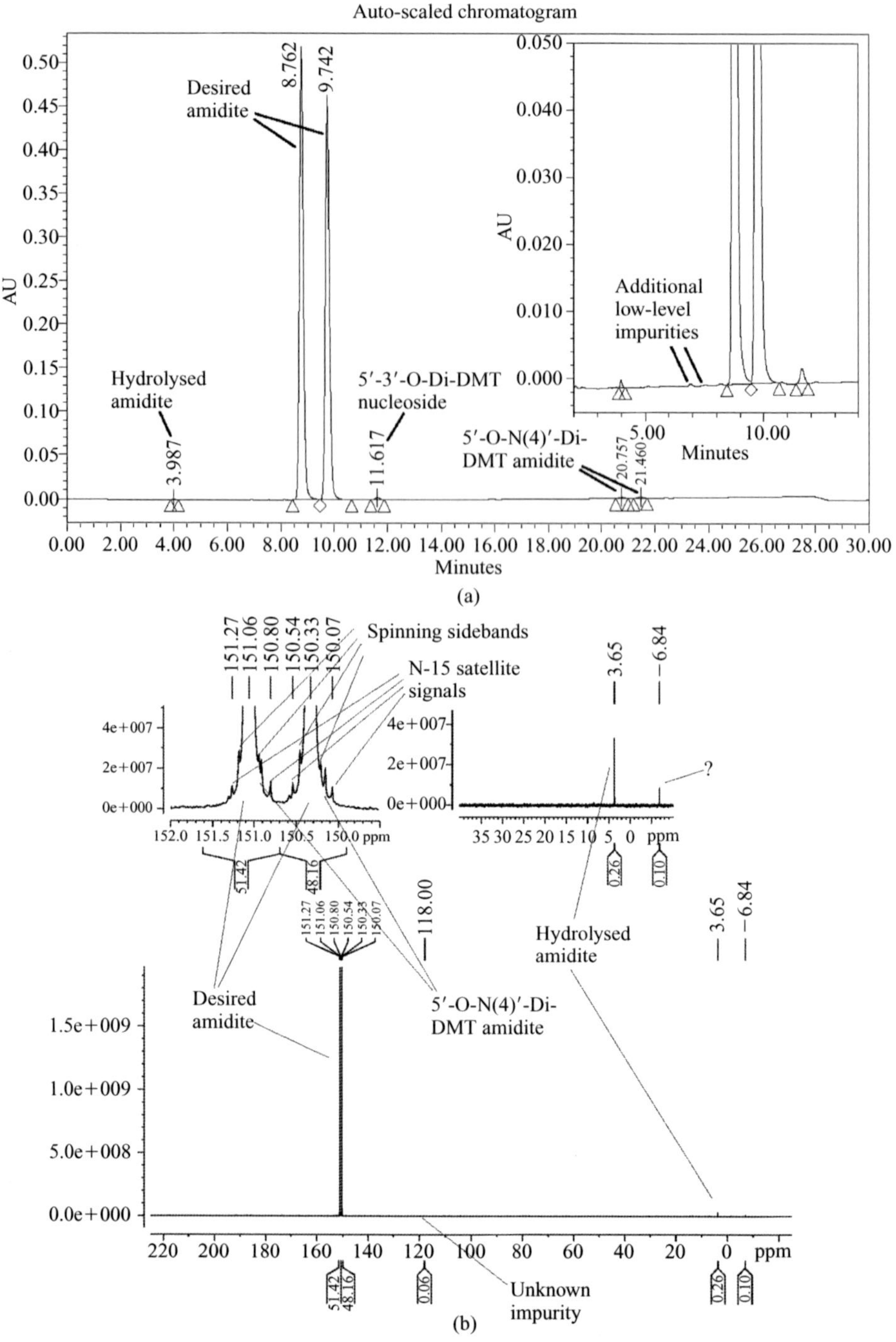

Figure 2.11 2′-*O*-methyl-A amidite analysed by (a) HPLC (99.3%) and (b) ^{31}P-NMR (99.5%).

development of solid supports utilised in oligonucleotide synthesis are summarised in Figure 2.12.

2.3.2.1 Nucleoside Loaded Supports

Traditionally, a nucleoside succinate is allowed to react with an amino or hydroxyl functionalised support to create a nucleoside loaded support [50]. The succinate linkage thus created between solid support and nucleoside is stable to conditions and reagents used during solid-phase synthesis. At the end of the synthesis, during basic C&D conditions the succinate linkage is hydrolysed, releasing the oligonucleotide chain from the support. Other more base labile linkers have been developed such as oxalate [51] and Q-linker [52, 53], but the succinate linkage is the most widely used one in large-scale synthesis. The stable succinate linkage allows selective removal of cyanoethyl phosphate protecting groups without releasing the oligonucleotide from the support using on-column base washing. This inhibits acrylonitrile adduct formation of the heterocyclic bases, in particular thymine [54].

CPG supports require a spacer in between the nucleoside succinate and the functional group of the support to ensure good coupling efficiencies. The amino or glyceryl functionalised CPG support is thereby derivatised with a long chain alkyl amine (LCAA) [55] (see Figure 2.12) or other similar linkers, as described by Sirna Therapeutics [56] or others [43] before attaching the nucleoside succinate. With polystyrene supports, functional (amino or hydroxyl) groups are reacted directly with the nucleoside succinate and no additional linker is required.

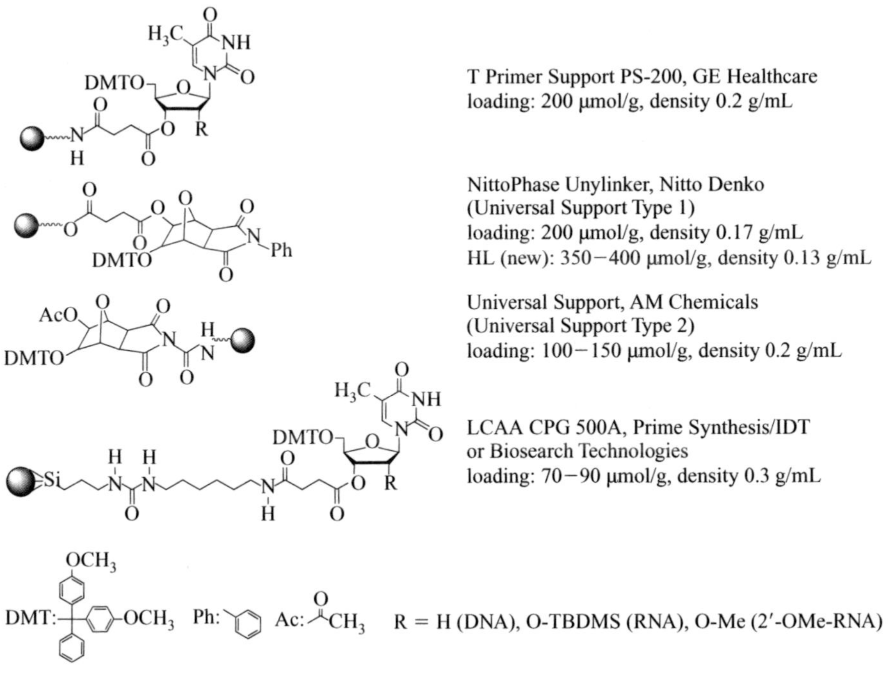

Figure 2.12 Common types of supports used in solid-phase oligonucleotide synthesis.

The coupling of a nucleoside-3′-*O*-succinate to the support can be performed with 2,2′-dithiobis(5-nitropyridine) and dimethylaminopyridine (DMAP), followed by addition of triphenylphosphine and LCAA-CPG [57]. Alternatively, the nucleoside-3′-*O*-succinate can be attached to LCAA-CPG using a variety of phosphonium or uronium coupling agents or soluble carbodiimides, namely diisopropylcarbodiimide [58], and DMAP [59, 60].

2.3.2.2 Formation of Branchmers

During manufacture or storage of a nucleoside loaded support, loss of protecting groups of the heterocyclic base can lead to impurities, called branchmers, in particular for cytidine (C) and adenine (A) nucleoside loaded supports. Free or unprotected exocyclic amino groups, when exposed to the coupling conditions, will also react with the incoming phosphoramidite resulting in branching and eventually leading to a species having almost twice the molecular weight ($2n - 1$) [61]. In RNA synthesis acidic triethylamine-trihydrogen fluoride (TEA-3HF) is commonly used for the removal of the 2′-*O* silyl protecting group. This acidic reagent reacts during desilylation with the amidate group between the exocyclic amino and the first nucleoside after the branching point, cleaving the chain and resulting in the formation of an $n - 1$ from the 3′-end carrying a 3′-phosphate group [62]. A capping step prior to the start of the synthesis would cap any free exocyclic amino moiety and thereby prevent 3′-terminal branching. However, any free 5′-hydroxyl group created by loss of the DMT group would be capped as well, resulting in yield losses.

2.3.2.3 Generation of $n - 1$ (PS)/$n - 1$ (PO) Impurities

$n - 1$ impurities containing a 3′-terminal PS/PO group are formed when any unprotected functional group left on the support is able to react with a nucleoside phosphoramidite. When using a nucleoside loaded support, this would lead to an oligonucleotide missing the 3′-base, (called $n - 1$) from the 3′-end, since the 3′-nucleoside is part of the support. The $n - 1$ would be connected to the support via a phosphodiester (PO, oxidation) or phosphorothioate bond (PS, sulfurisation). The cleavage of such a bond will typically result in an $n - 1$ impurity having a 3′-terminal phosphate or 3′-terminal phosphorothioate (3′-TPT) [63].

These impurities cannot easily be removed and therefore it is important that all functional groups remaining on the support are capped. Amino or hydroxyl groups of polystyrene supports usually react well with regular capping reagents. However, unreacted surface silanol groups of CPG are more difficult to cap and cannot be blocked with standard capping reagents. Silanol groups can be classified as 'free' or 'associated' [64]. The associated form consists of two adjacent silanols with hydrogen bound to each other (vicinal) or a silanol with two hydroxyls located on the same atom (geminal). Associated silanols are reactive to trimethylchlorosilane (TMCS), while free silanols are more reactive towards hexamethyldisilazane (HMDS) [65]. Repetitious silanisation (monolayer conditions) or single silanisation, followed by a dual silanol capping treatment with TMCS and HMDS can provide complete coverage of all silanol groups. The $n - 1$ impurity from the 3′-end used to be a common side product in oligonucleotide synthesis using CPG supports, because of the difficulties in capping free silanols. As standard capping reagents are unable to block free silanols, alternative means of capping using phosphorylation reagents, such as diethoxy *N,N*-diisopropyl [66] or bis-2-cyanoethyl-*N,N*-diisopropyl phosphoramidite, followed by oxidation [67] have been suggested.

Loss of capping of the support can also occur during storage. Capping loss during storage or incompletely capped support will have the same result; $3'(n-1)$ impurities in the final oligonucleotide. These $3'(n-1)$ impurities often contain a 3′-terminal thiophosphate or phosphate group. A capping step prior to the first synthesis cycle would prevent $n-1$ (3′-end) formation. While a simple capping step using acetic anhydride would efficiently cap free exocyclic amino groups of the heterocyclic base as mentioned above, it would be ineffective in capping free silanol groups of the CPG support. Also, any 5′-*O*-DMT group lost during storage would be capped as well, resulting in diminished yields.

$n-1$ impurities with 3′-terminal PS/PO are also formed when acid labile nucleosides are loaded onto the solid support. For example, deoxyadenosine and, to a lesser extent, deoxyguanosine supports are prone to depurination at prolonged acid treatments during synthesis resulting in n − 1 impurity (see process related impurity Section 2.3.1) [68–70]. The amount of *N*-glycosidic bond cleavage is highly dependent on the nature of the base protecting group. Classical protecting groups like benzoyl for adenosine and iso-butyryl for guanosine are more sensitive to depurination than fast deprotecting groups such as alkyl formamidine or phenoxyacetyl and p-substituted phenoxyacetyl groups, with formamidine being the least sensitive. Linking adenosine via the exocyclic amino group to the support has been suggested to minimise the $n-1$ impurity [68]. Then an adenine loss due to depurination would result in the loss of the oligonucleotide chain during synthesis, hence no depurinated impurity would remain in the product.

2.3.2.4 Universal Supports

Solid supports without pre-loaded nucleosides are termed universal support. These supports have several advantages over nucleoside loaded supports. They do not suffer from loss of base protecting groups during storage or preparation. The formation of branchmers, depurinated species, $n-1$ with 3′-terminal PS/PO as described for nucleoside loaded supports (see Sections 2.3.2.2 and 2.3.2.3) can be avoided. Additionally, same universal support can be used for oligodeoxynucleotide and oligoribonucleotide synthesis. Many different universal supports have been developed over the years [71–79]. Today the most widely used universal supports are UnyLinker [80, 81] (Figure 2.12), and Universal A-Support [82–84] (Figure 2.13). Unylinker was developed by Guzaev and Manoharan [80] and its use was further refined by Ravikumar *et al.* [81].

Two different variants of UnyLinker are available. The UnyLinker bicyclic ring system can either be connected to the support via the nitrogen of the maleimide group (type I) or through one of the hydroxyl groups of the diol (type II) (see Figure 2.12). The oligonucleotide chain grows on the other hydroxyl group of the diol. In case of the type II support the nitrogen of the maleimide group is typically protected by a phenyl group [81]. Glen Research also offers a methyl-protected derivative called

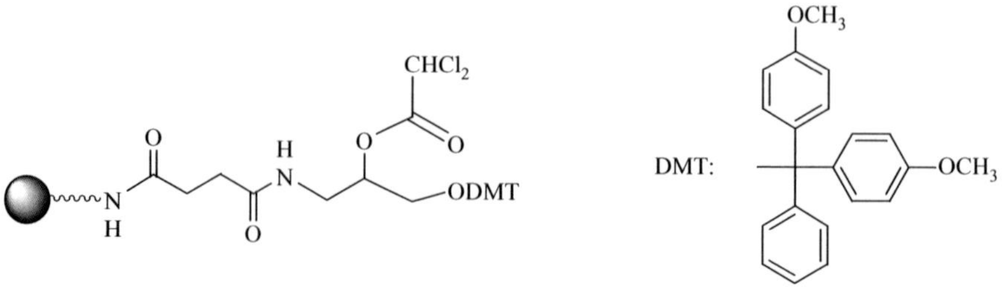

Universal A-Support

Loadings for plystyrene supports: 100 μmol/g, density: 0.2 g/mL

Figure 2.13 Structure of Universal A-Support.

UnySupport [85] for high-throughput oligonucleotide synthesis. During C&D the vicinal hydroxyl group, protected during the synthesis by an acetyl group (type I) or connected to the support via a succinate (type II), is liberated and attacks the phosphodiester (oxidation)/phosphorothioate (sulfurisation) group leading to the elimination of the 3′-OH oligonucleotide [81]. Incomplete cleavage of type II UnyLinker can lead to an impurity where the 3′-end of the oligonucleotide is still attached to the UnyLinker via a phosphodiester/phosphorothioate group (Figure 2.14). This resulting bicyclic impurity can be identified by mass spectrometric analysis as it has an additional mass of 263 (PO) or 279 Da (PS). This type of impurity can be avoided with the type II UnyLinker, in which the nitrogen of the maleimide group is connected to the support via a stable amide bond [86]. Incomplete cleavage of this type of

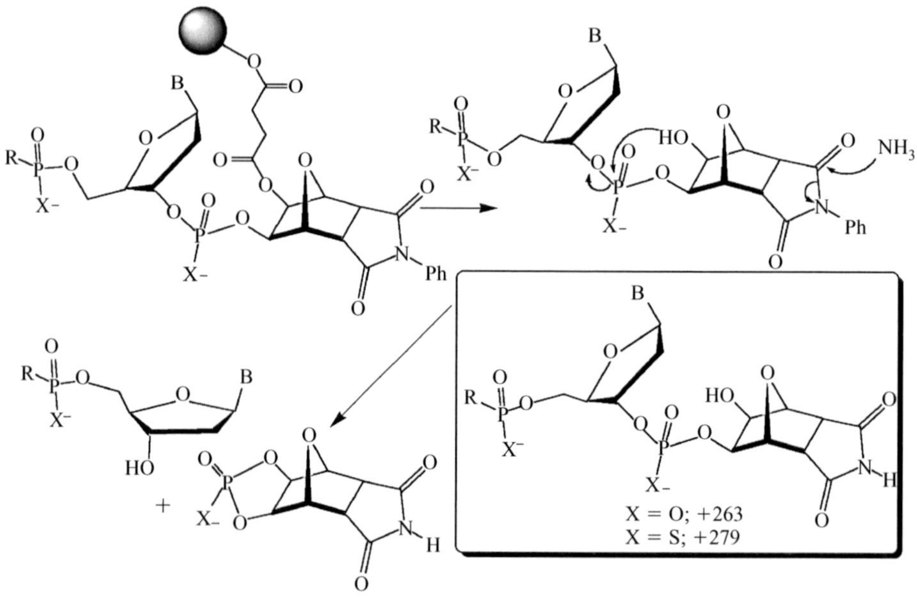

Figure 2.14 Proposed mechanism for formation of bicyclic impurity.

UnyLinker would only lead to some yield loss; however, extending the C&D time would minimise any loss in yield.

The hydroxyl group of the diol system required for oligonucleotide growth is typically protected with a DMT group. As a secondary hydroxyl group, it is less reactive and more sterically hindered than the 5′-hydroxyl group of a sugar. Hence, its removal requires a longer exposure time to detritylation reagent, dichloroacetic acid (DCA), during detritylation. Once the trityl group has been removed, the resultant secondary hydroxyl moiety reacts more slowly to the incoming phosphoramidite. Both of these phenomena can lead to the formation of $n - 1$ from the 3′-end, a common impurity found when using UnyLinker. However, these problems can be mitigated by extending the first detritylation step (or using 10% DCA solution) and extending the first coupling time.

The Universal A-Support was developed by Azhayev and co-workers and is available on both polystyrene and CPG supports [83, 84] (see Figure 2.13). The cleavage and 3′-dephosphorylation reaction is facilitated by a solution of 4 M anhydrous ammonia in methanol. The labile dichloroacetyl group is cleaved prior to the ß-elimination of the cyanoethyl protection group of the phosphate moiety. This is followed by the rapid intramolecular nucleophilic attack on the phosphotriester function by the hydroxyl group. This reaction is additionally assisted by the neighbouring amide function. For a quantitative release of the oligonucleotide from the support it is critical that the dichloroacetyl group is cleaved prior to the cyanoethyl phosphate protecting group closest to the spacer of the support. Once the cyanoethyl group is cleaved, the hydroxyl group is not able to attack the phosphorus since it now represents a negatively charged phosphodiester function and the oligonucleotide remains attached to the support. It therefore is important to use mildly basic conditions for the initial selective removal of the dichloroacetyl group. Universal A-Support is therefore not compatible with the base wash (e.g. diethylamine wash) typically performed in large-scale oligonucleotide manufacturing for the release of the phosphate protecting groups before initiating C&D, in order to minimise the formation of the N^3-cyanoethyl-thymidine (CNET) impurity.

After removing the solid support, further deprotection procedures typically employ addition of aqueous ammonium hydroxide to the solution of oligonucleotide in methanolic ammonia and overnight heating at 55°C. A more easily accessible Universal A-Support was recently introduced where the succinylamido linked support is replaced by an urea linked support [87, 88].

2.3.3 Reagent Related Impurities

2.3.3.1 Detritylation Reagent

Traditionally dilute solutions of trichloroacetic acid (TCA, pKa 0.8) or DCA (pKa 1.5) in dichloromethane (DCM) were used as detritylation reagent for oligonucleotide synthesis [69, 89–91]. Later it was recognised that a high concentration of a weak acid (15% DCA in DCM) is more effective at driving detritylation toward completion than either a low concentration of a weak acid (3% DCA in DCM), or a low concentration of a strong acid such as 3% TCA [89]. The high volatility (boiling point 40°C) of DCM and its high toxicity and carcinogenicity pose a health hazard for personnel and the environment in large-scale manufacturing of oligonucleotides. As a consequence, in large-scale synthesis, DCM has been replaced with toluene, without compromising yield or quality of the synthesised oligonucleotide [92–94]. Again, the amount of DCA in toluene appeared to be more effective at higher concentrations [95]. However, 3% DCA in toluene is still widely used in the industry as higher concentrations of DCA may be more corrosive to the synthesiser and its components, which are typically made of stainless steel.

Increased exposure of oligonucleotides to acids leads to depurination (see Section 2.4.1.2) and therefore the acid contact time needs to be controlled. The detritylation reaction is complex [69, 89–91] and the amount of acid needed depends on type of support used and increases with the oligonucleotide chain length and is therefore not constant. The trityl cation generated during detritylation has an intense yellow/orange colour and it has been used to determine the end-point of the detritylation reaction using spectrophotometric methods. Using this approach, the acid contact time can be minimised/controlled without affecting trityl removal. Incomplete removal of trityl can lead to the formation of $n - 1$ impurities (see Section 2.4.1). Commercial DCA has been found to contain trace amounts of chloral (trichloroacetaldehyde), which can react with free 5′-hydroxyl groups during the detritylation reaction, resulting in an adduct with an added mass of 147 Da [96] (see Section 2.4.1.2. for more details).

2.3.3.2 Activator

No impurities are known to result from impurities present in the activator solutions. Activators work through an acidic-based mechanism involving nucleophilic attack. Activators work not only as an acidic catalyst, thus making the di-alkyl amino group a better leaving group [97], but also replacing the amino group before reacting with the 5′-terminal hydroxyl [98–100]. Please refer to Section 2.4.2 for more details on activators.

2.3.3.3 Thiolation/Oxidation

After coupling, the P(III) triester linkage formed is unstable towards acidic reagent; and is therefore converted to a stable P(V) triester linkage before further chain elongation. The most prevalent oxidation reagent used in oligonucleotide synthesis is a water-containing iodine solution. The composition of the oxidation reagent used in large-scale manufacturing is typically 50 mM iodine in 90% pyridine and 10% water [101]. Concentrations of iodine used in oligonucleotide synthesis may vary from 20 to 100 mM and oxidation reagent for small-scale synthesisers typically includes tetrahydrofuran (THF) [16] as a solvent. Sometimes lutidine is used in place of pyridine [102]. THF is not compatible with some of the components of the pumps and valves of large-scale oligonucleotide synthesisers, prohibiting the use of THF in large-scale synthesis. Lower concentrations of iodine are used for oxidation in case of sensitive base or backbone modifications. However, some modifications have been found to be unstable to oxidation with iodine-containing solutions, regardless of iodine concentration. Also, in phosphodiester (PO), phosphorothioate (PS) gapmers iodine tends to convert some of the PS bonds into the more chemically stable PO bonds, resulting in an oligonucleotide with a mass of 16 Da less than the desired product.

As the coupling step in oligonucleotide synthesis is very moisture sensitive, a non-aqueous oxidation reagent would be preferred to maintain a water-free environment during each step of the synthesis cycle. Low-water iodine mixtures have been suggested for the synthesis of methylphosphonates to avoid side product formation [103]. One non-aqueous oxidiser which has been used for both DNA and RNA synthesis is t-butyl hydroperoxide [104, 105]. Peroxides, however, are unstable compounds and not suited for large-scale manufacturing. A very mild oxidation reagent useful for the synthesis of modified oligonucleotides has been recommended, which is 0.1–0.5M (1S)-(+)-(10-camphorsulfonyl) oxaziridine in anhydrous acetonitrile (ACN) [106–109]. Recently, several other non-aqueous and non-iodine oxidation reagents have been described [67, 110–113]. Besides the disadvantages mentioned, iodine remains the oxidation reagent of choice for oligonucleotide synthesis, as other non-aqueous reagents are either expensive or unstable and therefore not suited and/or economical for large-scale synthesis.

During oligonucleotide synthesis by the phosphoramidite approach, a phosphorothioate is introduced by oxidising the phosphite triester using a sulfur transfer reagent. The first described sulfur reagent was elemental sulfur [114]. However, elemental sulfur is difficult to keep fully dissolved and is therefore not suitable for automated synthesisers. With phosphorothioate oligonucleotides being one of the most important modifications, a large number of efficient sulfurising reagents have been developed over the years. These include tetraethylthiuram disulfide (TETD) [115], bis(*O,O*-diisopropoxy phosphinothioyl) disulfide (Stec's reagent) [116], 3H-1,2-benzodithiol-3-one-1,1,-dioxide (Beaucage reagent) [117], phenylacetyl disulfide (PADS) [118–120], 3-ethoxy-1,2,4-dithiazoline-5-one (EDITH) [121–123], 1,2-dithiazole-5-thione (xanthane hydride or ADTT) [124–126], 3-((dimethylamino-methylidene) amino)-3*H*-1,2,4-dithiazole-3-thione (DDTT) [127, 128], dimethylthiuram disulfide (DTD) [129, 130], 3-phenyl-1,2,4-dithiazoline-5-one (PolyOrg Sulfa or POS) [131] and various sulfonic acid derivatives [132].

Initially TETD was the most prevalent sulfurisation agent. Later TETD was replaced by Beaucage reagent owing to TETD's slow reactivity, in particular for RNA synthesis. While Beaucage reagent was more efficient with faster reaction kinetics, it was not very suitable for large-scale applications owing to its limited stability in solution and also to formation of phosphate diester impurity during thiolation. PADS and xanthane hydride were found to be inexpensive, efficient and fast-reacting sulfurising reagents. Therefore, these two sulfurising reagents prevail in large-scale manufacturing today. Reagents like DDTT and POS offer better stability and/or solubility and are currently being used in small-scale synthesis. Please refer to Section 2.4.3 for more details on sulfurisation reagents.

2.3.3.4 Capping Reagent

The capping reaction is a key step in oligonucleotide synthesis and serves in preventing unwanted chain elongation. Hence, control of the capping conditions is vital to limit impurity levels. Acetic anhydride is the most common capping reagent. Historically it was used as a mixture with 2,6-lutidine, THF and 4-*N*,*N*-dimethylamino pyridine (DMAP), which acts as an acetylation catalyst [16]. Recent studies [133] have shown that DMAP converts dG and dG-containing oligomers into fluorescent derivatives and in the presence of ammonium hydroxide convert further to 2,6 diaminopurine deoxyribonucleoside (2,6-DAP), a potentially mutagenic nucleoside analogue. *N*-Methylimidazole (NMI) has been suggested as an alternative to DMAP, which does not undergo the same side-reactions [133]. Today NMI has replaced DMAP as capping catalyst in routine oligonucleotide synthesis.

The protection of guanosine by the phenoxyacetyl group precludes the use of acetic anhydride as the capping agent. An exchange reaction occurs between the guanine phenoxyacetyl group and the acetyl group of acetic anhydride during the capping reaction. This impurity is quite stable towards standard C&D and can be detected by HPLC-MS and results in a mass of +42 Da. To avoid formation of this adduct, acetic anhydride can be replaced by phenoxyacetic anhydride [134] or trimethylacetic anhydride [135]. A similar adduct formation was observed for iso-butyryl-protected guanosine when using acetic anhydride as capping reagent. Isobutyric anhydride was suggested to replace acetic anhydride for capping [136, 137].

During synthesis of phosphorothioate oligonucleotides capping needs to be performed after sulfurisation. If this reaction occurs after capping, higher levels of the phosphodiester impurity can be observed [115]. For phosphodiester oligonucleotide synthesis, capping may be performed before or after oxidation.

Acetic anhydride cannot be combined with the acetylation catalyst prior to capping. Therefore, two capping reagents (Cap A and Cap B) are delivered during the synthesis from two different bottle positions, which are being mixed inline. One of the capping reagents contains NMI (or DMAP) the other contains acetic anhydride. In addition, one of either reagent contains a base, either pyridine or lutidine. THF has been replaced with ACN in large-scale synthesis. However, neither NMI nor acetic anhydride is stable in combination with base in the presence of ACN. Therefore, in large-scale synthesis three capping reagents are used: Cap A, containing 20% NMI in ACN; Cap B1 containing 40% acetic anhydride in ACN and Cap B2 containing 60% lutidine in ACN. Typically Cap B1 and Cap B2 are mixed together prior to use.

Capping efficiency is another important factor and, if uncontrolled, can lead to low yields and increased impurity levels. The efficiency of capping is dependent on the nature and amount of the activating agent, e.g. NMI or DMAP [138] in the capping mixture (see Table 2.2). DMAP appears to be more effective as a capping activator relative to NMI, although it has fallen out of favour owing to its tendency to lead to guanosine modifications (vide supra). To overcome these, Glen Research has introduced a phosphoramidite-based capping reagent called UniCap (see Figure 2.15), which is also well suited for H-phosphonate chemistry.

The activating agent, NMI, has been shown to contain low levels of trimethylhexahydrotriazine, which can react with nucleotides to form a +85 Da impurity [139]. The mechanism of formation of this impurity is thought to involve a Mannich type reaction between the enolate anion of acetic anhydride and the imine formed from trimethylhexahydrotriazine (Figure 2.16).

Table 2.2 Comparison of capping mixtures (Glen Research).

Synthesiser	Cap A solution	Cap B solution	Capping efficiency (%)
Expedite	THF/Ac_2O (9:1)	10% NMI/THF/Py (8:1)	90.5 ±1.9
ABI 394	$THF/Pyr/Ac_2O$ (8:1:1)	10% NMI/THF	88.8 ± 2.5
ABI 394	$THF/Lut/Ac_2O$ (8:1:1)	10% NMI/THF	89.1 ± 2.0
ABI 394	$THF/Lut/Ac_2O$ (8:1:1)	16% NMI/THF	96.6 ± 1.4
ABI 394	$THF/Lut/Ac_2O$ (8:1:1)	6.5% DMAP/THF	99.4 ± 0.3
All	UniCap	Not applicable	98.6 ± 0.4

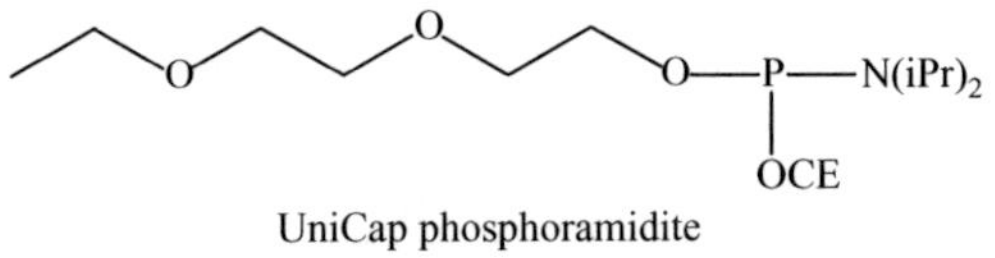

Figure 2.15 Structure of phosphoramidite capping reagent, UniCap.

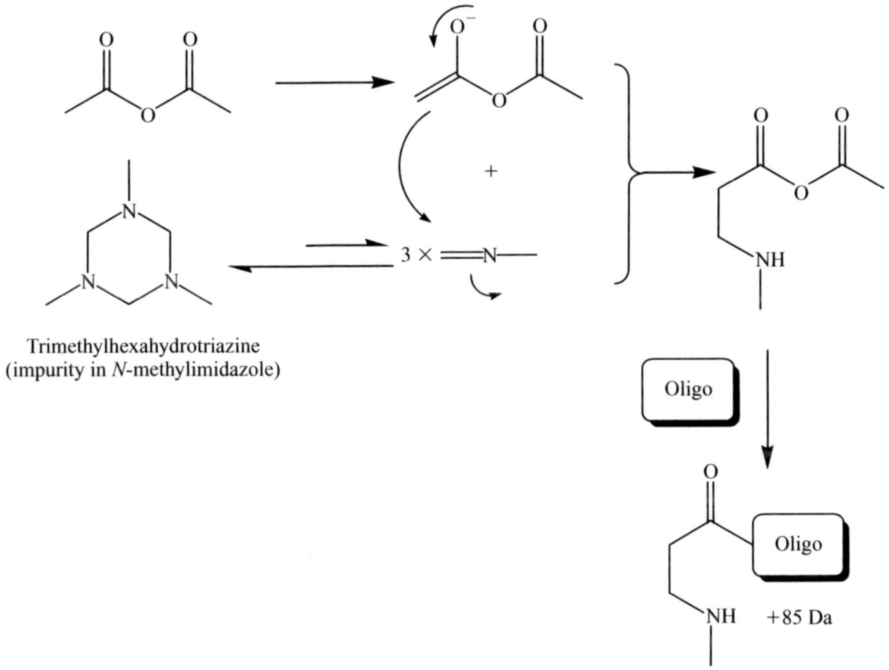

Figure 2.16 Postulated mechanism for the generation of NMI-based +85 Da impurity.

As discussed here and in Section 2.4.4, capping based on derivatives of acetic anhydride appears to be more complicated and gives rise to more impurities than would be expected. Therefore, capping using a terminating phosphoramidite like UniCap (Figure 2.15) was suggested not only for the capping of the support, as discussed in Section 2.3.2.3 [66, 67], but also for capping after each coupling step [138].

2.4 Process Related Impurities

2.4.1 Detritylation Step

2.4.1.1 Chloral Adduct Formation

It has been described in Section 2.3.3.1 that commercial DCA solution contains trichloroacetal, which may form adducts in the oligonucleotides chain. The combination of HPLC-MS, HPLC, synthesis of model compounds and NMR techniques led to the elucidation of a full-length product (FLP)+147 adduct associated with the incorporation of chloral (trichloroacetaldehyde) [96]. The product of the reaction of chloral with the 5′-hydroxyl is a hemiacetal which may further react with an incoming phosphoramidite through standard activation chemistry to provide structures of the type shown in Figure 2.17.

The link between chloral and the observed impurities was demonstrated unequivocally by synthesising the target oligonucleotide in the presence of 3% DCA in toluene containing 300 ppm of chloral hydrate, resulting in approximately 2% of the FLP+147 impurity for a phosphorothioate oligonucleotide. Reducing the chloral

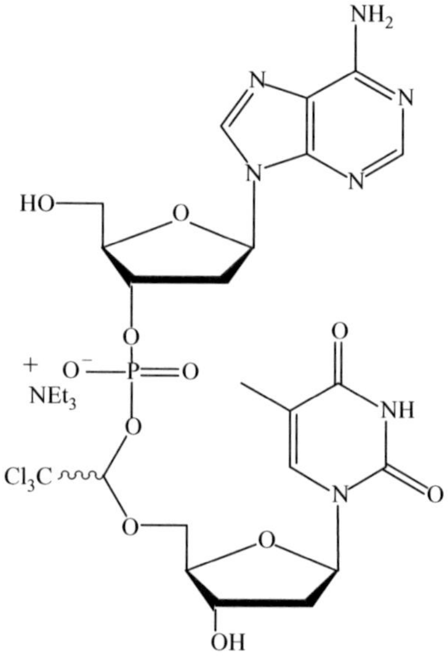

Figure 2.17 Chloral-modified d(ApT).

content to 100 ppm resulted in a decrease of this impurity to less than 1%. The most effective means for eliminating this impurity is by simple distillation of DCA, use of higher percentage DCA (10% DCA/toluene for example), to minimise contact time, or setting a specification to limit the amount of chloral found in DCA. However, an accurate quantification of chloral in DCA at such low levels remains challenging.

2.4.1.2 Depurination

The cleavage of the glycosidic bond connecting the sugar to the nucleosidic base in deoxyadenosine and deoxyguanosine is referred to as depurination. Because depurination is related to acid exposure [140, 141], very long sequences [142] or those having deoxyadenosine and, to a lesser extent, deoxyguanosine, at the 3′-end are particularly susceptible to undergoing depurination.

This phenomenon results in the excision of the purine base from the oligonucleotide and subsequent replacement with water to form the hydrolysis product. As shown in Figure 2.18, protonation at the N-7 position in adenine (purine) converts the base into a good leaving group. The expulsion of the nucleoside base is facilitated by the

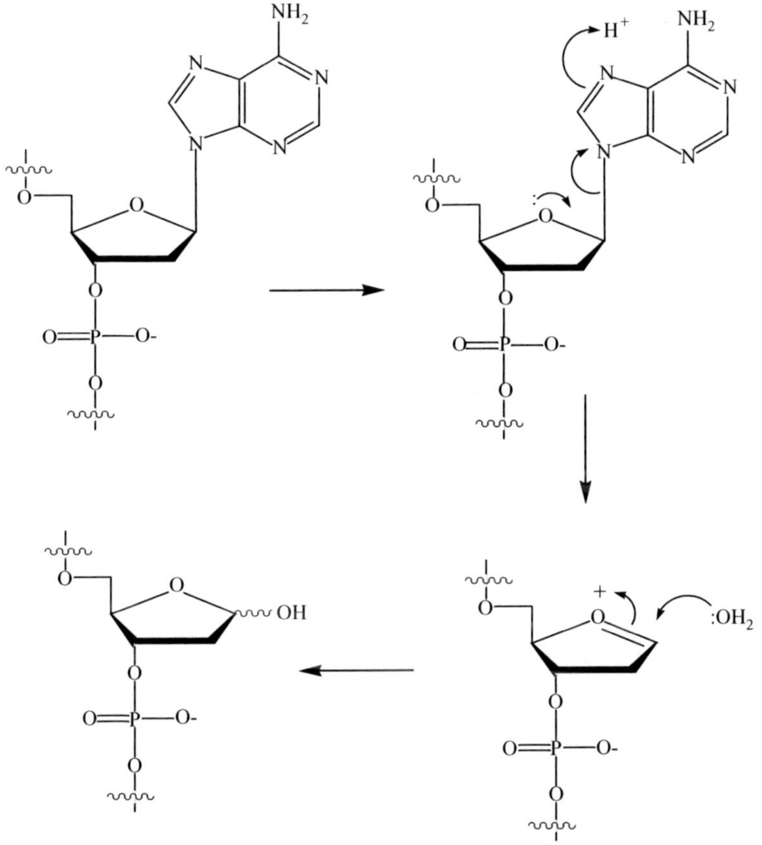

Figure 2.18 Postulated mechanism of acid-induced depurination.

anomeric oxygen atom of the deoxyribose. Adventitious water is responsible for quenching the oxonium ion to provide the cyclic acetal.

The glycosidic bond is known to be destabilised by electron-withdrawing groups at N-6 [143]. To counter this effect, the usual acyl-type protecting groups may be replaced by the electron-donating amidine (formamide) derivatives to provide enhanced stability toward depurination. The t-butoxyphenylacetyl protecting group has also been shown to curb depurination of 3′-terminal dA relative to its N-dmf and N-Bz analogues [144]. Varying the type and concentration of acid also displays a marked effect on depurination kinetics. Septak reported a kinetic study on the relationship between depurination and acid concentration and found that depurination can be minimised by employing high-concentration acid (15% DCA) relative to low-concentration acid (3%) [69]. The process of detritylation presents a complicated picture, which includes a binding equilibrium of acid/oligonucleotide and acid/ACN in addition to the detritylation reaction. It follows that the optimal detritylation approach is to use the deblock reagent that most effectively minimises such lag time while supporting rapid detritylation, namely 15% DCA.

The depurination as an artefact is also observed in the electrospray ionisation in the HPLC-MS instrument [145, 146]. This type of depurination is caused by collisions between analyte ions and the background gas. The level of depurination in the gas phase is proportional to the voltage used to initiate ionisation. As this species is generated in the absence of solvent, the water addition step observed under the acid step (as in real depurination) does not occur, hence the mass of dupurination fragments is typically lower by 18 Da. Figure 2.19 shows a representative depurination reaction scheme of deoxyadenosine under acid- and instrument-induced conditions, which can be detected as molecular ions M-117 Da and M-135 Da respectively. It should also be noted that a potential complication may arise in analysis of oligonucleotide sequences containing both adenosine and guanosine. Because of the similarity in molecular weight between adenosine and guanosine, instrument-induced depurination of adenosine (−135.13 Da) and chemical depurination of guanosine (−133.11 Da) fall within the mass accuracy of many HPLC-MS instruments (~2 Da).

Ring opening of the furanose sugar ring under acidic conditions exposes a reactive aldehyde which may undergo β-elimination, resulting in fragmentation to give a 3′-phosphate fragment. Elimination at the γ-position releases the 3′-phosphate fragment as shown in Figure 2.20.

Forcing oxidation conditions are reported to generate a ribolactone, which may undergo elimination of the 3′-terminal fragment to give the corresponding α,β-unsaturated ribolactone [147]. The authors suggested a mechanism as proposed in Figure 2.21 for its formation.

2.4.1.3 DMT-C Phosphonate Esters

DMT-C phosphonate esters have been reported to form when inadequate amounts of thiolation/oxidation reagent or insufficient contact time are provided [142] during oxidation. The normal reaction pathway of the coupling product, the P(III) triester **1**, is the oxidation at the phosphorus centre to give the stable product, the P(V) triester **2**, as shown in Figure 2.22. Owing to incomplete thiolation/oxidation, the intermedi-

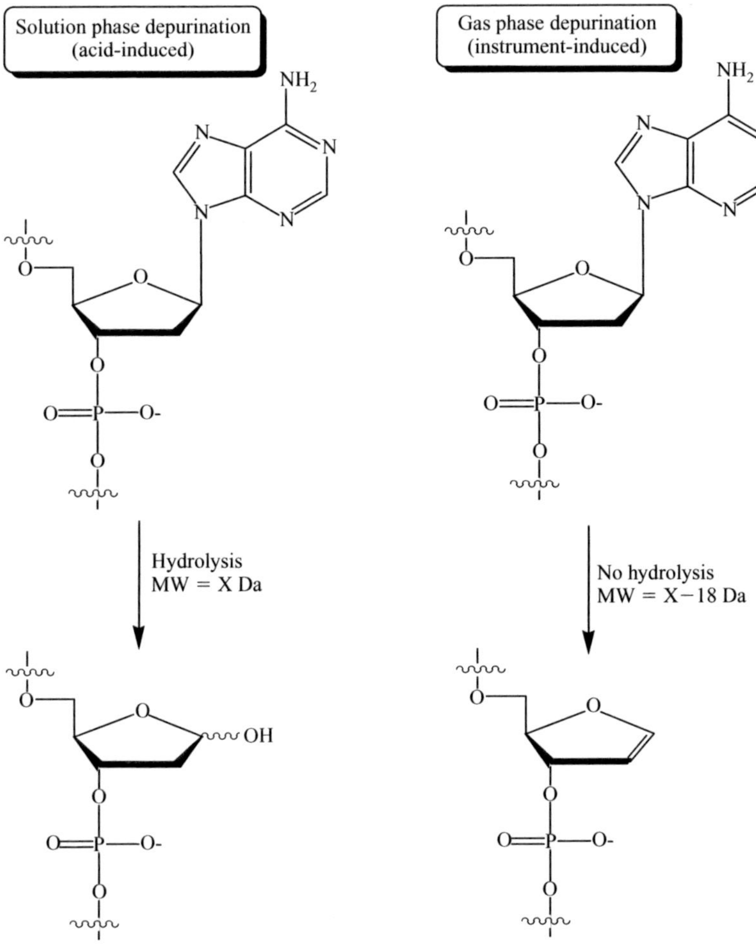

Figure 2.19 Depurination in solution versus gas phase.

ate phosphonium ion **3** [148], however, results from nucleophilic reaction of the reactive phosphite P(III) with the DMT cation liberated from the detritylation step. Decomposition of the phosphonium species to provide the corresponding esters is shown in Figure 2.23. Krotz and co-workers [148] reported the formation of DMT-C-phosphonate monoesters and diesters when only 1.7 equivalent of PADS was used to carry out thiolation. The products, which elute at slightly longer retention times in RP ion pairing (IP)-HPLC, were identified through the use of HPLC-MS.

The intermediate DMT-C phosphonium ion 3 represented in Figure 2.23 can undergo three separate bond scissions to follow Path A, to form a $n-1$ with a terminal DMT C-phosphonate as a result of the nucleoside cleaved at the 5′-end; to follow Path B to provide the $n-1$ product as a result of 3′-cleavage; and finally, to follow Path C, to result into an Arbuzov-type elimination of the cyanoethyl protecting group to form the full-length DMT-C phosphonate.

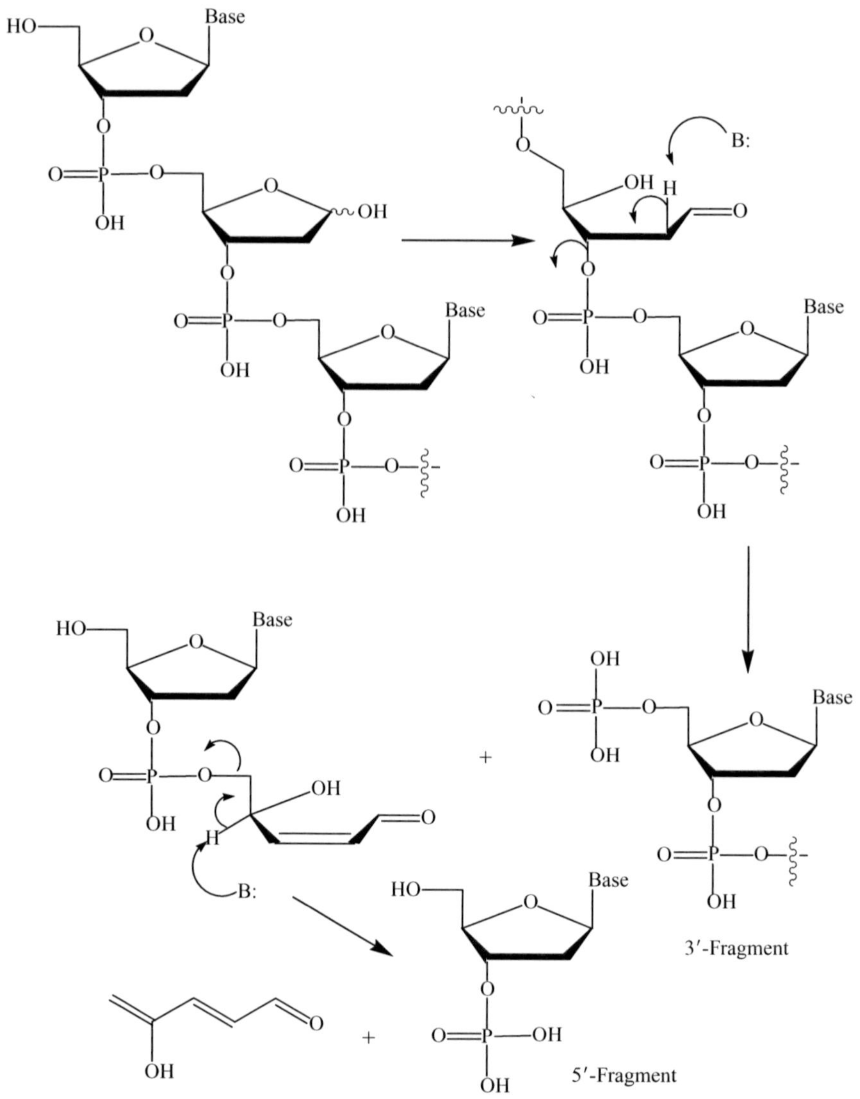

Figure 2.20 Fragmentation via β-elimination following depurination.

2.4.2 Coupling Step

2.4.2.1 $n - x$

The most commonly encountered impurity (and one of the more difficult to remove by purification) is the $n - 1$ impurity. It is also known as a shortmer, deletion or failure sequence. The $n - 1$ represents a class of impurities, rather than a specific impurity,

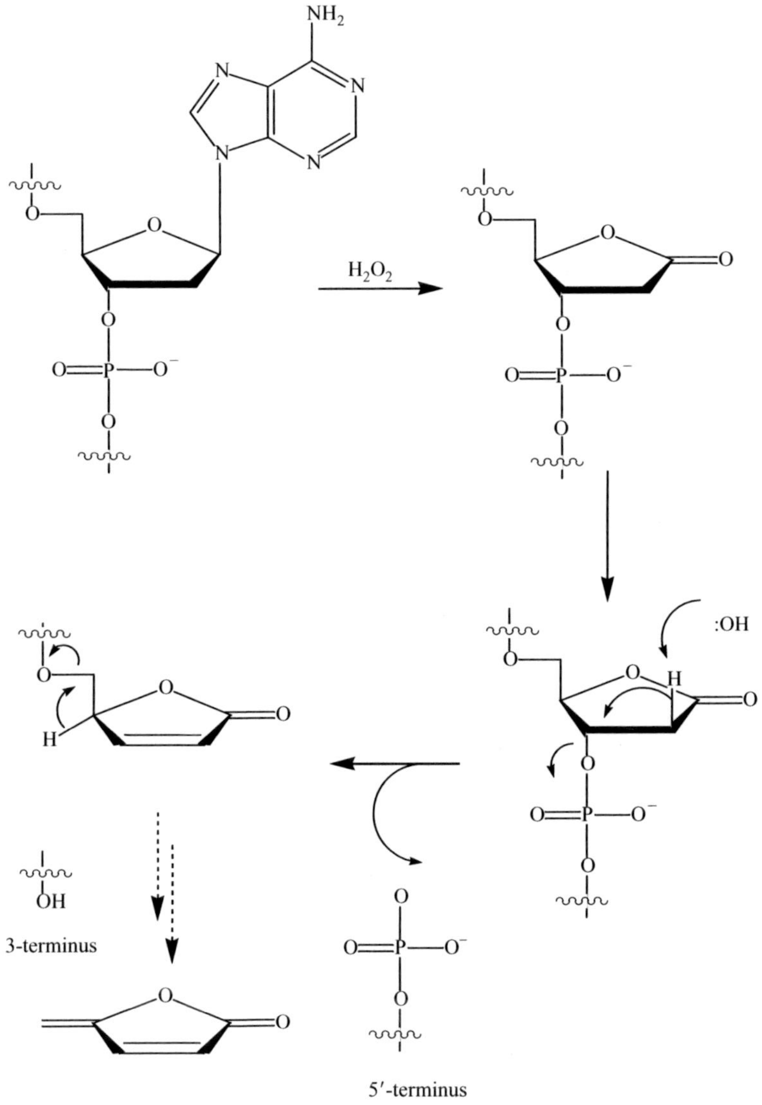

Figure 2.21 Degradation under oxidising conditions.

because it is shown to comprise all possible $n-1$ combinations. As many of the early oligonucleotide analyses were carried out using PAGE, which is more sensitive to size, rather than the nature of the sequence, the $n-1$ failures represent a single band in the gel. Agrawal and co-workers [149] elegantly proved the $n-1$ to be a heterogeneous mixture by excising the $n-1$ band from the gel of a 25-mer DNA phosphodiester synthesis product, tagging the $n-1$ band with dA tail and annealing it to a complementary plasmid. By transforming the plasmid in competent bacteria and

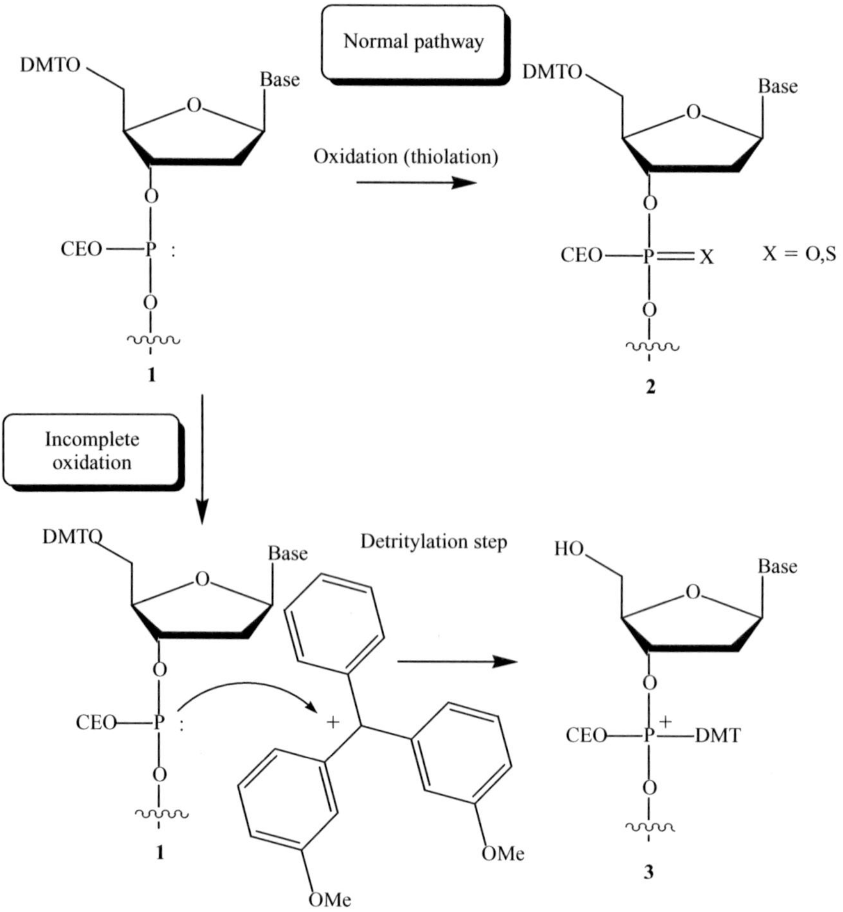

Figure 2.22 Formation of DMT-C phosphonate as a result of improper sulfurisation.

sequencing the expressed clones, they showed that the $n - 1$ deletion sequence was distributed among 70 different clones. Moreover, the highest frequency of clones was characterised by deletion at the 3′-terminus of the oligonucleotide.

Fearon and co-workers [150] utilised MS analysis of the PAGE-purified $n - 1$ deletion band to reveal the $n - 1$ impurity as a composite of random $n - 1$ failure sequences. The deletion sequences are thought to originate from low-frequency defects at each stage of the synthesis cycle – incomplete detritylation, coupling, oxidation (or sulfurisation) and capping, or a combination thereof. For example, a specific $n - 1$ can arise from a coupling failure followed by a failure to cap the free 5′-hydroxyl. Figure 2.24 outlines the origin of the $n - 1$ deletion sequences. The paragraphs below further provide a description of the impact on each synthetic step on the origin of the $n - 1$ impurities.

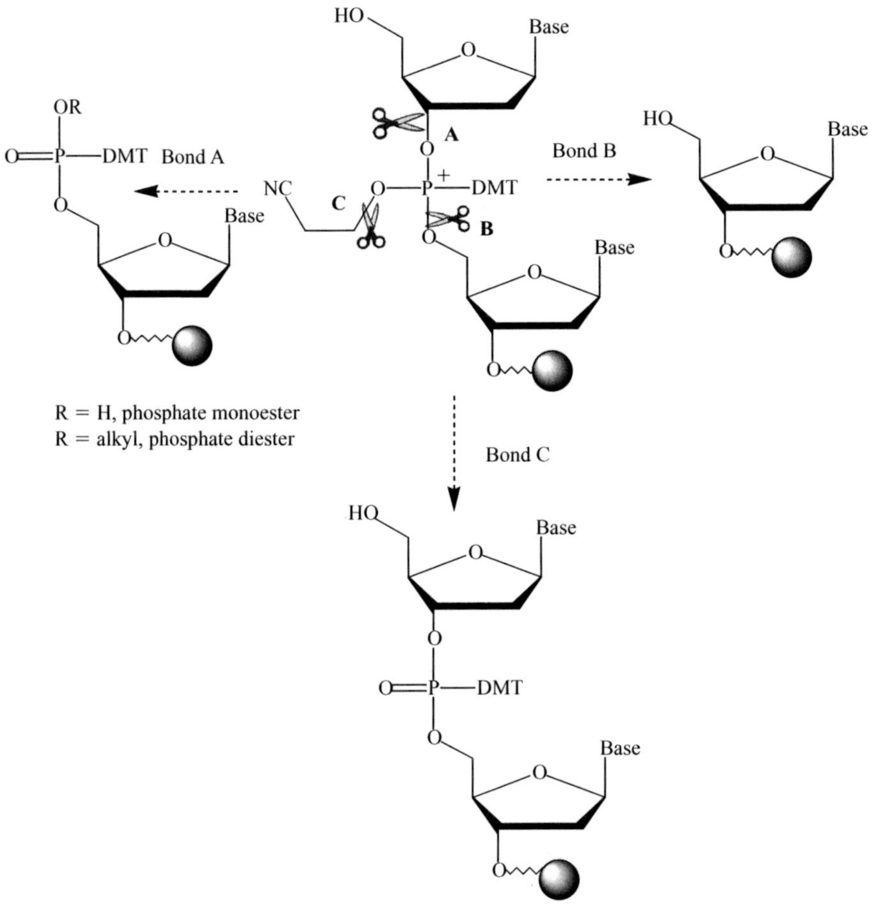

Figure 2.23 Reactive degradation pathways of the intermediate phosphonium species.

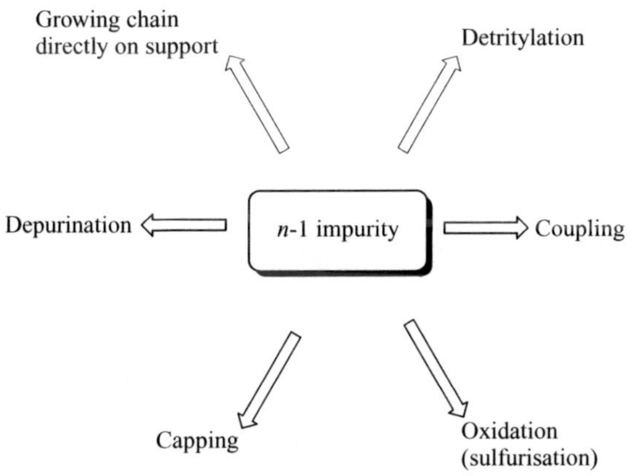

Figure 2.24 Origin of $n - 1$ deletion sequences.

Detritylation-Derived n – 1 Impurities

The synthesis cycle, shown in Figure 2.1, begins with detritylation, in which an acid (typically DCA) is responsible for removing the terminal DMT protecting group to expose the reactive 5′-hydroxyl functionality, which is then coupled with the incoming phosphoramidite nucleoside. An incomplete detritylation results in a mixture of full-length oligonucleotide having an exposed 5′-hydroxyl and its protected counterpart. After the chain elongation, oxidation and capping, the mixture of full-length product and its (now) $n-1$ trityl-on analogue enter the subsequent detritylation cycle. When the mixture is exposed to acid in the subsequent detritylation cycle, the free hydroxyl of the full-length product and its $n-1$ analogue will be liberated, resulting in the propagation of the $n-1$ species through the synthesis.

Coupling-Derived n – 1 Impurities

The coupling efficiency for solid-phase oligonucleotide synthesis is extremely efficient, usually approaching 99%. Despite the high efficiency, the iterative nature of oligonucleotide synthesis highly impacts the overall yield. An oligonucleotide of n bases possesses $n-1$ linkages and the overall synthesis yield expressed as: [coupling efficiency]$^{n-1}$. The percent yield for a synthesis of a 20-mer oligonucleotide at 99% coupling efficiency would then be calculated as $[0.99]^{19}$, or 82.6%. If the efficiency were to fall to 98%, the resulting yield would erode to 68%! The remainder of the product corresponds to shorter failure sequences. Inefficient coupling results in sequences having a free 5′-hydroxyl group, which are normally capped at the end of the synthesis cycle. A failed coupling results in $n-1$ only if the subsequent capping is not effective.

Support-Derived n – 1 Impurities

The $n-1$ may also originate from growing the oligonucleotide to improperly capped support or from detritylation of the first base on the support. These alternative mechanisms are discussed in Section 2.2.2. Chapter 4 of this book describes the use of HPLC-MS in the identification of oligonucleotide impurities.

Capping-Derived n – 1 Impurities

Following coupling, any unreacted 5′-hydroxyl remaining is capped, typically through reaction with acetic anhydride and an amine base, such as pyridine, picoline or lutidine. Ineffective capping leads to an exposed 5′-hydroxyl which can enter the subsequent coupling reaction. The result is the generation of $n-1$ species.

2.4.2.2 *n + x* Formation of High-Molecular-Weight Impurities

Another commonly encountered class of oligonucleotide impurity are the $n+x$ analogues, sometimes called longmers or extension sequences. As the formation of shortmers ($n-1$, $n-2$, etc.) would occur through failure of coupling reactions, longmers [151] ($n+1$, $n+2$) would originate in over-condensation during coupling cycles. Coupling requires the reaction of the free 5′-hydroxyl of the oligonucleotide on solid support with phosphoramidite facilitated by an activator, typically mixed inline and delivered to the column. Most activators are weak acids capable of protonating the phosphorus centre on the incoming phosphoramidite. The diisopropylamino group is then replaced by the conjugate base of the activator in the rate-determining step of the coupling cycle [98, 99]. The activator, in turn, is displaced by the free 5′-hydroxyl group of the growing oligonucleotide chain and chain extension is complete. The acidic nature of the activator may cause partial removal of the DMT protecting group on either the incoming phosphoramidite or the growing chain. The formation of longmers can occur by two possible scenarios. The first possibility is that during coupling the DMT-protecting group on the 5′-terminus is removed to transform the protected 5′-hydroxyl position to a second reactive site capable of undergoing an additional coupling.

The second scenario involves the removal of the DMT-protecting group of an amidite which would render it reactive toward a second equivalent of amidite, forming a dimer in situ, which is ultimately coupled to the growing chain on solid support. Either scenario results in the formation of an n + 1 species. The extent of over-coupling is dependent on the sequence, incoming phosphoramidite (in particular guanosine amidite) and the type of activator. The structures and acidities of commonly used activators are listed in Table 2.3.

Krotz and co-workers have carried out studies on the synthesis of deoxyoligonucleotide sequences of the form $(TTT)_6T$, $(TGG)_6T$, $(TCC)_6T$, and $(TAA)_6T$ promoted by 0.45M *S*-ethylthiotetrazole (ETT) and tetrazole, respectively [151]. They reported that the extent of longmer formation is highly dependent on the composition of the sequence. The order parallels the rate of -DMT removal: dG > dA > dC > T. Also, the quantity of longmers increases significantly when ETT (pKa = 4.28) is used relative to tetrazole (pKa = 4.89). Tetrazole generates a lower amount of longmers; however, owing to its explosive nature, its use has been restricted and highly regulated.

More acidic activators commonly used for RNA synthesis are *S*-ETT and *S*-benzylthio tetrazole. The use of dicyanoimidazole (DCI) [152] (pKa = 5.2) is almost 10 times less acidic than ETT, but possesses even greater nucleophilicity over its tetrazole-based competitors. DCI has a special advantage over tetrazole in the synthesis of long oligonucleotides and at larger scales where minimising monomer equivalency is essential. A pyridinium-based activator (pyridinium trifluoroacetate/*N*-methylimidazole) which shows similar reactivity and comparatively lower percentage of over-condensation products relative to tetrazole has also been developed [153]. Sinha and co-workers have developed a saccharin-based activator for replacement of tetrazole in large-scale manufacturing [154]. The formation n+mers using the saccharin activator are reportedly significantly reduced compared to ETT [154]. Page and co-workers carried out extensive NMR studies [155] toward elucidation of the mechanism of phosphoramidite activation by saccharin *N*-methylimidazole.

Chemists at Sigma-Aldrich have developed Activator 42 [156], which is reported to be one of the most efficient activators on the market. Although it is quite acidic (pKa 3.4), it is believed that Activator 42 is a poor activator of dimers, hence its resistance to form over-condensation products despite its highly acidic nature [157].

Table 2.3 Some commonly used activators and their respective pKa values.

Activator	Structure	pKa	Solubility in ACN [M]	Typical activator concentration	Explosive class
1-*H*-tetrazole	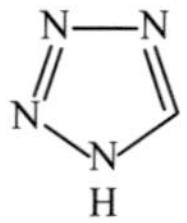	4.89	0.5	0.45M	A
S-ethylthio tetrazole	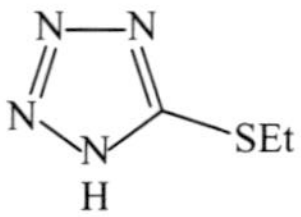	4.28	1.5	0.25M (DNA) 0.6M (RNA)	C
S-benzylthio tetrazole	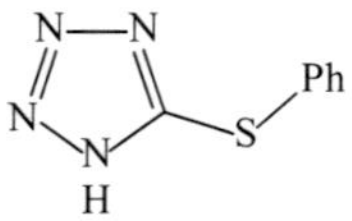	4.08	0.4	0.25–0.3M with or without NMI (RNA)	None
5-(3,5-Bistrifluoromethyl-phenyl)-1H-tetrazole (activator 42)	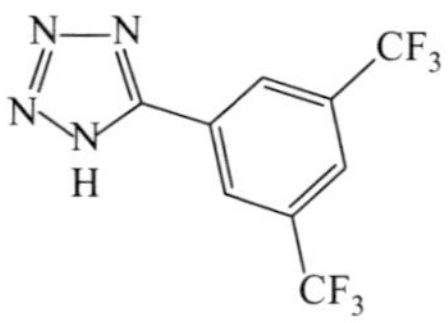	3.4	0.9		None

(*continued*)

Table 2.3 (*continued*)

Activator	Structure	pKa	Solubility in ACN [M]	Typical activator concentration	Explosive class
4,5-dicyanoimidazole	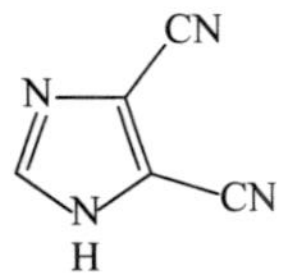	5.28	1.1	0.7M with 0.1M NMI	None
Saccharin 1-methylimazole	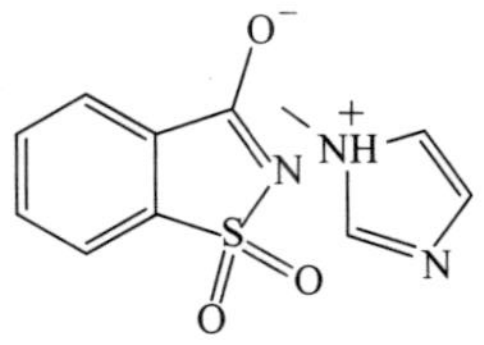	ND	0.25	0.2M	None

2.4.3 Thiolation/Oxidation Step

2.4.3.1 Sulfur Loss (PS to PO Exchange)

The susceptibility of phosphodiester oligonucleotides to nuclease-mediated degradation has spurred the development of a variety of backbone analogues with enhanced resistance to nuclease activity. Among them, the sulfur congener, or phosphorothioate (see Figure 2.25) is popular because of its enhanced stability, ease of synthesis and compatibility with a variety of oligonucleotide chemistries, that is siRNA, antisense DNA, aptamers, etc. Oxidative thiolation following coupling is an extremely efficient reaction.

A number of different reagents have been developed to effect sulfur transfer to the intermediate phosphite P(III) ester following coupling. These reagents are presented with their original references in Table 2.4. The DTD reagent offers the unique advantage of combining thiolation and capping in a single operation. PADS and xanthane hydride have been most extensively employed in large-scale manufacturing. The sulfur transfer reaction is extremely efficient; however, no chemical reaction is quantitative and the sulfurisation of phosphite esters is no exception.

Ineffective sulfurisation generally results in two families of impurities. The first and most prevalent is the phosphate diester, which originates from oxygen transfer to the phosphorus centre. The second class of impurities stems from nucleophilic reaction of the trityl cation with the phosphite triester; this sub-class is described in Section 2.3.1.3. The third class of impurities results from H-phosphonate formation followed by base catalysed strand cleavage during deprotection (see Section 2.4.5.6).

The Beaucage reagent [117], one of the earlier sulfur transfer reagents, is known to form a mixture of phosphorothioate and phosphate diesters because the byproduct of the Beaucage reagent (formed during sulfurisation), acts as a good oxygen donor, as shown in Figure 2.27.

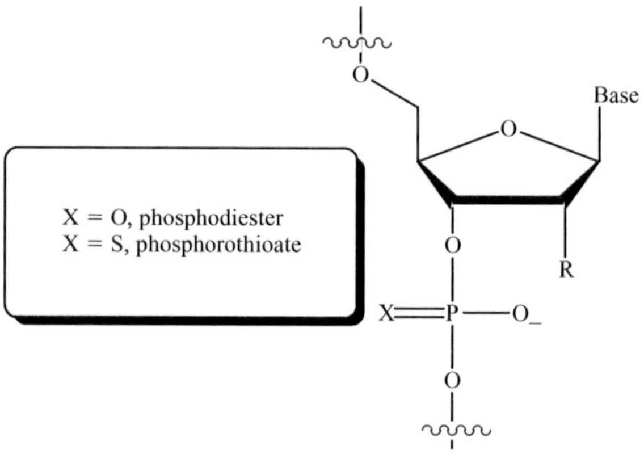

Figure 2.25 Structural comparison of phosphothioate and phosphodiester.

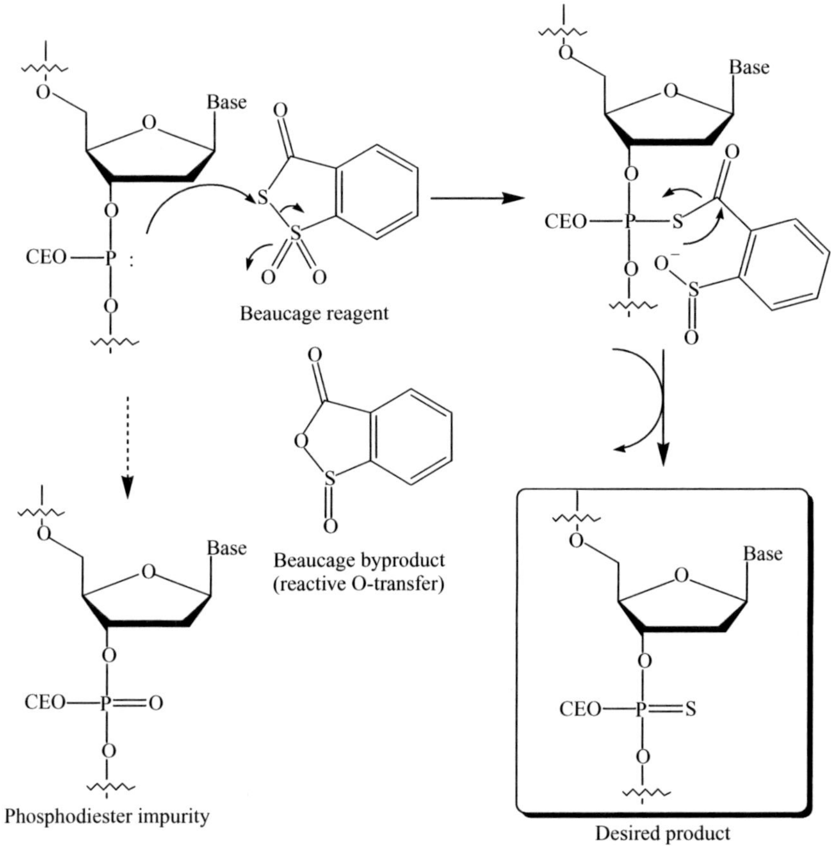

Figure 2.26 Proposed mechanism of sulfur transfer reaction using Beaucage reagent.

The high cost of the Beaucage reagent, coupled with its instability and tendency to form phosphodiester impurities (see Figure 2.27), prompted the investigation for low-cost, scalable sulfur transfer reagents (see Table 2.4).

One of the most commonly used reagents is PADS, originally introduced by van Boom and co-workers [120]. The PADS reagent is amenable to scale-up and is extremely efficient, often producing over 99.5% efficiency per linkage. A thorough study of the critical parameters found that the time between reagent preparation and its use is critical. Optimal thiolations were conducted after ageing a 0.2 M solution of PADS in a 1:1 mixture of ACN/3-picoline for 24–30 h [118].

2.4.3.2 Loss of Sulfur in Mixed PS/PO Backbones

It is common practice to enhance the nuclease stability of therapeutic oligonucleotides by flanking a central PO backbone with several phosphorothioate linkages at both the 5′- and 3′-termini. Introduction of a phosphorothioate linkage at the 3′-end increases

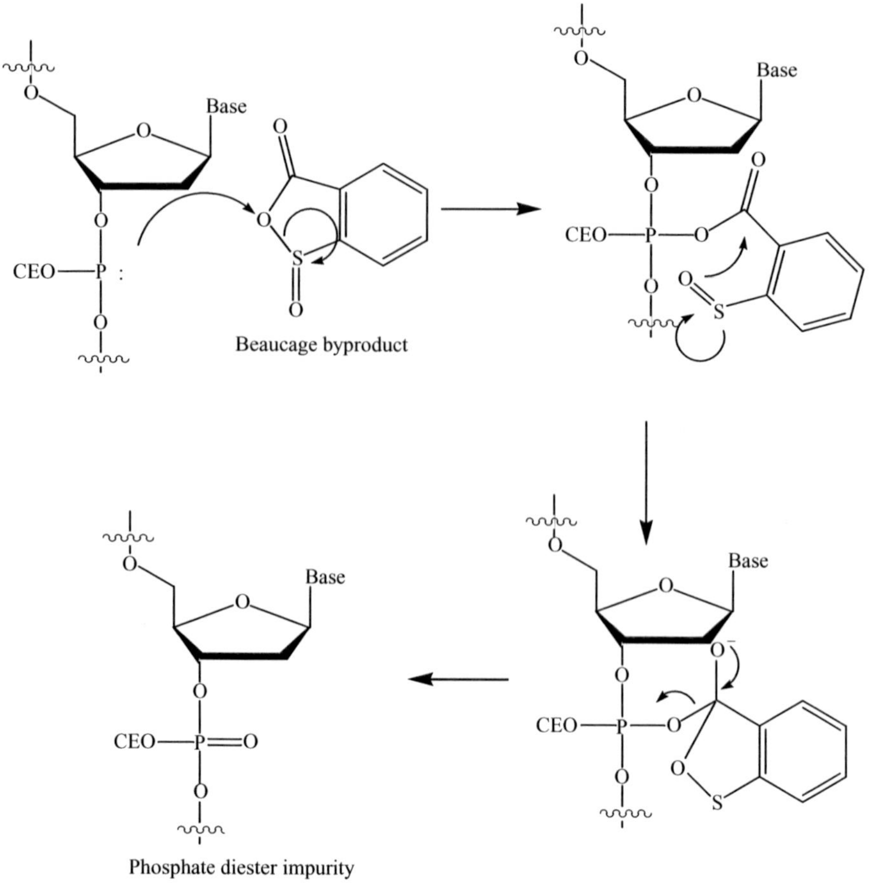

Figure 2.27 Proposed mechanism for undesired oxygen transfer using Beaucage reagent.

the risk of partial oxidation of the PS linkage to its more stable PO congener through successive oxidation steps in I_2, water and pyridine.

2.4.4 Capping Step

2.4.4.1 Capping After Premature Trityl Group Loss

The oxidation reagent typically used in oligonucleotide synthesis is composed of a 0.05M iodine solution in 90% pyridine in water. In aqueous solution, however, elemental iodine is in equilibrium with hydriodic acid, which can lead to partial detritylation of the product. Guanosine is especially prone to detritylation (see Section 3.2.2) [151]. When the standard coupling cycle is employed (detritylation, coupling, oxidation and capping), partial detritylation in the oxidation step exposes the free 5′-hydroxyl, which is subsequently capped. This leads to losses in the overall yield.

Table 2.4 Commonly used sulfurisation reagents.

Common name or acronym	Chemical name	Structure	Commonly used concentrations	Reference
TETD	Tetraethylthiuram disulfide	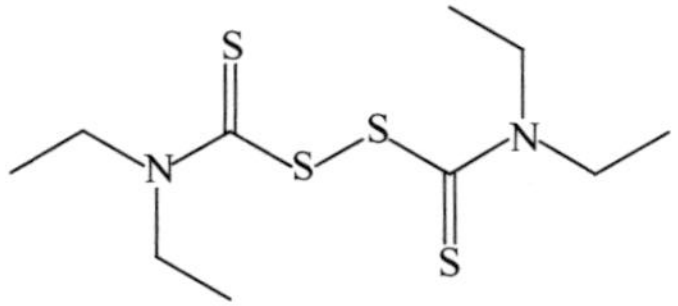	0.5M in ACN	[115]
Stec	Bis(*O*,*O*-diisopropoxy phosphinothioyl) disulfide	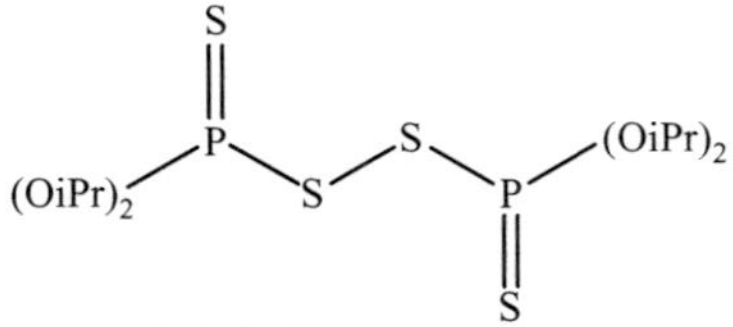	0.05M in ACN	[116]
Beaucage	3H-1,2-benzodithiol-3-one-1,1,-dioxide	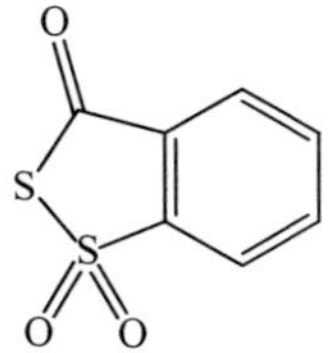	0.2M in ACN (limited long-term stability)	[117]
PADS	Phenylacetyl disulfide	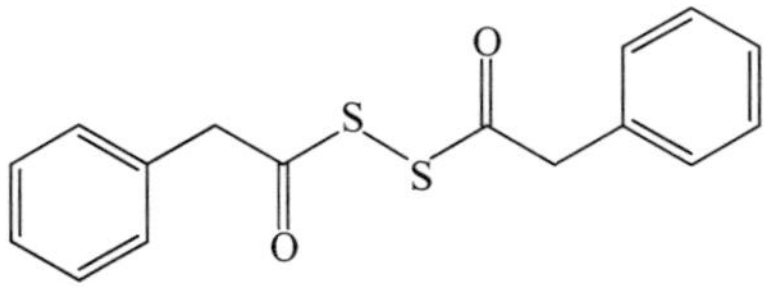	0.2M in 1:1: 3-picoline/ACN (limited long-term stability)	[118–120, 158, 159]

(*continued*)

Table 2.4 (*continued*)

Common name or acronym	Chemical name	Structure	Commonly used concentrations	Reference
EDITH	3-ethoxy-1,2,4-dithiazoline-5-one	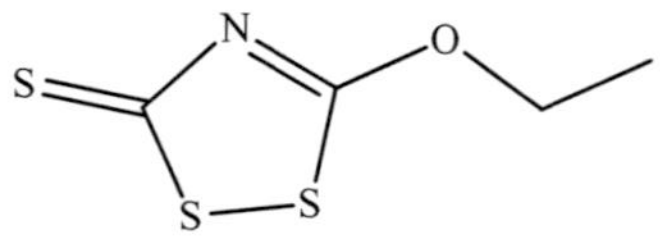	0.05M–0.1M in ACN	[121–123]
Xanthane hydride (ADTT)	1,2-dithiazole-5-thione	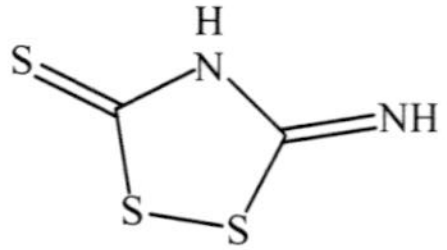	0.01M in ACN or 0.02M ACN-pyridine (9:1), 0.2M in pyridine	[124–126]
DDTT	3-((dimethylamino-methylidene)amino)-3*H*-1,2,4-dithiazole-3-thione	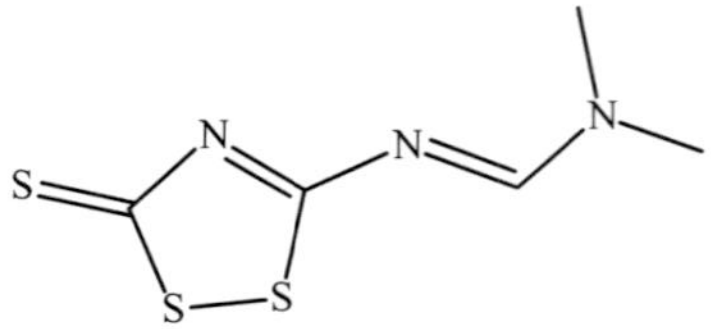	0.05M in ACN/ pyridine (60:40), 0.2M in pyridine	[127, 128]

(*continued*)

Table 2.4 (*continued*)

Common name or acronym	Chemical name	Structure	Commonly used concentrations	Reference
DTD	dimethylthiuram disulfide	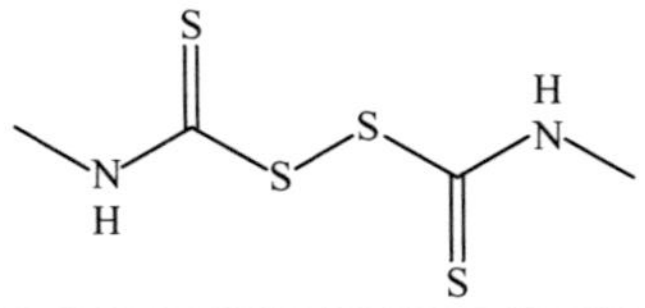		[129, 130]
PolyOrg Sulfa (POS)	3-phenyl-1,2,4-dithiazoline-5-one	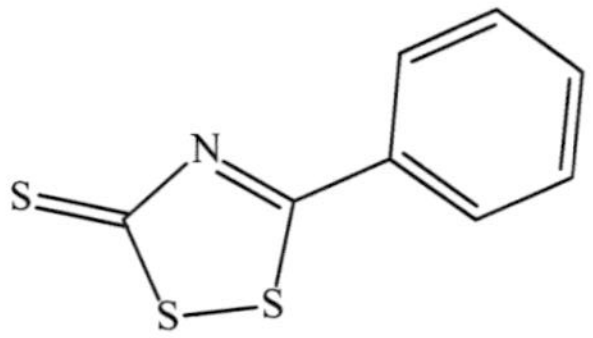	0.1M ACN	[131]
Not applicable	Sulfonic acid derivatives	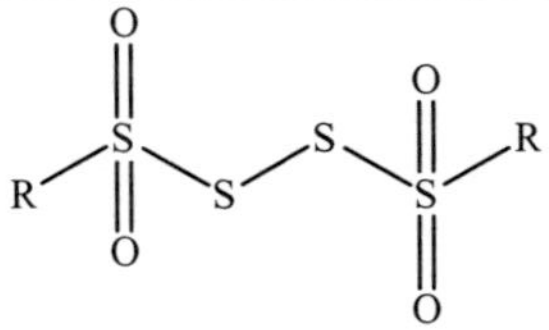	100% ACN	[132]

2.4.4.2 Modification of Guanosine

The correct order of capping and oxidation is a matter of debate as the impurities formed may impact the quality of oligonucleotides synthesis. It was reported by Pon *et al.* in 1985 [160] that modification of deoxyguanosine would take place during synthesis with phosphoramidites, leading to poor synthesis using a couple/oxidation/capping cycle. Chemists at Applied Biosystems reported the formation of a fluorescent intermediate of guanosine, which was further converted to 2,6-diaminopurine nucleoside base during high-temperature ammonia deprotection [133]. They undertook a systematic study of the chemical steps involved in oligonucleotide synthesis to determine which set of conditions was responsible for the formation of low levels of modified guanosine. They proposed that reaction of the incoming phosphoramidite with O-6 of the support bound guanosine leads to intermediate **2** in Figure 2.28.

Capping followed by oxidation gives the stable O-6 phosphate triester **3**. A second capping operation exposes the phosphate triester to DMAP, which replaces the 5′-nucleoside, giving stable, fluorescent intermediate **4** (see Figure 2.28). This intermediate can be isolated when treated with ammonia at ambient temperature. Prolonged heating in the presence of ammonia allows the concomitant removal of the isobutyryl protecting group on guanosine and the displacement of DMAP at C-6 to provide 2,6-diaminopurine (see Figure 2.29).

The formation of 2,6-diaminopurine was found to be reduced by nearly one third when a couple/capping/oxidation procedure was employed rather than a couple/oxidation/capping protocol. This phenomenon may be explained by a presumed destruction of the *O*-6 guanine phosphite intermediate by acetate ion prior to oxidation of the phosphite to its stable phosphate form [161]. The formation of 2,6-diaminopurine was eliminated or suppressed, when DMAP (4-*N*,*N*-dimethylaminopyridine) in the capping mixture was replaced by *N*-methylimidazole [133], irrespective of whether capping was performed prior to oxidation or not.

2.4.4.3 Guanosine Protecting Group Exchange

A well-known transamidation pathway occurs via exchange of the isobutyryl protecting group on guanosine for its acetyl equivalent [136]. Acetic anhydride, typically used in the capping step, serves as the acetylation reagent and repeated capping steps can result in the introduction of the acetyl group at the exocyclic 2-amino position in guanine. The resulting acetylated guanosine is quite stable under standard deprotection conditions, leading to a series of impurities having FLP+42 Da (or integral amounts thereof) as evidenced in the HPLC-MS spectrum. The use of isobutyryl anhydride in the capping step will eliminate the formation of the stable acetamide impurity.

2.4.4.4 Order of Capping

Early protocols suggested the use of a coupling/capping/oxidation cycle [162]. As was described above, employing a capping step prior to the oxidation step was shown to reduce the level of impurity of the modified guanosine. The argument against the original coupling/capping/oxidation sequence was that placing the capping step after the oxidation further reduces the level of moisture introduced from oxidation solution,

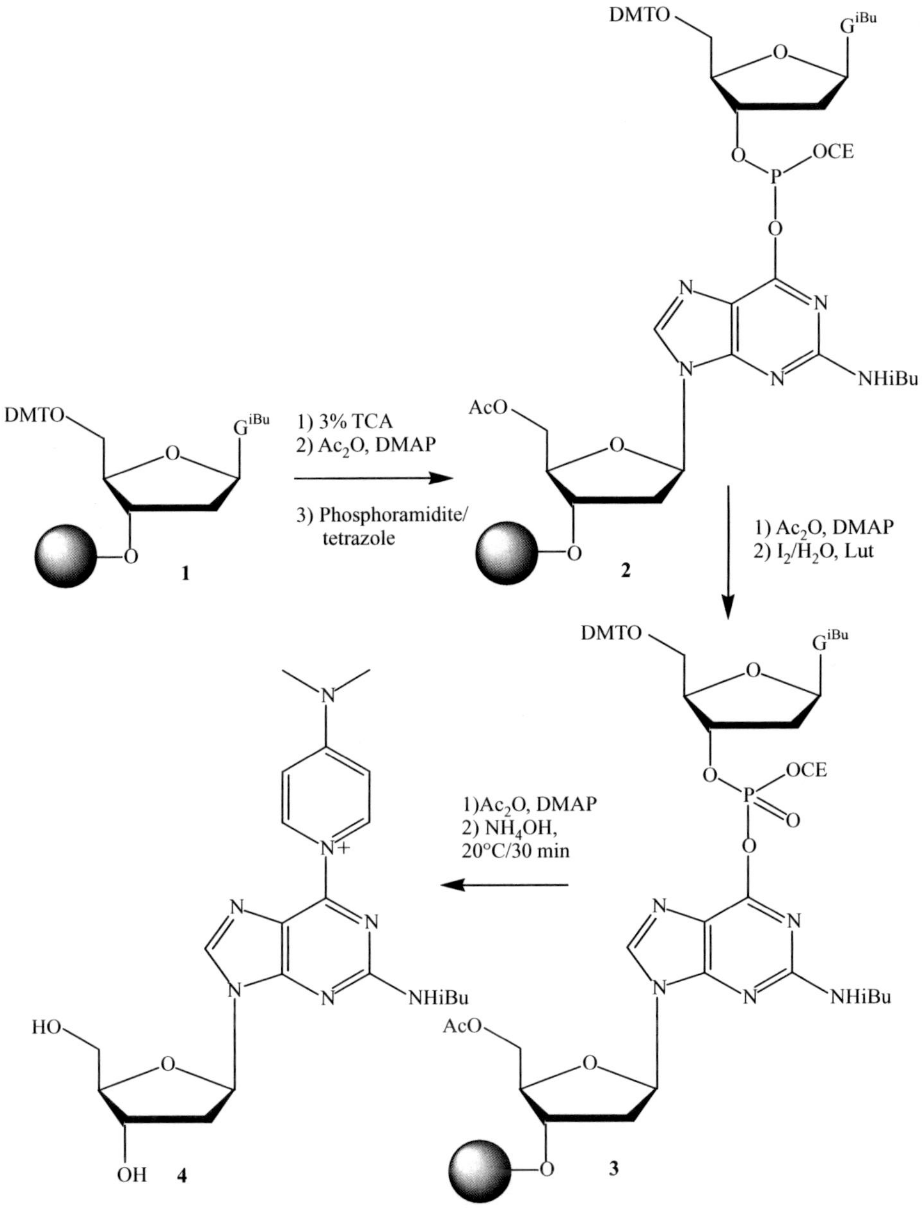

Figure 2.28 Proposed mechanism of the formation of fluorescent intermediate **4**.

which contains 10% water. In this manner, the support could be more effectively 'dried' in preparation for the subsequent coupling cycle. Incorporating both of these arguments would suggest a cycle having the following order: detritylation/coupling/capping/oxidation/capping. In fact, Sinha *et al.* [144] demonstrated that adapting

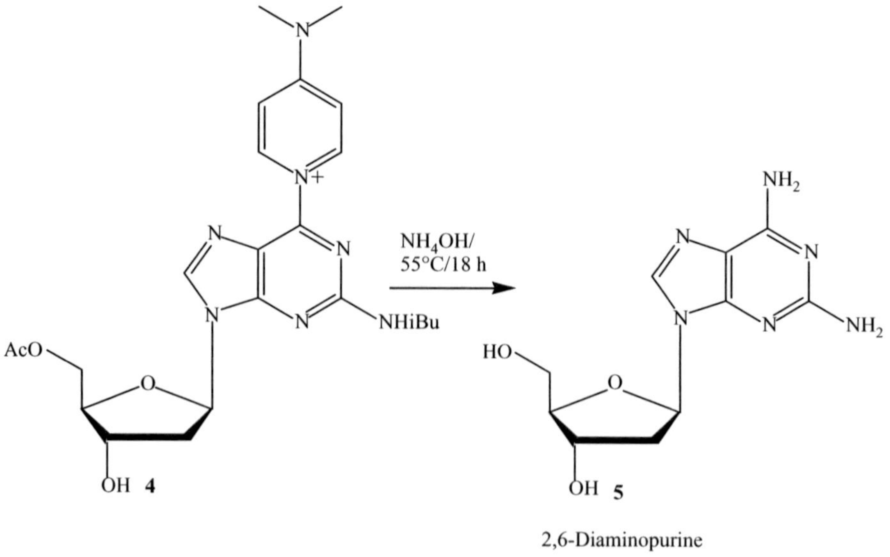

Figure 2.29 Formation of 2,6-diaminopurine **5**.

capping before and after oxidation (capping/oxidation/capping) improved the quality of synthesis of oligonucleotides using *t*-butylphenoxyacetyl-*N*-protected phosphoramidites. This protocol has eventually been adopted for the synthesis of many phosphodiester oligonucleotides and particularly for long RNAs [142, 163]. Phosphorothioate oligonucleotides are traditionally prepared by a coupling/thiolation/capping sequence; however, in this case no modification of guanosine has been observed.

2.4.5 C&D Step

2.4.5.1 Loss of Sulfur During C&D

Aside from incomplete sulfurisation during the thiolation step, phosphodiester contamination can result from the C&D step. The standard practice of heating the crude product in ammonia at 55°C to release the product from solid support and with simultaneous removal of the base-protecting groups is known to generate phosphodiester contamination. Several theories have been put forward to explain this phenomenon. It has been suggested that the conversion of phosphorothioate to phosphodiester backbone is catalysed by trace transition metals such as Fe^{+2}, Ni^{+2} and Cr^{+3}, which are contaminants in ammonia and may result from reagent transfer [164]. Ammonia solution containing 0.01M ethylenediaminetetraacetic acid (EDTA) resulted in a reduction of the PO content. Reese and Song [165] compared the ammonia deprotection of 0.05M solutions of d[Tp(s)T] at 50°C/15 h in the following mixtures: (case 1) ammonia with no treatment, (case 2) ammonia with 0.01M EDTA and finally, (case 3) ammonia containing 9:1 (v/v) 2-mercaptoethanol. Analysis of each crude mixture by ^{31}P NMR gave 0.8% and 0.4% for case 1 and case 2, respectively. The mixture containing mercaptoethanol contained no detectable PO. Turney *et al.* [166] compared an electropolished stainless steel vessel with a glass container for the C&D reaction of a phosphorothioate oligonucleotide. The PO content was identical for both containers, showing that the PO content is independent of the material of construction. However, the addition of stainless steel shavings led to an increase in PO content over time [166].

An alternative theory suggests that the introduction of the phosphodiester impurity results from an effective transfer of the cyanoethyl group to sulfur followed by a hydrolysis of the resulting phosphorylthioether in the presence of base during the C&D step. A possible mechanism is depicted in Figure 2.30.

2.4.5.2 Incomplete Removal of Protecting Groups

Following synthesis, the crude oligonucleotide on solid support is treated with a base, traditionally aqueous ammonia, to effect cleavage of the material from the solid support (fast reaction) with concomitant removal of the nucleoside base protecting

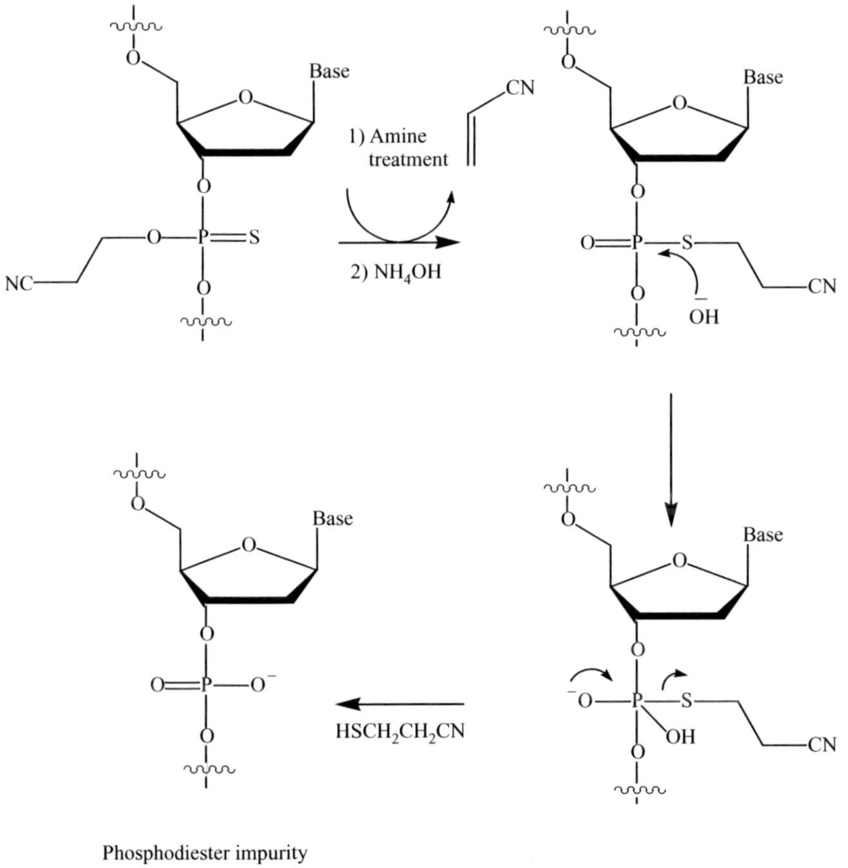

Figure 2.30 Proposed mechanism for loss of sulfur during cleavage and deprotection.

groups (slower reaction) in a single operation. While the cleavage of crude oligonucleotide from the resin generally is complete within 1 h at 55°C in aqueous ammonia, the removal of protecting groups, especially the isobutyryl protecting group on guanosine, can be the time-limiting step, often requiring as much as 17 h at 55°C or 120 h at room temperature [92]. Moreover, the time needed to carry out the deprotection is sequence dependent. Sequences having multiple guanosine residues, particularly in close proximity, can be especially troublesome. The use of HPLC-MS is used to detect the characteristic FLP+70 Da peak associated with a single guanosine residue containing its isobutyryl protecting group. The synthesis of DNA phosphorothioates can be difficult, because the high temperature necessary to effect protecting group removal may promote exchange of the phosphorothioate backbone linkage to its phosphodiester complement. In this case, the practitioner must find a balance between low-temperature deprotections, which minimise PS/PO exchange but may result in incomplete deprotection, and high-temperature conditions, which are necessary for complete deprotection but may favour PS/PO exchange.

2.4.5.3 CNET Adduct Formation

The release of the crude oligonucleotide from the solid support is typically effected by treatment with ammonia, which results in removal of the cyanoethylphosphate protecting groups with concomitant release of acrylonitrile. The highly reactive acrylonitrile is known to add to the N^3-position of thymidine [54] yielding an impurity with a characteristic mass of full-length product + 53 Da. The addition of acrylonitrile occurs preferentially to thymidine (T) as illustrated in Figure 2.31, although the reaction has been described for dG [167], dC, dA and U.

The unambiguous structural proof of cyanoethyl addition was performed by synthesising a $(T_{PS})_{18}T$ and treating the crude material on solid support with ammonium hydroxide to effect succinate cleavage and base deprotection. Removal of the 5′-DMT group was accomplished by treatment with aqueous acetic acid. The mixture was analysed by HPLC-MS and found to contain the characteristic (+53) Da peak. The material was desulfurised and digested with SVP and alkaline phosphatase. Confirmation of the alkylation at *N*-3 was obtained by HPLC co-injection of an independently synthesised sample of the modified nucleobase. Further studies on the stability of this adduct supported the relative stability of the akylated *N*-3 thymidine to the unblocking conditions.

The short half-life of acrylonitrile in ammonia solution can be exploited by treating the mixture with a scavenger to remove the cyanoethyl protecting groups and trap the so-formed acrylonitrile in a single operation [168, 169]. A number of methods have been described to curb the formation of deleterious CNET adducts using weak bases. The first such method was described by Eritja *et al.* [167], where cyanoethyl-protected oligonucleotides were treated with 40% triethylamine in pyridine for 3 h. Other approaches were described where the oligonucleotide was treated with diethylamine, *tert*-butylamine, morpholine or piperidine in organic solvents such as ACN, dioxane, or THF [169, 170]. Such a so-called ‘amine wash’ treatment is routinely used prior to releasing the oligonucleotide from the support when using concentrated ammonia solution for C&D.

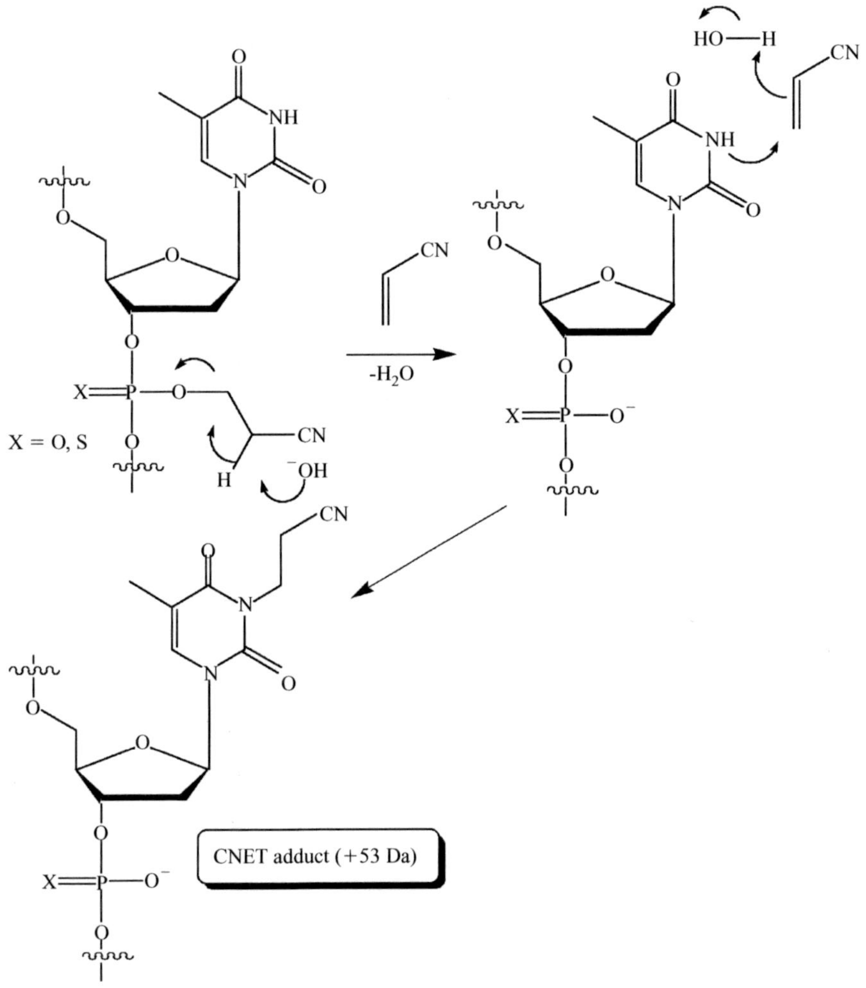

Figure 2.31 Postulated mechanism for CNET formation under base catalysis.

2.4.5.4 Transamidation of Amino Functionality

A second class of transamidation impurities results from the use of amino-linkers, typically tethered to either the 3′- or 5′-end of oligonucleotides as a linker for post-synthetic conjugation of polyethylene glycol (PEG) polymers, dyes, lipids and other small molecule conjugates and reporter groups. Such conjugates are widely employed as fluorescent probes or to increase lipophilicity for increased biodistribution. The linker is characterised by having a carbon chain of 3–12 carbon atoms or ethylene glycol units separating a terminal amine from its phosphoramidite functionality. When attachment at the 5′-end is required, the linker is introduced using standard phosphoramidite coupling techniques. The amine is commercially available in several different protecting group varieties, each possessing distinct advantages and disadvantages. A trifluoroacetamide protecting group is easily removed in base; however, it does not offer the lipophilic properties one may wish to exploit in a trityl-on purification approach and a free amine group is reactive under basic C&D conditions. A monomethoxytrityl (MMT) protecting group is more common and has the advantage of providing a good lipophilic handle for a trityl-on purification strategy. The MMT group, however, to some extent is partially lost under basic C&D conditions. The free amine, whether due to partial loss of MMT or removal of TFA protecting group, competes with ammonia in removing the acyl protecting group from the exocylic amine of the bases. This leads to transfer of the acyl moiety to the amine of the alkyl amine chain, rendering it inactive toward further manipulations (such as coupling with PEG-NHS (*N*-hydroxysuccinimide) or dyes). Strömberg and co-workers [171] report that elimination of the final capping step only marginally improves the ratio of unwanted acetylated impurity, further supporting the idea of transamidation, rather than direct amidation through the capping step. Hoffmann and co-workers at Noxxon suggest that treatment of the support with an *N,N*-dialkylamine, such as diethylamine, immediately following synthesis reduces the likelihood of internal transfer of acetyl to the terminal amine [163]. The impurity, which is usually characterised by having a slightly longer retention time than the deprotected amine in ion-pairing HPLC analysis, displays a characteristic +42, +72 and +104 Da molecular weight relative to its FLP for an acetyl, isobutyryl and benzoyl group, respectively [171]. The trifluoroacetamide-protected amine also suffers the same fate during ammonia-based C&D. The acyl protecting group, which is nearest to alkylamine, is transferred most and is the predominant impurity peak.

2.4.5.5 Transamination of Cytidine

Standard C&D protocol for oligonucleotides has traditionally been treatment with aqueous ammonia at 55°C. This general set of conditions suffers the drawback of long reaction times and often incomplete removal of the isobutyryl protecting group on guanosine. While the early use of DNA as polymerase chain reaction primers did not require its complete deprotection, therapeutic use of oligonucleotides demands a higher level of purity. Two separate strategies were employed to address the problem of long reaction times and incomplete deprotection. Protecting groups capable of undergoing milder deprotection conditions were developed, such as dimethylformamidine [143, 172, 173], phenoxyacetyl [134, 174, 175], isopropyl-phenoxyacetyl [176] and *tert*-butylphenoxyacetyl [144, 177]. An alternative approach relies on the use of more nucleophilic amines, such as hydrazine or alkylamine. A very mild deprotection procedure using gaseous ammonia or methylamine was initially described by Kempe [178] and further evaluated by Beaucage and co-workers [179]. However, a deprotection process relying on gaseous amines will be challenging to implement in large-scale oligonucleotide manufacturing. Using a 50:50 mixture of aqueous methylamine and ammonia, most oligonucleotides can be cleaved from the solid support and the base protecting groups (including isobutyryl from guanosine) can be removed by heating the mixture at 65°C for 10 min. The drawback to the use of alkylamines is their propensity to undergo transamination at C^4 of cytidine when the benzoyl protecting group is used, resulting into more than 10% of N^4-Me-C (see Figure 2.32), which can

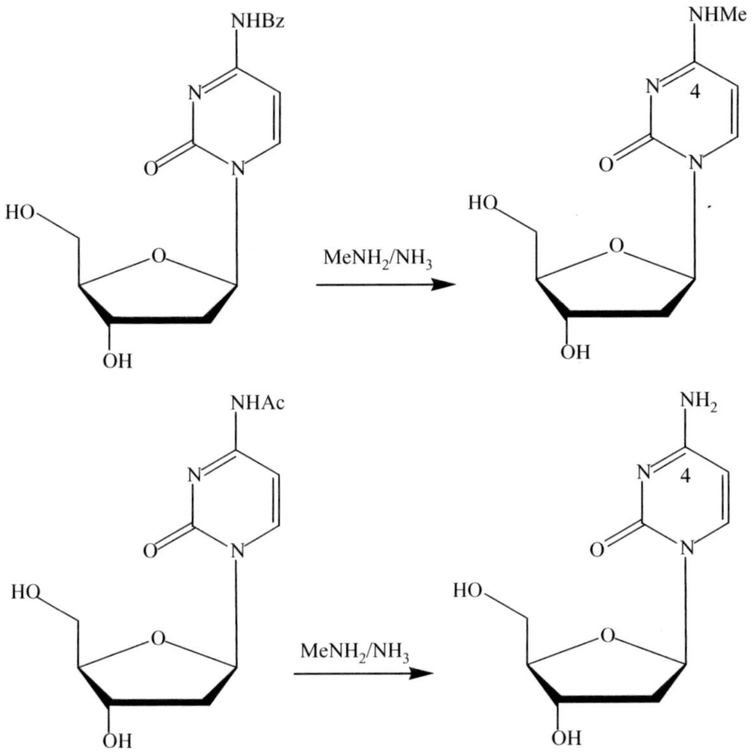

Figure 2.32 Treatment of N-Bz protected deoxycytidine with methylamine leads to transamination at C-4 (top); the use of acetyl-protected cytidine avoids transamination.

be detected in the MS by an additional mass of 14 Da. Prior to MS, such modifications were detected only through base digestion and HPLC analysis of the mixture. Reddy *et al.* [180] have described the acetyl group for protection of cytidine as an alternative which can minimise the introduction of transamination impurities. Acetyl protection has also been extended to methylphosphonates, which are particularly susceptible to degradation at high temperature with ammonia. In this case, complete removal of acetyl was seen without attendant transamination by treatment at room temperature with ethylenediamine/ethanol (1:1) for 7 h at room temperature [181].

The transamination at C^4 was also reported by Guzaev and co-workers [182] as a set of impurities associated with the cleavage of alkylphosphate tethers from CPG support with α, ω-alkanediamines. Their strategy relied on the use of a spacer between the 3′-terminal nucleoside and the CPG support which would be cleaved by a diamine for the purpose of introducing a reactive amine at the 3′-terminus, as depicted in Figure 2.33.

Cleavage promoted by the diamine only led to increased levels of transamination at dC, whereas pre-treatment with hydrazine liberated the benzoyl protecting group on deoxycytidine, thereby rendering it unreactive toward the diamine used. It should be

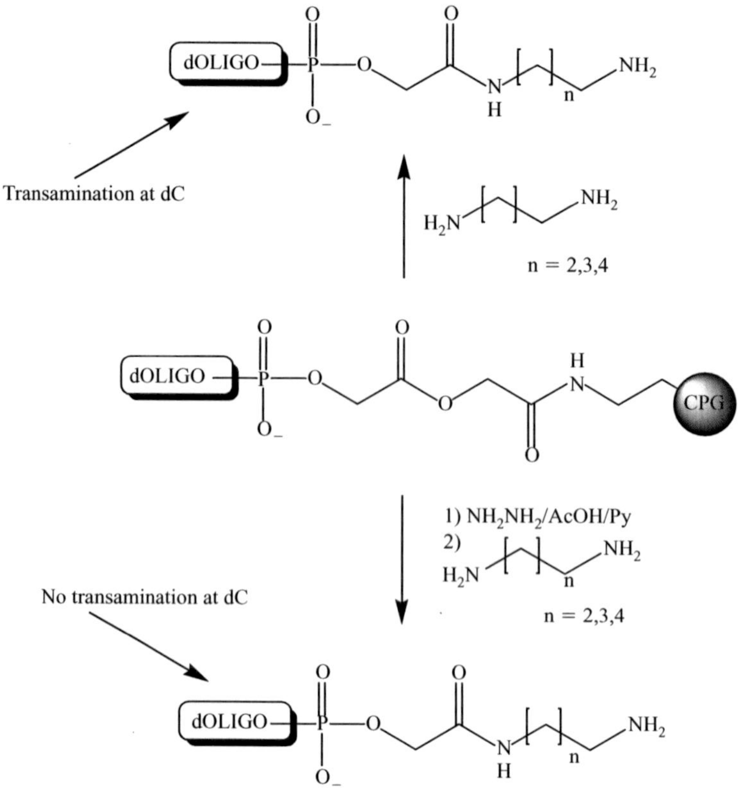

Figure 2.33 Observation of transamidation using α,ω-diamines for cleavage.

noted that only tethers of a certain length are observed to be stable toward hydrazine acetate.

2.4.5.6 Depurination and Resultant Strand Breakage During C&D

Depurination does not interfere with chain elongation steps; however, treatment of the product with ammonia in the cleavage step results in fragmentation of the oligonucleotide related to the acid-induced detritylation step as shown in Figure 2.34. The acetal is in equilibrium with its open-chain aldehyde tautomer, which is extremely susceptible to β-elimination in the presence of base. The elimination usually results in strand breakage [147].

The instability of the acetal tautomer in basic conditions also presents itself in trityl-on purification. The trityl-on purification mode, in which the 5′-trityl group is retained following the synthesis, is an attractive option for sequences in which the FLP is particularly difficult to separate from the $n-1$ and $n-2$ failure sequences by conventional anion-exchange or RP chromatographic techniques. The trityl-on mode, which is compatible with either anion exchange or RP chromatography, requires that the 5-trityl group be left intact at the end of the synthesis. The trityl-on crude

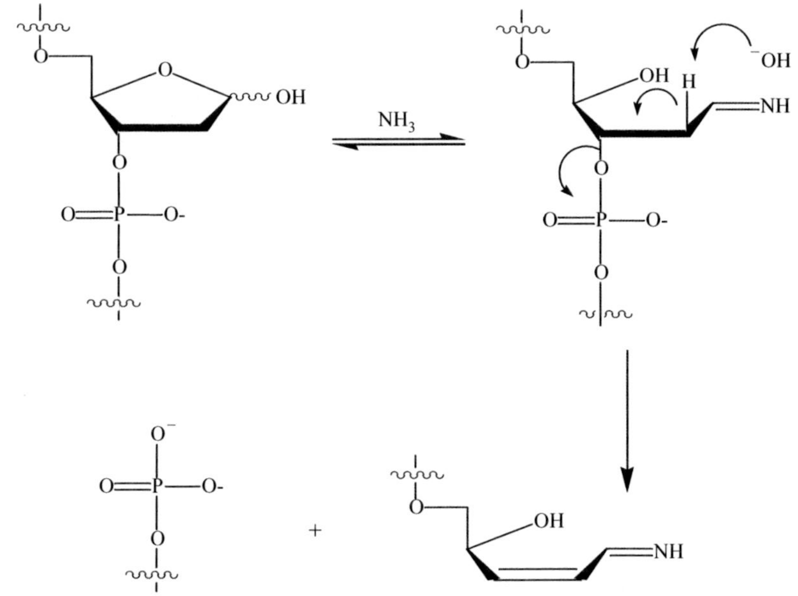

Figure 2.34 Postulated mechanism for strand breakage via depurination and ammonia treatment.

oligonucleotide is loaded onto the column and the attendant failure sequences, which are much less lipophilic, are washed away in preference to the more hydrophobic trityl-bearing oligonucleotide. The trityl group is removed on-column by acid treatment (typically acetic acid or DCA in water) and the parent oligonucleotide eluted with a salt gradient (anion exchange) or organic gradient (RP). Degradation owing to strand breakage often results from on-column acid treatment followed by exposure of the so-formed acetal to high-pH purification buffers during anion exchange trityl-on purification.

2.4.5.7 Strand Breakage via H-Phosphonate

Still another mechanism for strand breakage impurities may be analogous to the formation of the C-DMT phosphate ester due to incomplete oxidation or thiolation (see earlier). The failed sulfurisation/oxidation of the phosphite triester (shown in Figure 2.35), will undergo capping and ultimately be exposed to the subsequent detritylation reaction, where the phosphorus centre will undergo protonation by the DCA solution. The resultant H-phosphonate will continue to undergo chain elongation; however, the H-phosphonate will be degraded in the presence of ammonium hydroxide at high temperature during the C&D reaction, resulting in what would appear as coupling failure sequences.

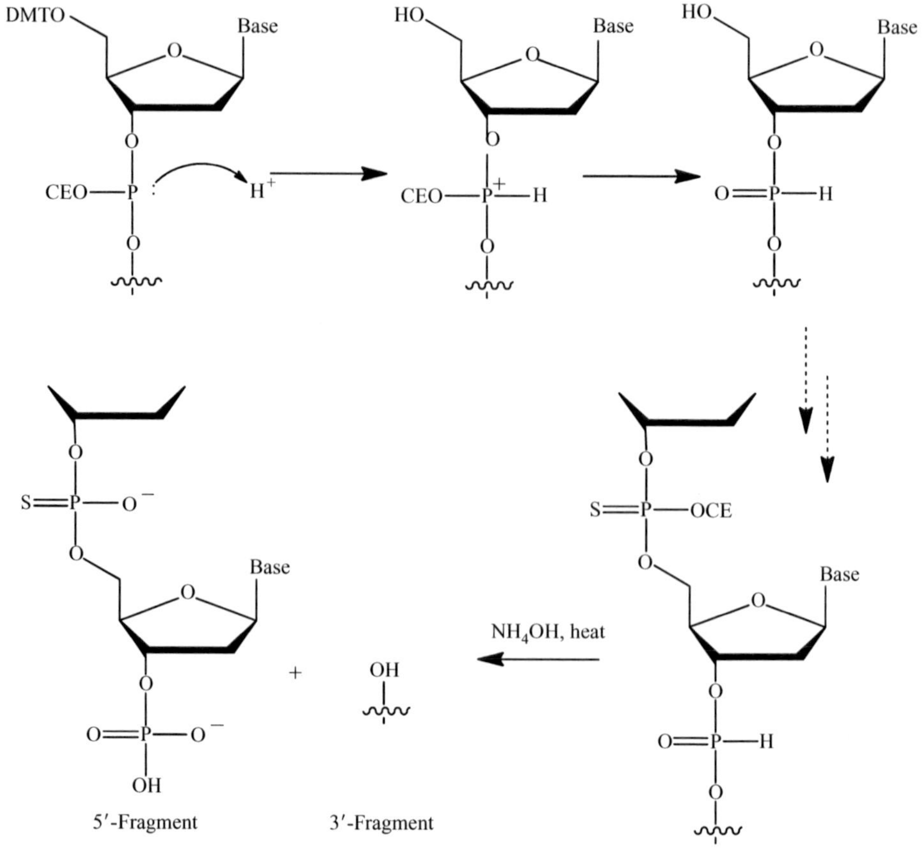

Figure 2.35 Proposed pathway for strand breakage resulting from failed sulfurisation.

2.4.5.8 Deamination

Deamination describes the replacement of the exocyclic amine moiety N^6 on adenine or N^4 cytidine with carbonyl functionality. The reaction is known to take place under physiological conditions and is catalysed by a class of enzymes termed deaminases [183, 184]. The principal catalyst for non-enzymatic deamination is the hydroxide ion and the reaction is accelerated by heat and base. In an attempt to determine the rate constant for the non-enzymatically catalysed reaction, Wolfenden [185] studied the deamination of adenine and cytidine under a variety of conditions. Table 2.5 summarises commonly encountered deaminated products and references to the original work. The product of deamination of adenine under these forcing conditions is inosine (entry 1) along with detectable amounts of adenine and hypoxanthine (entry 2). Under the same conditions, or in the presence of mercaptoethanol, cytidine is deaminated to urindine (entry 3). The heat-induced deamination of 5-methyl-deoxycytidine, deoxycytidine and N^4-methyldeoxycytidine at physiological pH between 70 and 105°C was investigated by Ehrlich *et al.* [186] (entries 4, 5). They report that the

Table 2.5 Commonly encountered de-aminated species.

Entry	Starting material	Conditions	Impurity	Reference
1	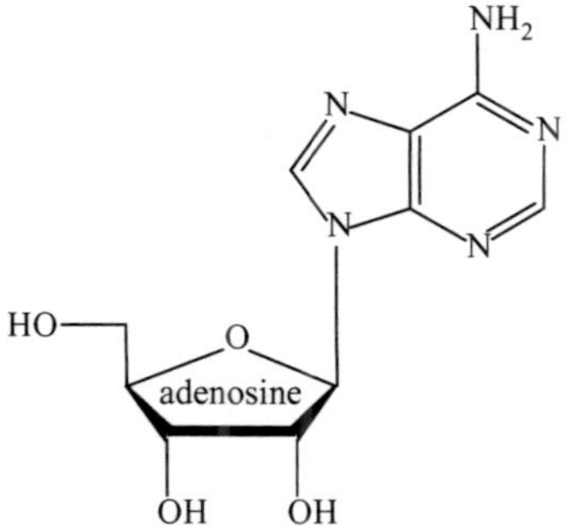	0.1M KOH at 85°C, 4 h	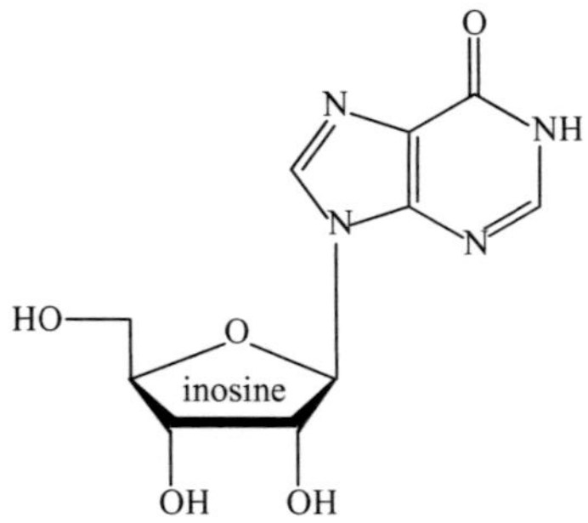	[182]
2	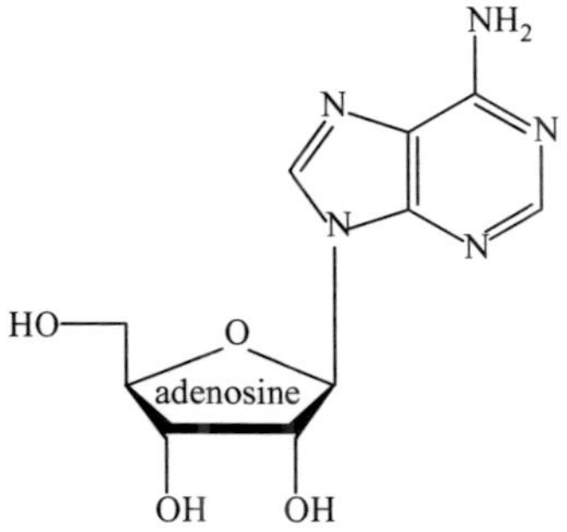	0.1M KOH at 85°C, 4 h	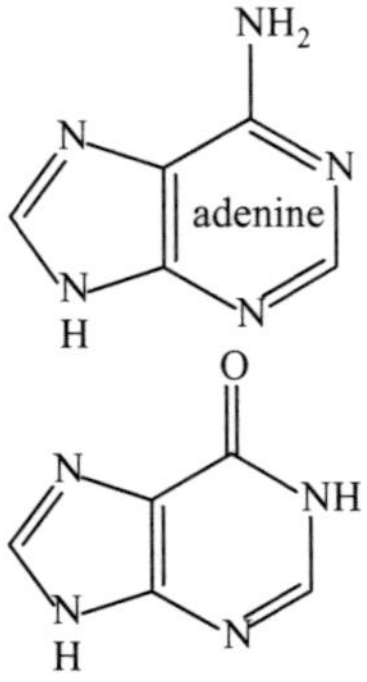 hypoxanthine	[182]

(*continued*)

Table 2.5 (*continued*)

Entry	Starting material	Conditions	Impurity	Reference
3	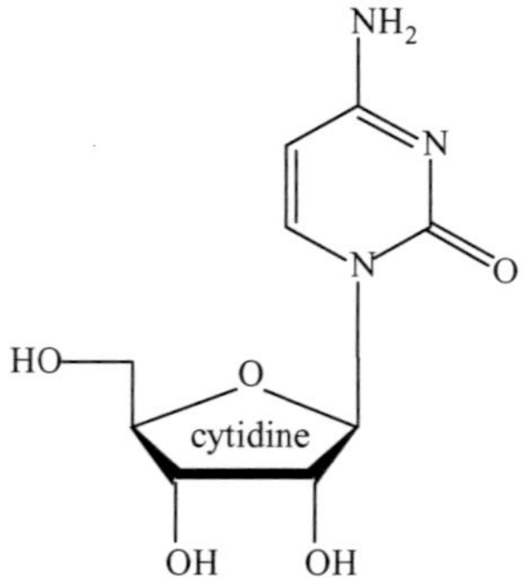	0.1M KOH at 85°C, 4 h or Phosphate dianion, 85°C, 4 h or Mercaptoethanol, 85°C, 4 h or Thermal only (slow)	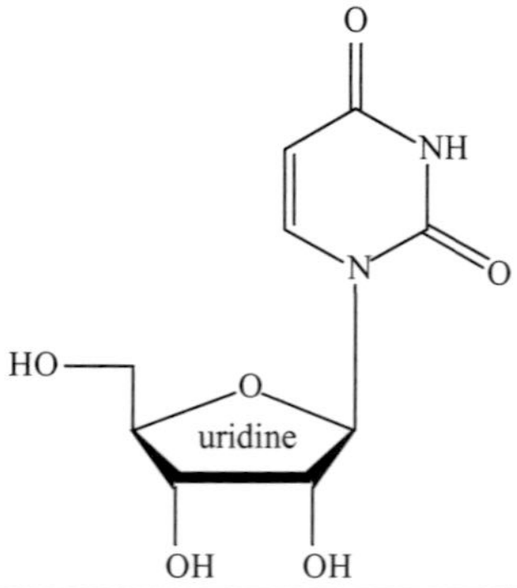	[182]
4	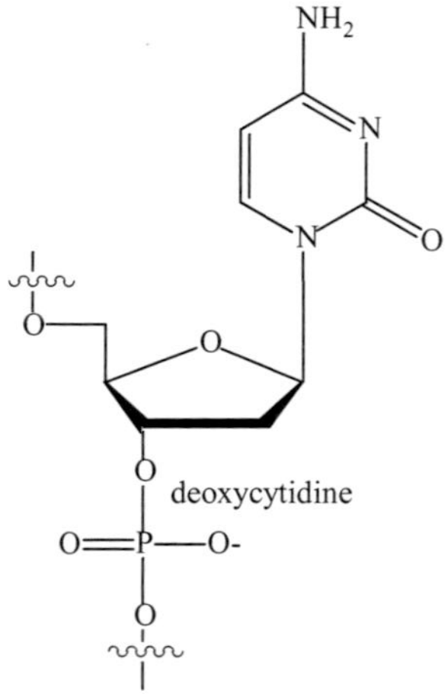	Heat, pH 7.4	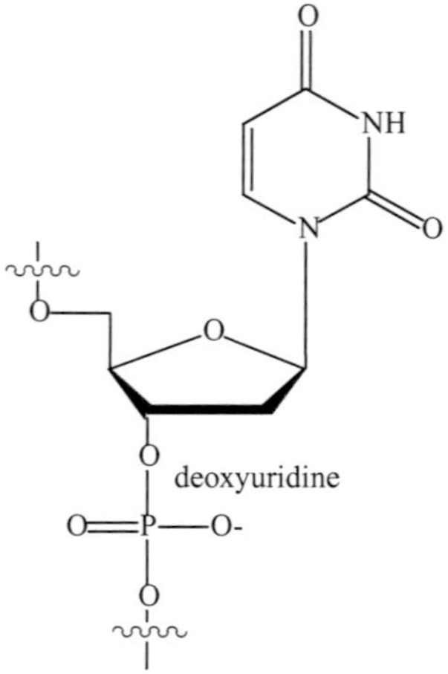	[186]

(*continued*)

Table 2.5 (*continued*)

Entry	Starting material	Conditions	Impurity	Reference
5	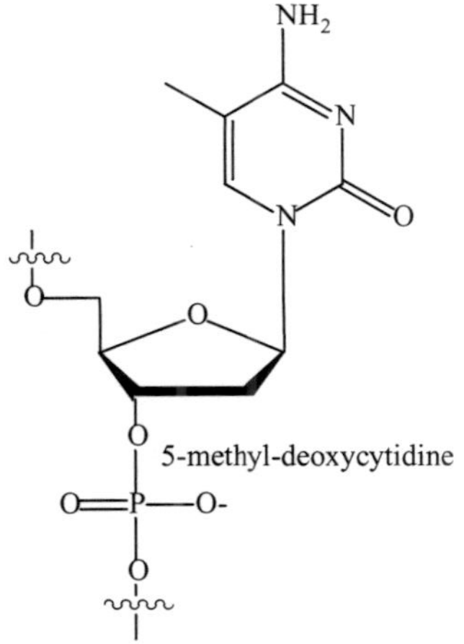	Heat, pH 7.4	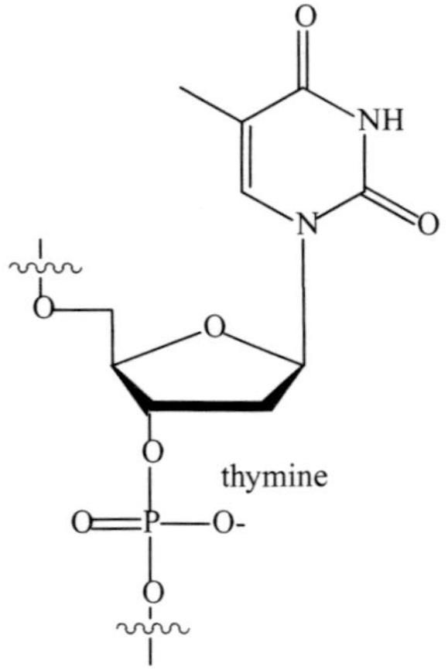	[186]

order of deamination proceeds: Me^5-dC ≫ dC > N^4-dC. Adventitious deamination may be introduced by the C&D reaction, which is typically done under basic conditions at high temperature [186, 187].

2.5 Chemistry-Specific Impurities

2.5.1 *Tert*-butyldimethylsilyl-RNA

The most commonly used protecting group for the 2′-hydroxyl group in the synthesis of RNA oligonucleotides remains the *tert*-butyldimethylsilyl (TBDMS) group [188–191]. The TBDMS group is compatible with standard phosphoramidite chemistry and is orthogonal to the base-labile nucleoside protecting groups. The standard procedure for RNA is: step 1: synthesis; step 2: release of oligonucleotide from support with concomitant removal of base-protecting groups (ammonia or methylamine-ammonia treatment); step 3: removal of *tert*-butyldimethylsilyl group using fluoride ions to liberate the 2′-hydroxyl group. Fluoride salts, such as tetrabutylammonium fluoride [191], TEA-HF [192–194] or pyridinium hydrogen fluoride [195, 196] are used to this end. Failure to completely remove the protecting groups will result in the appearance of late-eluting impurities in the RP HPLC. Each residual TBDMS group will add an integral value of 114 Da to the target oligonucleotide.

Careful control of the pH and temperature during 2′-hydroxyl deprotection of RNA is critically important. 'One-pot' deprotection protocols are common [197–199] and can be described as a process in which the base protecting groups are removed by treatment with methylamine, ammonia or a mixture of them, before direct addition of a desilylation reagent to the basic mixture at low temperature. Thus, two orthogonal protecting groups are removed in a single operation without isolation of intermediates. Ideally, the excess amine base will be neutralised by the acidic desilylation reagent (fast reaction) before the 2′-hydroxyl groups are liberated (slow reaction). If the 2′-hydroxyl group is exposed before the pH is neutralised, strand breakage is likely to occur, resulting in the formation of cyclic phosphates and 3′-truncated sequences, as shown in Figure 2.36 [147, 200].

2.5.2 2′-3′-Isomerisation of RNA

As early as the early 1950s, it was discovered that the alkaline hydrolysis of RNA yielded each mononucleotide as a pair of isomers. The pioneering work of Carter [201] and Cohn [202] in separation of these isomers by paper and anion exchange chromatography, respectively, was instrumental in the elucidation of the connectivity of the phosphate linkage. The structure of the adenyl isomers was shown to be identical with the product of phosphorylation of the 5′-DMT nucleoside [203]. Having narrowed the phosphate linkage to either the 2′-hydroxyl or the 3′-hydroxyl, Todd and Brown ultimately (and correctly) concluded that the natural

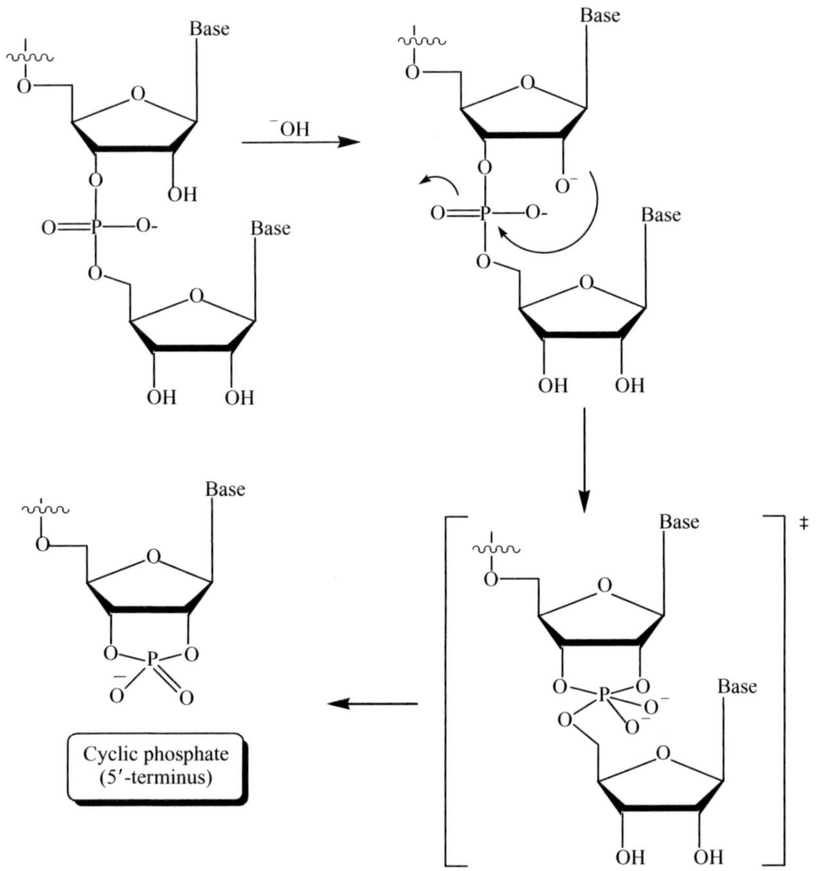

Figure 2.36 Proposed pathway for generation of the cyclic phosphate intermediate and intrastrand cleavage.

linkage is through the 3′-hydroxyl because it was the 3′-isomer which proved to be a metabolite for pancreatic ribonuclease [204]. Conversion of the isomers could be effected under acidic conditions, but not under basic catalysis. Under basic conditions, attack of the 2′-hydroxyl onto the neighbouring phosphate through a penta coordinated oxyphosphorane transition state results in loss of the 3′-terminal nucleoside to generate a cyclic phosphate intermediate (Figure 2.36) [200].

Attack of the cyclic phosphate by water or base yields the regioisomeric mixture of 2′- and 3′-phosphates described by Cohn and Carter (see Figure 2.37).

Acidic conditions favour the 3′- to 2′-isomerisation of RNA relative to intrastrand cleavage through the cyclic phosphate. A detailed treatment of the mechanism and kinetics of the cleavage and isomerisation of the phosphate bond has been published by Lönnberg and co-workers [200]. The basic mechanism is depicted in Figure 2.38.

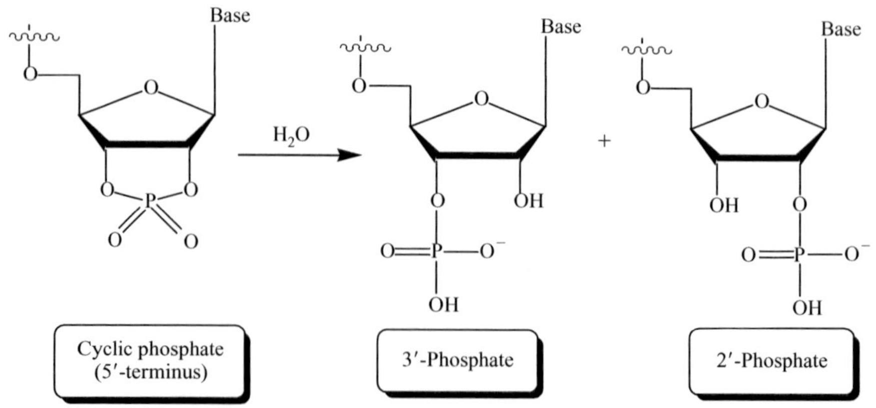

Figure 2.37 Base-catalysed hydrolysis of the cyclic phosphate.

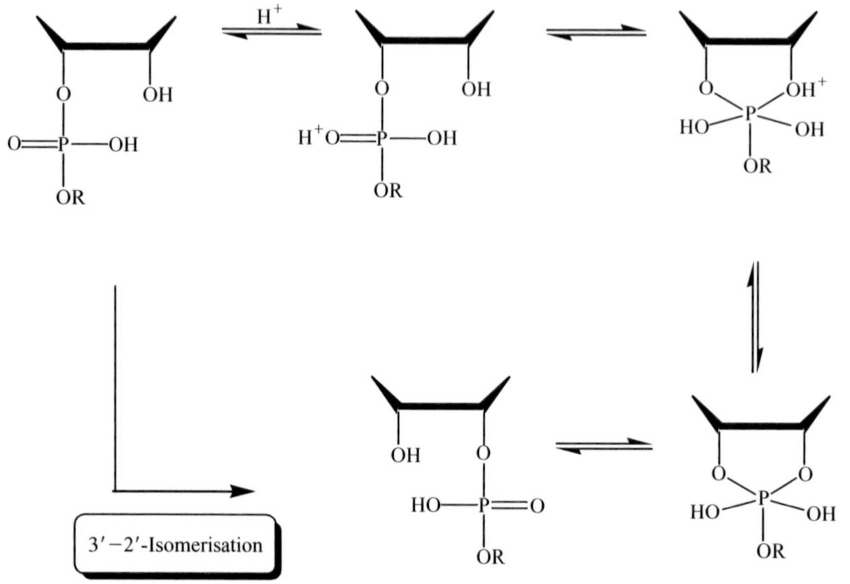

Figure 2.38 Proposed mechanism for 3′- to 2′-isomerisation.

2.5.3 Depyrimidation of Cytidine and Uridine

The amine-based deprotection protocol used to remove the base protecting groups can cause depryrimidation in cytidine and uridine (see Figure 2.39). The phenomenon is generally associated with methylamine-based deprotection conditions, although examples of these degradation products are also known to arise from ammonia treatment. The extent of degradation and relative distribution of the products is dependent on the sequence and deprotection conditions.

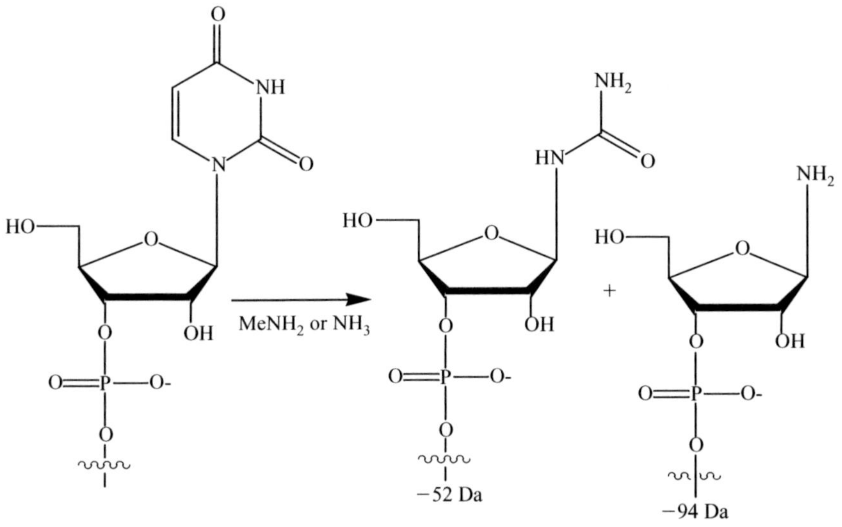

Figure 2.39 Detected impurities resulting from depyrimidation of uridine.

A proposed mechanism for depyrimidation of uridine is shown in Figure 2.40 [205]. In aqueous methylamine or mixtures of methylamine and ammonia, Michael addition of methylamine at the β-unsaturated position gives the corresponding enolate, which collapses to fragment the pyrimidine ring. Transamidation of newly formed α,β-unsaturated amide releases the nucleosidic urea, corresponding to a mass of 52 Da less than the parent peak. Further attack of methylamine at the urea is assumed to release the α-amino furan, accounting for a mass difference of 94 Da.

2.5.4 2′-Fluoro Modifications

There is a growing interest in the use of 2′-F modification [206, 207] and that has been demonstrated with the US Food and Drug Administration (FDA) approval of aptamer Macugen (R) (chimeric oligonucleotides containing 2′-F-nucleosides). Pallan and co-workers [206, 207] have also utilised 2′-F motif nucleosides, particularly in siRNA oligonucleotides, because of the enhanced nuclease resistance and binding stability associated with this feature. The dominant effect of the 2′-F results from locking the sugar in the C3′-*endo* conformation and thus leading to increased target affinity [208]. However, the lability of the 2′-F modification must be taken into consideration when working with mixed sequences consisting of RNA and 2′-F RNA. The 2′-F modification is subject to depyrimidation (especially with methylamine or ammonia/methylamine deprotection) and elimination and hydrolysis as shown in Figures 2.41 and 2.42 respectively. In the case of 2′-F-uridine, shown in Figure 2.41, a postulated mechanism begins with abstraction of the 2′-β-oriented proton with concomitant loss of the nucleoside base. Proton re-capture at the 2′-position occurs through assistance of the anomeric oxygen. The resultant positive charge is quenched by addition of water at

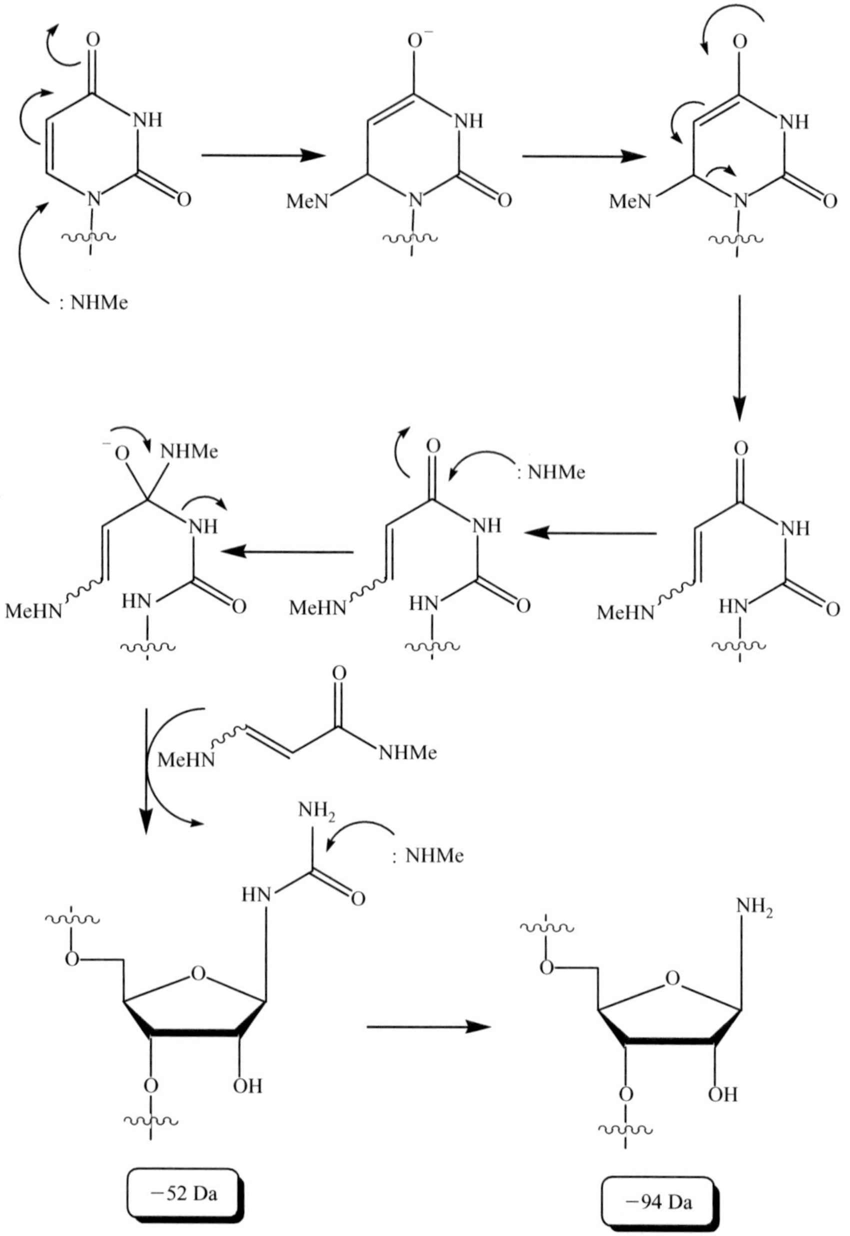

Figure 2.40 Proposed mechanism for depryimidation of uridine in the presence of ammonia and methylamine. For clarity, only the nucleobase is depicted throughout the mechanism.

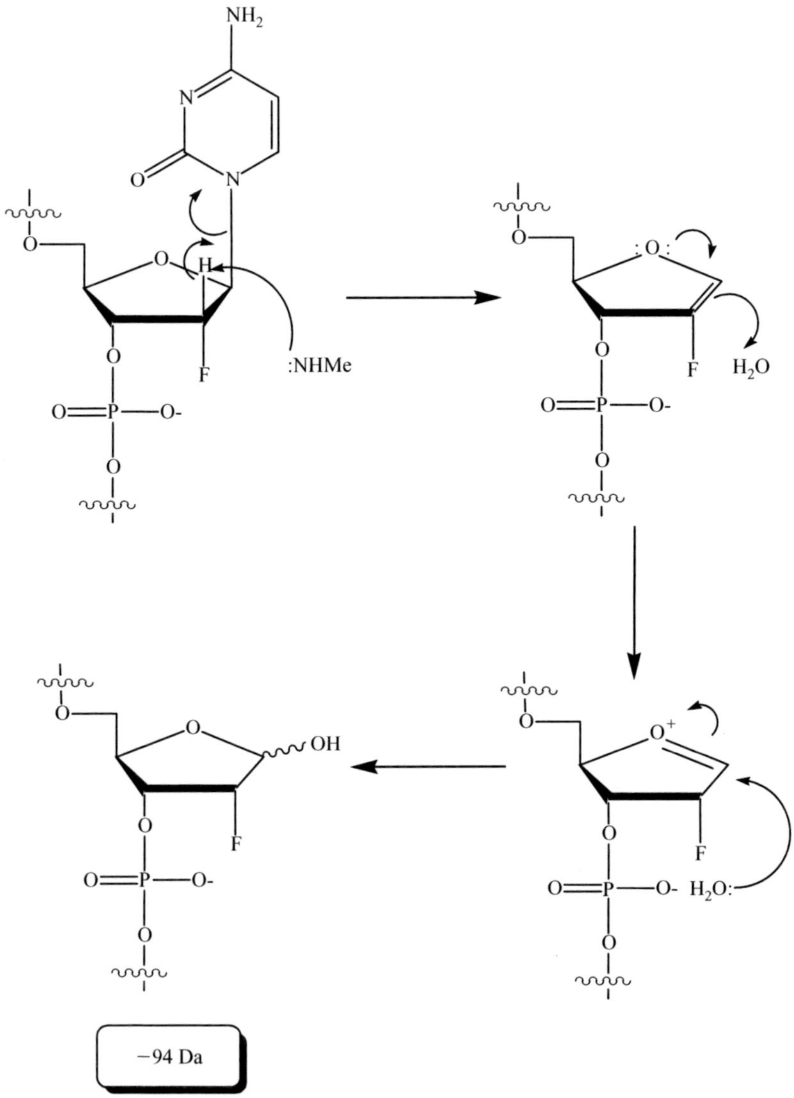

Figure 2.41 Postulated mechanism for depyrimidation of 2′-F-uridine.

the anomeric centre to give a hemiacetal having a net loss of 94 Da relative to the starting material.

The result is replacement of the 2′-fluorine with a hydroxyl group with the inversion of stereochemistry at the 2′-position. While elimination through anchiomeric assistance is possible in both 2′-F-2-deoxycytidine and 2′-F-2-deoxyurinidine, the rate of hydrolysis of cytidine is approximately 29 times that of uridine. The reaction proceeds under thermal and basic conditions [147, 209–211]. These side

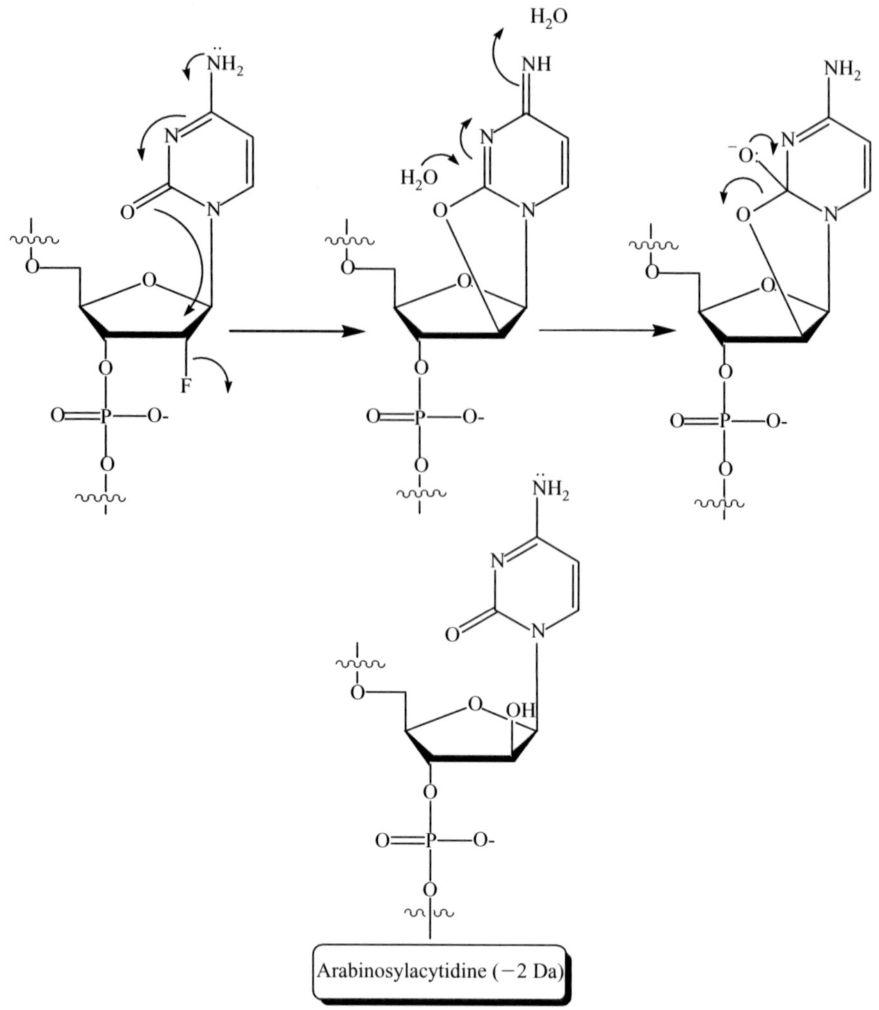

Figure 2.42 Proposed mechanism for elimination of 2′-F moiety.

reactions resulting from methylamine or methylamine/ammonium hydroxide-based C&D can often be minimised by using ethanolic ammonia solution (7M) for C&D.

2.5.5 Locked Nucleic Acids (LNAs)/Bicyclic Nucleic Acids (BNAs)

BNA [212, 213] and LNA [214, 215] are conformationally restricted synthetic nucleosides where the 2′-hydroxy group and 4′-carbon atom of the ribose ring are connected with a methylene bridge. Incorporation of a methylene bridge in ribose creates an additional five-member ring, which generates a constrained bicyclic structure with ideal conformation for an A-type duplex. Additionally, limited substitution of standard nucleosides in oligonucleotides with LNA/BNA provides better metabolic stability and higher binding affinity to its target. Owing to higher binding affinity, shorter sequences can be used in therapeutic investigation compared to standard oligonucleotides. Presently three oligonucleotides containing LNA/BNA modification are undergoing clinical trials.

Large-scale syntheses of LNA-phosphoramidites are currently performed by commercial manufacturers. The impurities present in LNA-phosphoramidite are very similar to standard phosphoramides (as described in Section 2.2.1). The quality of LNA-monomers (modified nucleosides) is controlled by ^{1}H NMR and HPLC.

2.6 Table of Impurities and Average Masses

The previous sections provided a survey of potential impurities found in oligonucleotides with origins in the raw materials and reagents and the process synthetic process. Especially in the case of chemistry-specific impurities, knowledge of the process is an essential part of understanding and controlling impurity formation. Table 2.6 displays a number of commonly encountered oligonucleotide impurities found in the MS spectra. The table is meant to be used by the MS practitioner to identify impurities based on potential chemical modifications at specific locations in the molecule.

2.7 Summary

Synthetic oligonucleotides have been explored for therapeutic and diagnostic purposes for more than two decades. These affords have presently culminated in FDA approving two drug products. Many oligonucleotides (ca. 100) are currently undergoing clinical trials and are at various stages of investigation. The drug development requires, apart from successful clinical trials, understanding of the quality of drug substance/drug product and proper characterisation of any impurities it may contain. In this chapter, an attempt has been made to describe the potential sources and measures that can be taken to control these impurities. In solid-phase oligonucleotide synthesis, based on phosphoramidite chemistry, sources of impurities have been subdivided into three categories: building block related (nucleoside-phosphoramidites), synthesis reagent related (solid supports and ancillary reagents) and process related (side reactions during synthesis).

The structures of nucleoside-phosphoramidite related impurities have been provided for clear understanding. Of the three types of impurities present in nucleoside-phosphoramidite, only reactive and critical components are incorporated in oligonucleotides synthesis and contribute to impurities in the drug substance. The nature of these potential impurities has been elucidated as far as possible. Impurities in synthesis reagents including various solid supports have been described. Potential impurities generated in the oligonucleotide due to impurities found in reagents and

Table 2.6 Average and monoisotopic mass differences relative to the full-length product for common impurities.

Problem	Affected locations	Mass difference (ave)
+PS (phosphothioate)	Backbone	+96.05
+PO (phosphate)	Backbone	+79.98
PS–PO conversion	Backbone	−16.07
Depurination	A base	−117.12
Depurination	A base	−135.13
Depurination	G base	−133.11
Depurination	G base	−151.13
Deamination	C, A, G bases	−0.98
Reaction with methylamine	C base	+14.03
+Acetyl	G base or amino linker	+42.04
+*β*-*N*-MAM	Unknown	+85.10
+Cyanoethyl	Heterocyclic bases, predominantly T	+53.06
+Isobutyryl	G base	+70.08
+Chloral	Backbone	+147.39
+*Tert*-butyl dimethylsilyl	rA, rG, rC, U	+114.26
+Benzoyl	A base	+104.11
Modified C	C base	+80.08
+DMT	5′*O* in all nucleosides	+302.37
Depyrimidation	U base	−52.03
Depyrimidation	U base	−94.07
-dA (PO)	A	−313.21
-dG (PO)	G	−329.21
-dC (PO)	C	−289.18
-dT (PO)	T	−304.19
-rA (PO)	rA	−329.21
-rG (PO)	rG	−345.21
-rC (PO)	rC	−305.18
-U (PO)	U	−306.17
-2′*O*-Me-rA (PO)	2′*O*-Me-rA	−343.23
-2′*O*-Me-rG (PO)	2′*O*-Me-rG	−359.23
-2′*O*-Me-rC (PO)	2′*O*-Me-rC	−319.21
-2′*O*-Me-U (PO)	2′*O*-Me-rU	−320.19
-2′F	2-F′ in all nucleosides	−20.00
-2′F-dA (PO)	2′*O*-Me-dA	−331.20
-2′F-dG (PO)	2′*O*-Me-dG	−347.20
-2′F-dC(PO)	2′*O*-Me-dC	−307.17
-2′F-dU (PO)	2′*O*-Me-dU	−308.16
-LNA-A (PO)	LNA-A	−341.22
-LNA-G (PO)	LNA-G	−357.22
-LNA-C (PO)	LNA-C	−317.19
-LNA-U (PO)	LNA-U	−318.18

solid supports have been illustrated and means to alleviate and control these impurities have been suggested.

Oligonucleotide impurities originating from the synthesis process (detritylation, coupling, capping, oxidation or sulfurisation, release of the oligonucleotide from the support and removal of protecting groups) have been explained and possible mechanisms to their formation have been provided.

Finally, commonly observed impurities present in oligonucleotides drug substance have been presented in a tabular form as a general reference.

2.8 Outlook

In the past two decades, concurrent to oligonucleotide-based therapeutic investigations, significant developments have been made in the analysis of oligonucleotides. Analytical techniques such as strong anion exchange HPLC, matrix-assisted light desorption ionisation–time of flight (MALDI-TOF) MS, IP-RP HPLC, HPLC-MS and MS/MS have been instrumental in identifying and characterising various impurities.

Impurities detected and indentified in nucleoside-phosphoramidites and synthesis reagents provide a means for controlling the quality of oligonucleotides. Similarly, post-synthesis analysis and in-process control during synthesis and purification of oligonucleotides further allow minimisation of impurities, thereby building confidence in the quality of oligonucleotides drugs. It is expected that these tools will be further utilised for maintaining the highest possible quality of therapeutic oligonucleotides in the future.

Acknowledgement

The authors thank Gary Burt from Girindus for providing the ^{31}P-NMR, HPLC and mass data of phosphoramidites released by the QC department of Girindus.

References

1. Michelson, A.M., Todd, A.R., Nucleotides part XXXII. Synthesis of a dithymidine dinucleotide containing a 3′: 5′-internucleotidic linkage, *J. Chem. Soc.*, 1955, 2632–2638.
2. Schaller, H., Weimann, G., Lerch, B., Khorana, H.G., Studies on polynucleotides. XXIV.1 The stepwise synthesis of specific deoxyribopolynucleotides (4).2 protected derivatives of deoxyribonucleosides and new syntheses of deoxyribonucleoside-3″ phosphates, *J. Am. Chem. Soc.*, 1963, 85, 3821–3827.
3. Gilham, P.T., Khorana, H.G., Studies on polynucleotides. I. A new and general method for the chemical synthesis of the C5″-C3″ internucleotidic linkage. Syntheses of deoxyribo-dinucleotides, *J. Am. Chem. Soc.*, 1958, 80, 6212–6222.
4. Khorana, H.G., *Some Recent Developments in the Chemistry of Phosphate Esters of Biological Interest*, 1961, John Wiley, London.
5. Khorana, H.G., Total synthesis of a gene, *Science*, 1979, 203, 614–625.

6. Letsinger, R.L., Mahadevan, V., Oligonucleotide synthesis on a polymer support, *J. Am. Chem. Soc.*, 1965, 87, 3526–3527.
7. Reese, C.B., The chemical synthesis of oligo- and poly-nucleotides by the phosphotriester approach, *Tetrahedron*, 1978, 34, 3143–3179.
8. Itakura, K., Bahl, C.P., Katagiri, N., Michiniewicz, J., Wightman, R.H., Narang, S.A., A modified triester method for the synthesis of deoxyribopolynucleotides, *Can. J. Chem.*, 1973, 51, 3649–3651.
9. Itakura, K., Katagiri, N., Bahl, C.P., Wightman, R.H., Narang, S.A., Improved triester approach for the synthesis of pentadecathymidylic acid, *J. Am. Chem. Soc.*, 1975, 97, 7327–7332.
10. Narang, S.A., DNA synthesis, *Tetrahedron*, 1983, 39, 3–22.
11. Catlin, J.C., Cramer, F., Deoxy oligonucleotide synthesis via the triester method, *J. Org. Chem.*, 1973, 38, 245–250.
12. Zamecnik, P.C., Stephenson, M.L., Inhibition of Rous sarcoma virus replication and cell transformation by a specific oligodeoxynucleotide, *Proc. Natl Acad. Sci. USA*, 1978, 75, 280–284.
13. Letsinger, R.L., Lunsford, W.B., Synthesis of thymidine oligonucleotides by phosphite triester intermediates, *J. Am. Chem. Soc.*, 1976, 98, 3655–3661.
14. Beaucage, S.L., Caruthers, M.H., Deoxynucleoside phosphoramidites – A new class of key intermediates for deoxypolynucleotide synthesis, *Tetrahedron Lett.*, 1981, 22, 1859–1862.
15. McBride, L.J., Caruthers, M.H., An investigation of several deoxynucleoside phosphoramidites useful for synthesizing deoxyoligonucleotides, *Tetrahedron Lett.*, 1983, 24, 245–248.
16. Sinha, N.D., Biernat, J., McManus, J., Koster, H., Polymer support oligonucleotide synthesis XVIII: use of beta-cyanoethyl-N,N-dialkylamino-/N-morpholino phosphoramidite of deoxynucleosides for the synthesis of DNA fragments simplifying deprotection and isolation of the final product, *Nucleic Acids Res.*, 1984, 12, 4539–4557.
17. Sinha, N.D., Biernat, J., Koster, H., β-Cyanoethyl N,N-dialkylamino/N-morpholinomonochloro phosphoamidites, new phosphitylating agents facilitating ease of deprotection and work-up of synthesized oligonucleotides, *Tetrahedron Lett.*, 1983, 24, 5843–5846.
18. Garegg, P.J., Regberg, T., Stawinski, J., Stromberg, R., Nucleoside hydrogenphosphonates in oligonucleotide synthesis, *Chem. Scripta*, 1986, 26, 59–62.
19. Froehler, B.C., Ng, P.G., Matteucci, M.D., Synthesis of DNA via deoxynucleoside H-phosphonate intermediates, *Nucleic Acids Res.*, 1986, 14, 5399–5407.
20. Hall, R.H., Todd, A., Webb, R.F., Nucleotides. Part XLI. Mixed anhydrides as intermediates in the synthesis of dinucleoside phosphates, *J. Chem. Soc.*, 1957, 3291–3296.
21. Ogilvie, K.K., Sadana, K.L., Thompson, E.A., Quilliam, M.A., Westmore, J.B., The use of silyl groups in protecting the hydroxyl functions of ribonucleosides, *Tetrahedron Lett.*, 1974, 15, 2861–2863.
22. Engels, J.W., Uhlmann, E., Gene synthesis, *Angew. Chem. Int. Ed. Engl.*, 1989, 28, 716–734.
23. Engels, J.W., Sprunkel, B., Uhlmann, E., DNA synthesis. In *Biotechnology*, 2nd edn, Rehm, H.J., Reed, G., Pühler, A., Stadler, P. (Eds), 1993, VCH, Weinheim.
24. Beaucage, S., Iyer, R.P., Advances in the synthesis of oligonucleotides by the phosphoramidite approach, *Tetrahedron*, 1992, 48, 2223–2311.
25. Fritz, H.-J., Eick, D., Werr, W., Analysis of oligodeoxynucleotides. In *Chemical and Enzymatic Synthesis of Gene Fragments: A Laboratory Manual*, Gassen, H.G., Lang, A. (Eds), 1982, Verlag Chemie, Weinheim.
26. Eadie, J.S., McBride, L.J., Efcavitch, J.W., Hoff, L.B., Cathcart, R., High-performance liquid chromatographic analysis of oligodeoxyribonucleotide base composition, *Analyt. Biochem.*, 1987, 165, 442–447.
27. Zon, G., Thompson, J.A., A review of high-performance liquid chromatography in nucleic acids research: II. Isolation, purification and analysis of oligodeoxyribonucleotides, *Biochromatography*, 1986, 1, 22–32.
28. Jay, E., Bambara, R., Padmanabhan, R., Wu, R., DNA sequence analysis: a general, simple and rapid method for sequencing large oligodeoxyribonucleotide fragments by mapping, *Nucleic Acids Res.*, 1974, 1, 331–353.

29. Tu, C.P., Jay, E., Bahl, C.P., Wu, R., A reliable mapping method for sequence determination of oligodeoxyribonucleotides by mobility shift analysis, *Analyt. Biochem.*, 1976, 74, 73–93.
30. Bonilla, J., Srivatsa, G.S., *Handbook of Analysis of Oligonucleotides and Related Products*, 2011, CRC Press, Boca Raton, FL.
31. Sanghvi, Y.S., A roadmap to the assembly of synthetic DNA from raw materials. In *Working Papers for Synthetic Genomics: Risks and Benefits for Science and Society*, Garfinkel, M.S., Endy, D., Epstein, G.L., Friedman, R.M. (Eds), 2005.
32. Brown, P., Will, R.G., Bradley, R., Asher, D.M., Detwiler, L., Bovine spongiform encephalopathy and variant Creutzfeldt-Jakob disease: background, evolution, and current concerns, *Emerg. Infect. Dis.*, 2001, 7, 6–16.
33. Sanghvi, Y.S., A status update of modified oligonucleotides for chemotherapeutics applications, *Curr. Protoc. Nucleic Acid Chem.*, 2011, Chapter 4, Unit 4 1, 1–22.
34. He, K., Hasan, A., Classification and characterization of impurities in phosphoramidites used in making therapeutic oligonucleotides: risk mitigation strategies for entering clinical phases, *Technical Note MK0010D01*, 2009, Thermo Fisher Scientific, Milwaukee, WI.
35. Ravikumar, V.T., Kumar, R.K., Stereoselective synthesis of alkylphosphonates: a facile rearrangement of cyanoethyl-protected nucleoside phosphoramidites, *Org. Proc. Res. Dev.*, 2004, 8, 603–608.
36. Krotz, A.H., Rentel, C., Gorman, D., Olsen, P., Gaus, H.J., McCardle, J.V., Scozzari, A.N., Solution stability and degradation pathway of deoxyribonucleoside phosphoramidites in acetonitrile, *Nucleosides Nucleotides Nucleic Acids*, 2004, 23, 767–775.
37. Nielsen, J., Marugg, J.E., Van Boom, J.H., Honnens, J., Taagaard, M., Dahl, O., Thermal instability of some alkyl phosphorodiamidites, *J. Chem Res. (S)*, 1986, 26–27.
38. Pauling, L., The nature of the chemical bond. IV. The energy of single bonds and the relative electronegativity of atoms, *J. Am. Chem. Soc.*, 1932, 54, 3570–3582.
39. Allred, A.L., Electronegativity values from thermochemical data, *J. Inorg. Nucl. Chem.*, 1961, 17, 215–221.
40. Kupihar, Z., Timar, Z., Darula, Z., Dellinger, D.J., Caruthers, M.H., An electrospray mass spectrometric method for accurate mass determination of highly acid-sensitive phosphoramidites, *Rapid Communs Mass Spectrom.*, 2008, 22, 533–540.
41. Kupihar, Z., Timar, Z., Dellinger, D.J., Caruthers, M.H., Accurate mass analysis of phosphoramidites by electrospray mass spectrometry, *Nucleosides Nucleotides Nucleic Acids*, 2005, 24, 663–666.
42. Fujitake, M., Harusawa, S., Araki, L., Yamaguchi, M., Lilley, D. M. J., Zhaob, Z., Kurihara, T., Accurate molecular weight measurements of nucleoside phosphoramidites: a suitable matrix of mass spectrometry, *Tetrahedron*, 2005, 61, 4689–4699.
43. Katzhendler, J., Cohen, S., Rahamim, E., Weisz, M., Ringel, I., Deutsch, J., The effect of spacer, linkage and solid support on the synthesis of oligonucleotides, *Tetrahedron*, 1989, 45, 2777–2792.
44. Kozlov, I.A., Dang, M., Sikes, K., Kotseroglou, T., Barker, D.L., Zhao, C., Significant improvement of quality for long oligonucleotides by using controlled pore glass with large pores, *Nucleosides Nucleotides Nucleic Acids*, 2005, 24, 1037–1041.
45. Vu, H., McCollum, C., Lotys, C., Andrus, A., New reagents and solid support for automated oligonucleotide synthesis, *Nucleic Acids Symp. Ser.*, 1990, 63–64.
46. McCollum, C., Andrus, A., An optimized polystyrene support for rapid, efficient oligonucleotide synthesis, *Tetrahedron Lett.*, 1991, 32, 4069–4072.
47. Reddy, M.P., Michael, M.A., Farooqui, F., Girgis, S., New and efficient solid support for the synthesis of nucleic acids, *Tetrahedron Lett.*, 1994, 35, 5771–5774.
48. Pon, R.T., Solid-phase supports for oligonucleotide synthesis, *Methods Mol. Biol.*, 1993, 20, 465–496.
49. Pon, R.T., Solid-phase supports for oligonucleotide synthesis, *Curr. Protoc. Nucleic Acid Chem.*, 2001, Chapter 3, Unit 3.1.
50. Yip, K.F., Tsou, K.C., A new polymer-support method for the synthesis of ribooligonucleotide, *J. Am. Chem. Soc.*, 1971, 93, 3272–3276.

51. Alul, R.H., Singman, C.N., Zhang, G.R., Letsinger, R.L., Oxalyl-CPG: a labile support for synthesis of sensitive oligonucleotide derivatives, *Nucleic Acids Res.*, 1991, 19, 1527–1532.
52. Pon, R.T., Yu, S., Hydroquinone-O,O′-diacetic acid (‘Q-linker’) as a replacement for succinyl and oxalyl linker arms in solid phase oligonucleotide synthesis, *Nucleic Acids Res.*, 1997, 25, 3629–3635.
53. Pon, R.T., Yu, S., Hydroquinone-O,O synthesise-diacetic acid as a more labile replacement for succinic acid linkers in solid-phase oligonucleotide synthesis, *Tetrahedron Lett.*, 1997, 38, 3327–3330.
54. Capaldi, D.C., Gaus, H., Krotz, A.H., Arnold, J., Carty, R.L., Moore, M.N., Scozzari, A.N., Lowery, K., Cole, D.L., Ravikumar, V.T., Synthesis of high-quality antisense drugs. Addition of acrylonitrile to phosphorothioate oligonucleotides: adduct characterization and avoidance, *Org. Process Res. Dev.*, 2003, 7, 832–838.
55. Gough, G.R., Brunden, M.J., Gilham, P.T., 2′(3′)-O-benzoyluridine 5′ linked to glass: an all-purpose support for solid phase synthesis of oligodeoxyribonucleotides, *Tetrahedron Lett.*, 1983, 24, 5321–5324.
56. Vargeese, C., Wang, W., Methods and reagents for oligonucleotide synthesis, 2007, Sirna Therapeutics, Inc., US Patent US7205399.
57. Kumar, P., Sharma, A.K., Sharma, P., Garg, B.S., Gupta, K.C., Express protocol for functionalization of polymer supports for oligonucleotide synthesis, *Nucleosides Nucleotides*, 1996, 15, 879–888.
58. Bhongle, N.N., Tang, J.Y., A convenient and practical method for derivatization of solid supports for nucleic acid synthesis, *Synth. Commun.*, 1995, 25, 3671–3679.
59. Pon, R.T., Yu, S., Rapid automated derivatization of solid-phase supports for oligonueleotide synthesis using uronium or phosphonium coupling reagents, *Tetrahedron Lett.*, 1997, 38, 3331–3334.
60. Pon, R.T., Yu, S., Sanghvi, Y.S., Rapid esterification of nucleosides to solid-phase supports for oligonucleotide synthesis using uronium and phosphonium coupling reagents, *Bioconjug. Chem.*, 1999, 10, 1051–1057.
61. Kurata, C., Bradley, K., Gaus, H., Luu, N., Cedillo, I., Ravikumar, V.T., Van Sooy, K., McArdle, J.V., Capaldi, D.C., Characterization of high molecular weight impurities in synthetic phosphorothioate oligonucleotides, *Bioorg. Med. Chem. Lett.*, 2006, 16, 607–614.
62. Cazenave, C., Bathany, K., Rayner, B., Formation of N-branched oligonucleotides as by-products in solid-phase oligonucleotide synthesis, *Oligonucleotides*, 2006, 16, 181–185.
63. Ravikumar, V.T., Capaldi, D.C., Lima, W.F., Lesnik, E., Turney, B., Cole, D.L., Antisense phosphorothioate oligodeoxyribonucleotide targeted against ICAM-1: synthetic and biological characterization of a process-related impurity formed during oligonucleotide synthesis, *Bioorg. Med. Chem.*, 2003, 11, 4673–4679.
64. Holik, M., Matejkova, B., Determination of hydroxyl groups and water content in silica by nuclear magnetic resonance spectroscopy, *J. Chromatogr.*, 1981, 213, 33–39.
65. Snyder, I.R., *Principles of Adsorption Chromatography*, 1968, Marcel Dekker, New York.
66. Yu, D., Tang, J.-Y., Iyer, R.P., Agrawal, S., Diethoxy N, N-diisopropyl phosphoramidite as an improved capping reagent in the synthesis of oligonucleotides using phosphoramidite chemistry, *Tetrahedron Lett.*, 1994, 35, 8565–8568.
67. Cvetovich, R.J., Hydrogen peroxide oxidation of phosphite triesters in oligonucleotide syntheses, *Org. Process Res. Dev.*, 2010, 14, 295–297.
68. Ravikumar, V.T., Kumar, R.K., Capaldi, D.C., Turney, B., Rentel, C., Cole, D.L., Antisense phosphorothioate oligodeoxyribonucleotide targeted against ICAM-1: Use of I-linker to eliminate 3′-terminal phosphorothioate monoester formation, *Org. Process Res. Dev.*, 2003, 7, 259–266.
69. Septak, M., Kinetic studies on depurination and detritylation of CPG-bound intermediates during oligonucleotide synthesis, *Nucleic Acids Res.*, 1996, 24, 3053–3058.
70. Suzuki, T., Ohsumi, S., Makino, K., Mechanistic studies on depurination and apurinic site chain breakage in oligodeoxyribonucleotides, *Nucleic Acids Res.*, 1994, 22, 4997–5003.

71. Nelson, P.S., Muthini, S., Vierra, M., Acosta, L., Smith, T.H., Rainbow Universal CPG: a versatile solid support for oligonucleotide synthesis, *Biotechniques*, 1997, 22, 752–756.
72. Lyttle, M.H., Dick, D.J., Hudson, D., Cook, R.M., A phosphate bound universal linker for DNA synthesis, *Nucleosides Nucleotides*, 1999, 18, 1809–1824.
73. Schcuer-Larsen, C., Rosenbohm, C., Jøgensen, T.J.D., Wengel, J., Introduction of a universal solid support for oligonucleotide synthesis, *Nucleosides Nucleotides*, 1997, 16, 67–80.
74. Scott, S., Hardy, P., Sheppard, R.C., McLean, M.J., A universal support for oligonucleotide synthesis. In *Innovations and Perspectives in Solid Phase Synthesis, 3rd International Symposium*, Epton, R. (Ed.), 1994, Mayflower Worldwide.
75. Kumar, P., Dhawan, G., Chandra, R., Gupta, K.C., Polyamine-assisted rapid and clean cleavage of oligonucleotides from cis-diol bearing universal support, *Nucleic Acids Res.*, 2002, 30, e130.
76. Kumar, P., Mahajan, S., Gupta, K.C., Universal reusable polymer support for oligonucleotide synthesis, *J. Org. Chem.*, 2004, 69, 6482–6485.
77. Anderson, E., Brown, T., Picken, D., Novel photocleavable universal support for oligonucleotide synthesis, *Nucleosides Nucleotides Nucleic Acids*, 2003, 22, 1403–1406.
78. Anderson, K.M., Jaquinod, L., Jensen, M.A., Ngo, N., Davis, R.W., A novel catechol-based universal support for oligonucleotide synthesis, *J. Org. Chem.*, 2007, 72, 9875–9880.
79. Morvan, F., Meyer, A., Vasseur, J.J., A universal and recyclable solid support for oligonucleotide synthesis, *Curr. Protoc. Nucleic Acid Chem.*, 2007, Chapter 3, Unit 3.16.
80. Guzaev, A.P., Manoharan, M., A conformationally preorganized universal solid support for efficient oligonucleotide synthesis, *J. Am. Chem. Soc.*, 2003, 125, 2380–2381.
81. Ravikumar, V.T., Kumar, R.K., Olsen, P., Moor, M.N., Carty, R.L., Andrade, M., Gorman, D., Zhu, X., Cedillo, I., Wang, Z., Mendez, L., Scozzari, A.N., Aguirre, G., Somanathan, R., Bernees, S., UnyLinker: An efficient and scaleable synthesis of oligonucleotides utilizing a universal linker molecule: a novel approach to enhance the purity of drugs, *Org. Process Res. Dev.*, 2008, 12, 399–410.
82. Azhayev, A.V., A new universal solid support for oligonucleotide synthesis, *Tetrahedron*, 1999, 55, 787.
83. Azhayev, A.V., Antopolsky, M.L., Amide group assisted 3′-dephosphorylation of oligonucleotides synthesized on universal A-supports, *Tetrahedron*, 2001, 57, 4977–4986.
84. Azhayev, A.V., Antopolsky, M.L., Tennilä, T.M.L., Mackie, H., Randolph, J.B., A comparative study of commercially available universal supports, *Genetic Engng News*, 2005, 25.
85. Glen Research Corporation, High load Glen UnySupport, *The Glen Report*, 2009, Vol. 21, No. 1, Glen Research Corporation, Sterling, VA.
86. Guzaev, A.P., *Guidelines for Selection and Use of Universal Solid Supports*. AM Chemicals, Oceanside, CA.
87. Glen Research Corporation, Improving universal support II for oligonucleotide synthesis, *The Glen Report*, 2008, Vol. 20, No. 1, Glen Research Corporation, Sterling, VA.
88. Yagodkin, A., Azhayev, A., Carbamoylation of amines, thiophenols, mercaptanes and phenols employing organic azides, 2010, US Patent US2010093981, Metkinen Chemistry, Finland.
89. Paul, C.H., Royappa, A.T., Acid binding and detritylation during oligonucleotide synthesis, *Nucleic Acids Res.*, 1996, 24, 3048–3052.
90. Caruthers, M.H., Barone, A.D., Beaucage, S.L., Dodds, D.R., Fisher, E.F., McBride, L.J., Matteucci, M., Stabinsky, Z., Tang, J.Y., Chemical synthesis of deoxyoligonucleotides by the phosphoramidite method, *Methods Enzymol.*, 1987, 154, 287–313.
91. Atkinson, T., Smith, M., Solid-phase synthesis of oligodeoxyribonucleotides by the phosphotriester method. In *Oligonucleotide Synthesis: A Practical Approach*, Gait, M.J. (Ed.), 1984, IRL Press, Oxford/Washington, DC.
92. Krotz, A.H., Carty, R.L., Moore, M.N., Scozzari, A.N., Cole, D.L., Ravikumar, V.T., Synthesis of antisense oligonucleotides using environmentally friendly and safe deprotection procedures, *Green Chem.*, 1999, 1, 277–281.
93. Krotz, A.H., Carty, R.L., Scozzari, A.N., Cole, D.L., Ravikumar, V.T., Large-scale synthesis of antisense oligonucleotides without chlorinated solvents, *Org. Process Res. Dev.*, 2000, 4, 190–193.

94. Krotz, A.H., Cole, D.L., Ravikumar, V.T., Synthesis of an antisense oligonucleotide targeted against C-raf kinase: efficient oligonucleotide synthesis without chlorinated solvents, *Bioorg. Med. Chem.*, 1999, 7, 435–439.
95. Cheruvallath, Z.S., Carty, R.L., Andrade, M., Moore, M.N., Song, Q., Rentel, C., Cole, D.L., Ravikumar, V.T., Efficient synthesis of antisense phosphorothioate oligonucleotides: evaluation of dichloroacetic acid at higher concentration to reduce cycle time, *Org. Process Res. Dev.*, 2003, 7, 917–920.
96. Gaus, H., Olsen, P., Sooy, K.V., Rentel, C., Turney, B., Walker, K.L., McArdle, J.V., Capaldi, D.C., Trichloroacetaldehyde modified oligonucleotides, *Bioorg. Med. Chem. Lett.*, 2005, 15, 4118–4124.
97. Nurminen, E.J., Mattinen, J.K., Lönnberg, H., Protonation of phosphoramidites. The effect on nucleophilic displacement, *J. Chem. Soc. Perkin Trans.*, 2000, 2, 2238–2240.
98. Dahl, B.H., Nielsen, J., Dahl, O., Mechanistic studies on the phosphoramidite coupling reaction in oligonucleotide synthesis. I. Evidence for nucleophilic catalysis by tetrazole, rate variations with the phosphorus substituents. *Nucleic Acids Res.*, 1987, 15, 1729–1743.
99. Berner, S., Muhlegger, K., Seliger, H., Studies on the role of tetrazole in the activation of phosphoramidites, *Nucleic Acids Res.*, 1989, 17, 853–864.
100. Nurminen, E.J., Mattinen, J.K., Lönnberg, H., Kinetics and mechanism of tetrazole-catalyzed phosphoramidite alcoholysis, *J. Chem. Soc. Perkin Trans. 2*, 1998, 1621–1628.
101. Beaucage, S.L., Caruthers, M.H., Synthetic strategies and parameters involved in the synthesis of oligodeoxyribonucleotides according to the phosphoramidite method, *Curr. Protoc. Nucleic Acid Chem.*, 2001, Chapter 3, Unit 3.3.
102. Dorman, M.A., Noble, S.A., McBride, L.J., Caruthers, M.H., Synthesis of oligodeoxynucleotides and oligodeoxynucleotide analogs using phosphoramidite intermediates, *Tetrahedron*, 1984, 40, 95–102.
103. Hogrefe, R.I., Reynolds, M.A., Vaghefi, M.M., Young, K.M., Riley, T.A., Klem, R.E., Arnold, L.J., Jr, An improved method for the synthesis and deprotection of methylphosphonate oligonucleotides. In *Protocols for Oligonucleotides and Analogs: Synthesis and Properties*, Agrawal, S. (Ed.), 1993, Humana Press, Totowa, NJ.
104. Hayakawa, Y., Uchiyama, M., Noyori, R., Nonaqueous oxidation of nucleoside phosphites to the phosphates, *Tetrahedron Lett.*, 1986, 27, 4191–4194.
105. Sproat, B., Colonna, F., Mullah, B., Tsou, D., Andrus, A., Hampel, A., Vinayak, R., An efficient method for the isolation and purification of oligoribonucleotides, *Nucleosides Nucleotides*, 1995, 14, 255–273.
106. Glen Research Corporation, Non-aqueous oxidation with 10-camphorsulfonyl-oxaziridine, *The Glen Report*, 1996, Vol. 9, No. 1, pp. 8–9, Glen Research Corporation, Sterling, VA.
107. Ugi, I., Jacob, P., Landgraf, B., Rupp, C., Lemmen, P., Verfuerth, U., Phosphite oxidation and the preparation of five-membered cyclic phosphorylating reagents via the phosphates, *Nucleosides Nucleotides*, 1988, 7, 605–608.
108. Dellinger, D.J., Sheehan, D.M., Christensen, N.K., Lindberg, J.G., Caruthers, M.H., Solid-phase chemical synthesis of phosphonoacetate and thiophosphonoacetate oligodeoxynucleotides, *J. Am. Chem. Soc.*, 2003, 125, 940–950.
109. Manoharan, M., Lu, Y., Casper, M.D., Ju, G., Allyl group as a protecting group for internucleotide phosphate and thiophosphate linkages in oligonucleotide synthesis: facile oxidation and deprotection conditions, *Org. Lett.*, 2000, 2, 243–246.
110. Sierzchala, A.B., Dellinger, D.J., Betley, J.R., Wyrzykiewicz, T.K., Yamada, C.M., Caruthers, M.H., Solid-phase oligodeoxynucleotide synthesis: a two-step cycle using peroxy anion deprotection, *J. Am. Chem. Soc.*, 2003, 125, 13427–13441.
111. Uzagare, M.C., Padiya, K.J., Salunkhe, M.M., Sanghvi, Y.S., NBS–DMSO as a nonaqueous nonbasic oxidation reagent for the synthesis of oligonucleotides, *Bioorg. Med. Chem. Lett.*, 2003, 13, 3537–3540.
112. Kataoka, M., Hattori, A., Okino, S., Hyodo, M., Asano, M., Kawai, R., Hayakawa, Y., Ethyl(methyl)dioxirane as an efficient reagent for the oxidation of nucleoside phosphites into phosphates under nonbasic anhydrous conditions, *Org. Lett.*, 2001, 3, 815–818.

113. Saneyoshi, H., Miyata, K., Seio, K., Sekine, M., 1,1-Dihydroperoxycyclododecane as a new, crystalline non-hygroscopic oxidizer for the chemical synthesis of oligodeoxyribonucleotides, *Tetrahedron Lett.*, 2006, 47, 8945–8947.
114. Burgers, P.M.J., Eckstein, F., Synthesis of dinucleoside monophosphorothioates via addition of sulphur to phosphite triesters, *Tetrahedron Lett.*, 1978, 19, 3835–3838.
115. Vu, H., Hirschbein, B.L., Internucleotide phosphite sulfurization with tetraethylthiuram disulfide. Phosphorothioate oligonucleotide synthesis via phosphoramidite chemistry, *Tetrahedron Lett.*, 1991, 32, 3005–3008
116. Stec, W.J., Uznanski, B., Wilk, A., Hirschbein, B.L., Fearon, K.L., Bergot, B.J., Bis(O,O-diisopropoxy phosphinothioyl) disulfide – a highly efficient sulfurizing reagent for cost-effective synthesis of oligo(nucleoside phosphorothioate)s, *Tetrahedron Lett.*, 1993, 34, 5317–5320.
117. Iyer, R.P., Phillips, L.R., Egan, W., Regan, J.B., Beaucage, S.L., The automated synthesis of sulfur-containing oligodeoxyribonucleotides using 3H-1,2-benzodithiol-3-one 1,1-dioxide as a sulfur-transfer reagent, *J. Org. Chem.*, 1990, 55, 4693–4699.
118. Krotz, A.H., Gorman, D., Mataruse, P., Foster, C., Godbout, J.D., Coffin, C.C., Scozzari, A.N., Phosphorothioate oligonucleotides with low phosphate diester content: greater than 99.9% sulfurization efficiency with 'aged' solutions of phenylacetyl disulfide (PADS), *Org. Process Res. Dev.*, 2004, 8, 852–858.
119. Cheruvallath, Z.S., Carty, R.L., Moore, M.N., Capaldi, D.C., Krotz, A.H., Wheeler, P.D., Turney, B.J., Craig, S.R., Gaus, H.J., Scozzari, A.N., Cole, D.L., Ravikumar, V.T., Synthesis of antisense oligonucleotides: replacement of 3H-1,2-benzodithiol-3-one 1,1-dioxide (Beaucage reagent) with phenylacetyl disulfide (PADS) as efficient sulfurization reagent: from bench to bulk manufacture of active pharmaceutical ingredient, *Org. Process Res. Dev.*, 2000, 4, 199–204.
120. Roelen, H.C.P.F., Kamer, P.C.J., van den Elst, H., van der Marel, G.A., van Boom, J.H., A study on the use of phenylacetyl disulfide in the solid-phase synthesis of oligodeoxynucleoside phosphorothioates, *Recl. Trav. Chim. Pays-Bas*, 1991, 110, 325–331.
121. Ma, M.Y., Dignam, J.C., Fong, G.W., Li, L., Gray, S.H., Jacob-Samuel, B., George, S.T., Evaluation of 3-ethoxy-1,2,4-dithiazoline-5-one (EDITH) as a new sulfurizing reagent in combination with labile exocyclic amino protecting groups for solid-phase oligonucleotide synthesis, *Nucleic Acids Res.*, 1997, 25, 3590–3593.
122. Xu, Q., Barany, G., Hammer, R.P., Musier-Forsyth, K., Efficient introduction of phosphorothioates into RNA oligonucleotides by 3-ethoxy-1,2,4-dithiazoline-5-one (EDITH), *Nucleic Acids Res.*, 1996, 24, 3643–3644.
123. Xu, Q., Musier-Forsyth, K., Hammer, R.P., Barany, G., Use of 1,2,4-dithiazolidine-3,5-dione (DtsNH) and 3-ethoxy-1,2,4-dithiazoline-5-one (EDITH) for synthesis of phosphorothioate-containing oligodeoxyribonucleotides, *Nucleic Acids Res.*, 1996, 24, 1602–1607.
124. Tang, J.-Y., Han, Y., Tang, J.X., Zhang, Z., Large-scale synthesis of oligonucleotide phosphorothioates using 3-amino-1,2,4-dithiazole-5-thione as an efficient sulfur-transfer reagent, *Org. Process Res. Dev.*, 2000, 4, 194–198.
125. Hanusek, J., Russell, M.A., Laws, A.P., Jansa, P., Atherton, J.H., Fettes, K., Page, M.I., Mechanism of the sulfurisation of phosphines and phosphites using 3-amino-1,2,4-dithiazole-5-thione (xanthane hydride), *Org. Biomol. Chem.*, 2007, 5, 478–484.
126. Han, Y., Tang, J., Zhang, Z., Tang, J.-Y., Sulfur transfer reagents for oligonucleotide synthesis, 2000, US Patent 6096881, Hybridon, Inc., USA.
127. Guzaev, A.P., Sulfur transfer reagents for oligonucleotide synthesis, 2010, US Patent 7723528, AM Chemicals, LLC, USA.
128. Guzaev, A.P., Reactivity of 3H-1,2,4-dithiazole-3-thiones and 3H-1,2-dithiole-3-thiones as sulfurizing agents for oligonucleotide synthesis, *Tetrahedron Lett.*, 2011, 52, 434–437.
129. Wang, Z., Song, Q., Sanghvi, Y.S., Dimethylthiarum disulfide: new sulfur transfer reagent in phosphorothioates oligonucleotide synthesis, *Methods Mol. Biol.*, 2005, 288, 51–64.
130. Song, Q., Wang, Z., Sanghvi, Y.S., A short, novel, and cheaper procedure for oligonucleotide synthesis using automated solid phase synthesizer, *Nucleosides Nucleotides Nucleic Acids*, 2003, 22, 629–633.

131. Roy, S.K., Tang, J.-Y., Sulfurizing reagent: 3-aryl-1,2,4-dithiazoline-5-ones, 2002, US Patent US6500944, Avecia Biotechnology, Inc., USA.
132. Efimov, V.A., Kalinkina, A.L., Chakhmakhcheva, O.G., Hill, T.S., Jayaraman, K., New efficient sulfurizing reagents for the preparation of oligodeoxyribonucleotide phosphorothioate analogues, *Nucleic Acids Res.*, 1995, 23, 4029–4033.
133. Eadie, J.S., Davidson, D.S., Guanine modification during chemical DNA synthesis, *Nucleic Acids Res.*, 1987, 15, 8333–8349.
134. Chaix, C., Molko, D., Teoule, R., The use of labile base protecting groups in oligoribonucleotide synthesis, *Tetrahedron Lett.*, 1989, 30, 71–74.
135. Zhu, Q., Delaney, M.O., Greenberg, M.M., Observation and elimination of N-acetylation of oligonucleotides prepared using fast-deprotecting phosphoramidites and ultra-mild deprotection, *Bioorg Med Chem Lett.*, 2001, 11, 1105–1107.
136. Sinha, N., Kuchimanchi, S.N., Miranda, G., Shaikh, S., Manufacture of therapeutic oligonucleotides: Development of new reagents and processes, *Indian J. Chem.*, 2006, 45B, 2297–2304.
137. Sinha, N.D., Michaud, D.P., Recent developments in the chemistry, analysis and control for the manufacture of therapeutic-grade synthetic oligonucleotides, *Curr. Opin. Drug Discov. Devel.*, 2007, 10, 807–818.
138. Glen Research Corporation, UniCap phosphoramidite, An alternative to acetic anhydride capping, *The Glen Report*, 2004, Vol. 17, No. 1, pp. 14–15, Glen Research Corporation, Sterling, VA.
139. Lorenz, S., Przybytek, J., Snoble, K., Synthesis of oligonucleotides or phosphorothioate oligonucleotide with a capping agent of N-methylimidazole free of 1,3,5-trimethylhexahydro-1,3,5-triazine, 2009, Patent WO2009052034, Honeywell International, Inc., USA.
140. Horn, T., Urdea, M.S., Solid supported hydrolysis of apurinic sites in synthetic oligonucleotides for rapid and efficient purification on reverse-phase cartridges, *Nucleic Acids Res.*, 1988, 16, 11559–11571.
141. Efcavitch, J.W., Heiner, C., Depurination as a yield decreasing mechanism in oligodeoxynucleotide synthesis, *Nucleosides Nucleotides*, 1985, 4, 267–267.
142. Glen Research Corporation, Technical brief – Synthesis of long oligonucleotides, *The Glen Report*, 2009, Vol. 21, No. 2, Glen Research Corporation, Sterling, VA.
143. Froehler, B.C., Matteucci, M.D., Dialkylformamidines: depurination resistant N6-protecting group for deoxyadenosine, *Nucleic Acids Res.*, 1983, 11, 8031–8036.
144. Sinha, N.D., Davis, P., Usman, N., Perez, J., Hodge, R., Kremsky, J., Casale, R., Labile exocyclic amine protection of nucleosides in DNA, RNA and oligonucleotide analog synthesis facilitating N-deacylation, minimizing depurination and chain degradation, *Biochimie*, 1993, 75, 13–23.
145. Gut, I.G., Depurination of DNA and matrix-assisted laser desorption/ionization mass spectrometry, *Int. J. Mass Spectrometry and Ion Processes*, 1997, 169/170, 313–323.
146. Pourshahian, S., McCarthy, S., Analysis of oligonucleotides by liquid chromatography and mass spectrometry. In *Handbook of Analysis of Oligonucleotides and Related Products*, Bonilla, J., Srivatsa, G.S. (Eds), 2011, CRC Press, Boca Raton, FL.
147. Calvitt, C.J., Levin, D.S., Shepperd, B.T., Gruenloh, C.J., Chemistry at the 2′ position of constituent nucleotides controls degradation pathways of highly modified oligonucleotide molecules, *Oligonucleotides*, 2010, 20, 239–251.
148. Capaldi, D.C., Gaus, H.J., Carty, R.L., Moore, M.N., Turney, B.J., Decottignies, S.D., McArdle, J.V., Scozzari, A.N., Ravikumar, V.T., Krotz, A.H., Formation of 4,4′-dimethoxytrityl-C-phosphonate oligonucleotides, *Bioorg. Med. Chem. Lett.*, 2004, 14, 4683–4690.
149. Temsamani, J., Kubert, M., Agrawal, S., Sequence identity of the n-1 product of a synthetic oligonucleotide, *Nucleic Acids Res.*, 1995, 23, 1841–1844.
150. Fearon, K.L., Stults, J.T., Bergot, B.J., Christensen, L.M., Raible, A.M., Investigation of the 'n – 1' impurity in phosphorothioate oligodeoxynucleotides synthesized by the solid-phase beta-cyanoethyl phosphoramidite method using stepwise sulfurization, *Nucleic Acids Res.*, 1995, 23, 2754–2761.
151. Krotz, A.H., Klopchin, P.G., Walker, K.L., Srivatsa, G.S., Cole, D.L., Ravikumar, V.T., On the formation of longmers in phosphorothioate oligodeoxyribonucleotide synthesis, *Tetrahedron Lett.*, 1997, 38, 3875–3878.

152. Vargeese, C., Carter, J., Pieken, W., Efficient activation of nucleoside phosphoramidites with 4,5-dicyanoimidazole during oligonucleotide synthesis, *Nucleic Acids Res.*, 1998, 26, 1046.
153. Eleuteri, A., Capaldi, D.C., Krotz, A.H., Cole, D.L., Ravikumar, V.T., Pyridinium trifluoroacetate/N-methylimidazole as an efficient activator for oligonucleotide synthesis via the phosphoramidite method, *Org. Process Res. Dev.*, 2000, 4, 182–189.
154. Sinha, N.D., Foster, P., Kuchimanchi, S.N., Miranda, G., Shaikh, S., Michaud, D., Highly effective non-explosive activators based on saccharin for the synthesis of oligonucleotides and phosphoramidites, *Nucleosides Nucleotides Nucleic Acids*, 2007, 26, 1615–1618.
155. Russell, M.A., Laws, A.P., Atherton, J.H., Page, M.I., The mechanism of the phosphoramidite synthesis of polynucleotides, *Org. Biomol. Chem.*, 2008, 6, 3270–3275.
156. Wolter, A., Leuck, M., Activators for oligonucleotide and phosphoramidite synthesis, 2011, US Patent US7897758, Sigma-Aldrich Co., USA.
157. Wolter, A., Personal communication, 2011, Sigma-Aldrich, Hamburg, Germany.
158. Kumar, R.K., Olsen, P., Ravikumar, V.T., An alternative advantageous protocol for efficient synthesis of phosphorothioate oligonucleotides utilizing phenylacetyl disulfide (PADS), *Nucleosides Nucleotides Nucleic Acids*, 2007, 26, 181–188.
159. Ravikumar, V.T., Andrade, M., Carty, R.L., Dan, A., Barone, S., Development of siRNA for therapeutics: efficient synthesis of phosphorothioate RNA utilizing phenylacetyl disulfide (PADS), 2006, *Bioorg. Med. Chem. Lett.*, 16, 2513–2517.
160. Pon, R.T., Damha, M.J., Ogilvie, K.K., Modification of guanine bases by nucleoside phosphoramidite reagents during the solid phase synthesis of oligonucleotides, *Nucleic Acids Res.*, 1985, 13, 6447–6465.
161. Pon, R.T., Usman, N., Damha, M.J., Ogilvie, K.K., Prevention of guanine modification and chain cleavage during the solid phase synthesis of oligonucleotides using phosphoramidite derivatives, *Nucleic Acids Res.*, 1986, 14, 6453–6470.
162. Brown, T., Brown, D.J.S., Modern machine-aided methods of oligodeoxyribonucleotide synthesis. In *Oligonucleotides and Analogues: A Practical Approach*, Eckstein, F. (Ed.), 1991, IRL Press, Oxford; New York.
163. Hoffmann, S., Hoos, J., Klussmann, S., Vonhoff, S., RNA aptamers and spiegelmers: synthesis, purification, and post-synthetic PEG conjugation, *Curr. Protoc. Nucleic Acid Chem.*, 2011, Chapter 4, Unit 4.46, 1–30.
164. Kodra, J.T., Kehler, J., Dahl, O., Stability of oligodeoxynucleoside phosphorodithioates and phosphorothioates in aqueous ammonia, *Nucleic Acids Res.*, 1995, 23, 3349–3350.
165. Reese, C. B., Song, Q., Avoidance of sulfur loss during ammonia treatment of oligonucleotide phosphorothioates, *Nucleic Acids Res.*, 1997, 25, 2943–2944.
166. Turney, B.J., Cheruvallath, Z.S., Andrade, M., Cole, D.L., Ravikumar, V.T., Stability of phosphorothioate oligonucleotides in aqueous ammonia in presence of stainless steel, *Nucleosides Nucleotides*, 1999, 18, 89–93.
167. Eritja, R., Robles, J., Avino, A., Albericio, F., Pedroso, E., A synthetic procedure for the preparation of oligonucleotides without using ammonia and its application for the synthesis of oligonucleotides containing O-4-alkyl thymidines, *Tetrahedron*, 1993, 48, 4171–4182.
168. Umemoto, T., Wada, T., Nitromethane as a scavenger of acrylonitrile in the deprotection of synthetic oligonucleotides, *Tetrahedron Lett.*, 2005, 46, 4251–4253.
169. Ravikumar, V.T., Manoharan, M., Capaldi, D.C., Krotz, A.H., Cole, D.L., Guzaev, A., Process for the synthesis of oligomeric compounds, 2002, US Patent 6465628, ISIS Pharmaceuticals, USA.
170. Sinha, N.D., Method of preventing modification of synthetic oligonucleotides, 2006, US Patent 7038027, Avecia Biotechnology, Inc., USA.
171. Zaramella, S., Yeheskiely, E., Strömberg, R., A method for solid-phase synthesis of oligonucleotide 5′-peptide-conjugates using acid-labile alpha-amino protections, *J. Am. Chem. Soc.*, 2004, 126, 14029–14035.
172. McBride, L.J., Kierzek, R., Beaucage, S.L., Caruthers, M.H., Amidine protecting groups for oligonucleotide synthesis, *J. Am. Chem. Soc.*, 1986, 108, 2040–2048.
173. Vu, H., McCollum, C., Jacobson, K., Theisen, P., Vfnayak, R., Spiess, E., Andrus, A., Fast

oligonucleotide deprotection phosphoramidite chemistry for DNA synthesis, *Tetrahedron Lett.*, 1990, 31, 7269–7272.
174. Schulhof, J.C., Molko, D., Teoule, R., The final deprotection step in oligonucleotide synthesis is reduced to a mild and rapid ammonia treatment by using labile base-protecting groups, *Nucleic Acids Res.*, 1987, 15, 397–416.
175. Wu, T., Ogilvie, K.K., Pon, R.T., N-Phenoxyacetylated guanosine and adenosine phosphoramidites in the solid phase synthesis of oligoribonucleotides: Synthesis of a ribozyme sequence, *Tetrahedron Lett.*, 1988, 29, 4249–4252.
176. Uznanski, B., Grajkowski, A., Wilk, A., The isopropoxyacetic group for convenient base protection during solid-support synthesis of oligodeoxyribonucleotides and their triester analogs, *Nucleic Acids Res.*, 1989, 17, 4863–4871.
177. Köster, H., Kulikowski, K., Liese, T., Heikens, W., Kohli, V., N-acyl protecting groups for deoxynucleosides : A quantitative and comparative study, *Tetrahedron*, 1981, 37, 363–369.
178. Kempe, T., Recovery of oligonucleotides by gas phase cleavage, 1998, US Patent 5738829, Barrskogen, Inc., USA.
179. Boal, J.H., Wilk, A., Harindranath, N., Max, E.E., Kempe, T., Beaucage, S.L., Cleavage of oligodeoxyribonucleotides from controlled-pore glass supports and their rapid deprotection by gaseous amines, *Nucleic Acids Res.*, 1996, 24, 3115–3117.
180. Reddy, M.P., Hanna, N.B., Farooqui, F., Fast cleavage and deprotection of oligonucleotides, *Tetrahedron Lett.*, 1994, 35, 4311–4314.
181. Reddy, M.P., Hanna, N.B., Farooqui, F., Elimination of transamination side product by the use of dCAc methylphosphonamidite in the synthesis of oligonucleoside methylphosphonates, *Tetrahedron Lett.*, 1996, 37, 8691–8694.
182. Hovinen, J., Guzaev, A., Azhayev, A., Lönnberg, H., Novel solid supports for the preparation of 3′-derivatized oligonucleotides: introduction of 3′-alkylphosphate tether groups bearing amino, carboxy, carboxamido, and mercapto functionalities, *Tetrahedron*, 1994, 50, 7203–7218.
183. Bass, B.L., RNA editing by adenosine deaminases that act on RNA, *Ann. Rev. Biochem.*, 2002, 71, 817–846.
184. Navaratnam, N., Sarwar, R., An overview of cytidine deaminases, *Int. J. Hematol.*, 2006, 83, 195–200.
185. Wolfenden, R., Are there limits to enzyme-inhibitor binding discrimination? Inferences from the behavior of nucleoside deaminases, *Pharmacol. Ther.*, 1993, 60, 235–244.
186. Ehrlich, M., Norris, K.F., Wang, R.Y., Kuo, K.C., Gehrke, C.W., DNA cytosine methylation and heat-induced deamination, *Biosci. Rep.*, 1986, 6, 387–393.
187. Frick, L., MacNeela, J.P., Wolfenden, R., Transition state stabilization by deaminases: rates of nonenzymatic hydrolysis of adenosine and cytidine, *Bioorg. Chem.*, 1987, 15, 100–108.
188. Ogilvie, K.K., Theriault, N., Sadana, K.L., Synthesis of oligoribonucleotides, *J. Am. Chem. Soc.*, 1977, 99, 7741–7743.
189. Ogilvie, K.K., Nemer, M.J., Theriault, N., Pon, R., Seifert, J.M., A complete procedure for the chemical synthesis of oligoribonucleotides, *Nucleic Acids Symp. Ser.*, 1980, 147–150.
190. Usman, N., Pon, R.T., Ogilvie, K.K., Preparation of ribonucleoside 3′-O-phosphoramidites and their application to the automated solid phase synthesis of oligonucleotides, *Tetrahedron Lett.*, 1985, 26, 4567–4570.
191. Usman, N., Ogilvie, K.K., Jiang, M.-Y., Cedergrens, R.J., Automated chemical synthesis of long oligoribonucleotides using 2′-O-silylated ribonucleoside 3′-O-phosphoramidites on a controlled-pore glass support: synthesis of a 43-nucleotide sequence similar to the 3′-half molecule of an *Escherichia coli* formylmethionine tRNA, *J. Am. Chem. Soc.*, 1986, 109, 7845–7854.
192. Duplaa, A.-M., Gasparutto, D., Livache, T., Molko, D., Teoule, R., Process for the synthesis of ribonucleic acid (RNA) using a novel deprotection reagent, 1996, US Patent 5552539, Commissariat a l'Energie Atomique and CIS BIO International, France.
193. Westman, E., Stromberg, R., Removal of t-butyldimethylsilyl protection in RNA-synthesis. Triethylamine trihydrofluoride (TEA, 3HF) is a more reliable alternative to tetrabutylammonium fluoride (TBAF), *Nucleic Acids Res.*, 1994, 22, 2430–2431.
194. Wincott, F., DiRenzo, A., Shaffer, C., Grimm, S., Tracz, D., Workman, C., Sweedler, D.,

Gonzalez, C., Scaringe, S., Usman, N., Synthesis, deprotection, analysis and purification of RNA and ribozymes, *Nucleic Acids Res.*, 1995, 23, 2677–2684.

195. Sekine, M., Iimura, S., Furusawa, K., Synthesis of a new class of 2′-phosphorylated oligoribonucleotides capable of conversion to oligoribonucleotides, *J. Org. Chem.*, 1993, 58, 3204–3208.
196. Tang, X.Q., Liao, X., Piccirilli, J.A., 2′-C-Branched ribonucleosides: synthesis of the phosphoramidite derivatives of 2′-C-beta-methylcytidine and their incorporation into oligonucleotides, *J. Org. Chem.*, 1999, 64, 747–754.
197. Bellon, L., Workman, C.T., Deprotection of RNA, 2001, US Patent 6303773, Ribozyme Pharmaceuticals, Inc., USA.
198. Bellon, L., Wincott, F., Oligonucleotide synthesis. In *Practical Solid-Phase Synthesis: A Book Companion*, Kates, S.A., Albericio, F. (Eds), 2000, Marcel Dekker, New York.
199. Vargeese, C., Deprotection and purification of oligonucleotides and their derivatives, 2006, US Patent 6989442, Sirna Therapeutics, Inc., USA.
200. Oivanen, M., Kuusela, S., Lönnberg, H., Kinetics and mechanisms for the cleavage and isomerization of the phosphodiester bonds of RNA by Bronsted acids and bases, *Chem. Rev.*, 1998, 98, 961–990.
201. Carter, C.E., Paper chromatography of purine and pyrimidine derivatives of yeast ribonucleic acid, *J. Am. Chem. Soc.*, 1950, 72, 1466–1471.
202. Cohn, W.E., The anion-exchange separation of ribonucleotides, *J. Am. Chem. Soc.*, 1950, 72, 1471–1478.
203. Brown, D.M., Todd, A.R., Nucleotides. Part X. Some observations on the structure and chemical behaviour of the nucleic acids, *J. Chem. Soc.*, 1952, 52–58.
204. Brown, D.M., Todd, A.R., Nucleotides, Part X XI: Action of ribonuclease on simple esters of mononucleotides nucleotides, *J. Chem. Soc.*, 1953, 2040–2049.
205. Hill, K.W., Understanding oligonucleotide production: identification of process-related impurities, *DIA 1st Oligonucleotide-based Therapeutics Conference, Washington, DC*, 2007.
206. Pallan, P.S., Greene, E.M., Jicman, P.A., Pandey, R.K., Manoharan, M., Rozners, E., Egli, M., Unexpected origins of the enhanced pairing affinity of 2′-fluoro-modified RNA, *Nucleic Acids Res.*, 2011, 39, 3482–3495.
207. Manoharan, M., Akinc, A., Pandey, R.K., Qin, J., Hadwiger, P., John, M., Mills, K., Charisse, K., Maier, M.A., Nechev, L., Greene, E.M., Pallan, P.S., Rozners, E., Rajeev, K.G., Egli, M., Unique gene-silencing and structural properties of 2′-fluoro-modified siRNAs, *Angew. Chem. Int. Ed. Engl.*, 2011, 50, 2284–2288.
208. Kawasaki, A.M., Casper, M.D., Freier, S.M., Lesnik, E.A., Zounes, M.C., Cummins, L.L., Gonzalez, C., Cook, P.D., Uniformly modified 2′-deoxy-2′-fluoro phosphorothioate oligonucleotides as nuclease-resistant antisense compounds with high affinity and specificity for RNA targets, *J. Med. Chem.*, 1993, 36, 831–841.
209. Krug, A., Oretskaya, T.S., Volkov, E.M., Cech, D., Shabarova, Z.A., Rosenthal, A., The behaviour of 2′-deoxy-2′-fluorouridine incorporated into oligonucleotides by the phosphoramidite approach, *Nucleosides Nucleotides*, 1989, 8, 1473–1483.
210. Doerr, I.L., Fox, J.J., Nucleosides. XXXIX. 2′-Deoxy-2′-fluorocytidine, 1-.beta.-D-arabinofuranosyl-2-amino-1,4(2H)-4-iminopyrimidine, and related derivatives, *J. Org. Chem.*, 1967, 32, 1462–1471.
211. Seiffert, S., Debelak, H., Hadwiger, P., Jahn-Hofmann, K., Roehl, I., Vornlocher, H.P., Noll, B., Characterization of side reactions during the annealing of small interfering RNAs, *Analyt. Biochem.*, 2011, 414, 47–57.
212. Takeshi, I., Satoshi, O., Bicyclonucleoside and oligonucleotide analogues, 2001, US Patent 6268490, Imanishi Takeshi, Japan.
213. Obika, S., Nanbu, D., Hari, Y., Morio, K.-i., In, Y., Ishida, T., Imanishi, T., Synthesis of 2′-O,4′-C-methyleneuridine and -cytidine. Novel bicyclic nucleosides having a fixed C3′-endo sugar puckering, *Tetrahedron Lett.*, 1997, 38, 8735–8738.
214. Koshkin, A.A., Singh, S.K., Nielsen, P., Rajwanshi, V.K., Kumar, R., Meldgaard, M., Olsen, C.E., Wengel, J., LNA (Locked nucleic acids): Synthesis of the adenine, cytosine, guanine, 5-

methylcytosine, thymine and uracil bicyclonucleoside monomers, oligomerisation, and unprecedented nucleic acid recognition, *Tetrahedron*, 1998, 54, 3607–3630.
215. Wengel, J., Nielsen, P., Oligonucleotide analogues, 1999, Patent WO9914226, Exiqon, A/S.

Separation of Oligonucleotides and Related Substances

3

Bernhard Noll and Ingo Roehl

3.1 Introduction

As oligonucleotides steadily gain importance in diagnostic and therapeutic applications, analytical techniques to analyse and control these molecules are becoming more powerful and versatile. Compared to most other synthetically manufactured molecules, oligonucleotides are large and complex entities with the propensity to participate in intra- and intermolecular interactions, which can lead to stable secondary and tertiary structures that may affect analysis.

The analytical methods described in this chapter cover a wide range of chemically modified single- and double-stranded deoxyribonucleic acid (DNA) and ribonucleic acid (RNA) molecules. Before introducing the general principles of chromatographic separation, a brief overview is given on physico-chemical properties of oligonucleotides that are relevant in the context of separation, as well as some basic information on the detection techniques and factors influencing the extinction coefficient. Other techniques used to determine the quality of an oligonucleotide, such as the determination of water content and salt content, are outside the scope of this chapter and are not described here.

This chapter is intended to give an overview of the most common chromatographic techniques that involve separation of the main product from impurities, namely anion exchange (AEX), ion pairing-reversed phase (IP-RP) and size related exclusion chromatography (SEC). For AEX and IP-RP, denaturing as well as non-denaturing methods are described. The former involve conditions intended to prevent non-covalent interactions, the latter are employed to characterise the molecules with possible non-covalent interactions intact. Before the different applications are described in detail, an introduction into the theoretical background of each of the

Analysis of Oligonucleotides and their Related Substances, edited by George Okafo, David Elder and Mike Webb.

separation principles is given and the impact of chromatographic parameters on separation is discussed. For every chromatographic technique, several examples and applications for a wide variety of oligonucleotides are provided.

Finally, a brief outlook into future trends in oligonucleotide analysis is given. Ultra-high-performance liquid chromatography (UPLC) technique is discussed and the combination of chromatographic techniques with mass spectrometric detection (liquid chromatography–mass spectrometry (LC-MS)) will be described. Table 3.1 gives a very general overview of a range of different chromatographic separation techniques, including their advantages and drawbacks. However, it has to be noted that every oligonucleotide analyte is a distinct entity with unique properties and may require a certain amount of method development to achieve the desired solution for a given separation task.

A general approach to identify the most important techniques and methods to characterise the analyte should start by investigating the individual properties of the molecule. For example, solubility tests, MS, ultraviolet (UV) spectroscopy and T_m analysis could be employed. It is also useful to perform initial chromatographic analysis using published methods. Ideally, a set of expected impurities could be synthesised and spiked to solutions of the analyte. The obtained samples can then be used to determine selectivity of the employed chromatographic techniques with

Table 3.1 Overview of separation techniques.

Separation method (denaturing)	**Try first for separations of:**	**Pros and cons**	**Comments**
AEX	PO impurities impurities of different charge	High capacity, scalable to preparative scale	Elution order often dependent on base composition
IP-RP	Impurities containing (lipophilic) modifications Shortmers Longmers	Very high resolution possible	Compatible with ESI-MS
Separation method (non-denaturing)	**Try first for separations of:**	**Pros and cons**	**Comments**
AEX	Multimers, duplexes and monomers	High capacity, scalable to preparative scale	Elution order primarily dependent on charge (neutral pH)
IP-RP	Duplexes and monomers, duplex variants	Very high resolution possible	Compatible with ESI-MS
SEC	Multimers, duplexes and monomers Unstructured monomers from hairpin structures	Physiological buffer conditions possible, low capacity	Elution according to hydrodynamic size

regard to the analyte and its impurities. In cases where the analyte is capable of forming secondary or tertiary structures, initial analysis could be performed to test the impact of the respective chromatographic media on the stability of the possible structures. A well-chosen set of initial experiments determining the characteristics of the analyte will facilitate method development and give important information on sample preparation.

3.2 Chromatographic Analysis of Oligonucleotides

3.2.1 Physico-chemical Properties of Nucleic Acids

Oligonucleotides are linear molecules built from smaller units/building blocks, the so-called nucleotides. If not indicated otherwise, the term oligonucleotides as used in this chapter will refer to molecules containing either one or both of the most common species of nucleotides, namely RNA and DNA. In an oligonucleotide, the individual nucleotides are linked *via* phosphate moieties that connect the 3′ and 5′ oxygen atoms of the ribose sugars. Sugar moieties and phosphate groups form the 'backbone' of the oligomer, the linear succession of the individual bases of the nucleotide forms the base sequence (see Chapter 1). The backbone is highly charged and hydrophilic. Conversely, the nitrogenous bases attached to C-1 of the sugar moieties are relatively hydrophobic and can participate in π-bond interactions (base stacking). A very good and detailed review of the structures, properties and functions of nucleic acids is available from University Science Books [1].

Oligonucleotides can be present in single-stranded (unpaired) form or, in the case of complementary base sequences, in double-stranded (or duplex) form. In the duplex form, strand association is mediated non-covalently by the formation of hydrogen bonds between the complementary purine and pyrimidine bases of two strands (base-pairing) and stacking interactions between the π-electron systems of the heterocyclic bases. The resulting three-dimensional helical structure is commonly referred to as the 'Watson–Crick double-helix', which was first published by James D. Watson and Francis Crick in 1953 [2] (Figure 3.1, see colour insert).

Single-stranded and duplex oligonucleotides differ significantly in their ability to interact with the surfaces of typical liquid chromatography columns. In general terms, a single strand may be considered a more flexible structure, where the hydrophilic and ionic sugar–phosphate backbone linkages may be free to rotate, allowing ionic interactions with stationary phase surfaces. At the same time, the π–electron systems of the bases may be accessible for hydrophobic interactions.

Compared to a single strand, a helical double strand is considered more rigid and rod-like in shape. The charged phosphates are positioned at the outside of the Watson–Crick helix, while the heterocyclic bases, connected *via* hydrogen bonds, form the hydrophobic core and may not be readily accessible for interactions with external surfaces. As a result, the freedom of the hydrophobic bases to interact with external surfaces is severely constrained in the double strand (Figure 3.1).

3.2.1.1 pH Effects

The pH of the surrounding medium can have a considerable influence on the charge state of the oligonucleotide in solution. At neutral or moderately basic pH values, the naturally occurring unmodified phosphate groups (and most common synthetic analogues) carry one negative charge per phosphate group, resulting in an overall negatively charged molecule. At pH values above approximately 10.5 the imino-proton at the guanine and thymine bases can dissociate, whereas at pH values below 6.5 the N3 nitrogen of the cytosine base and at approximately pH 4.5 the N1 nitrogen of the adenine can be protonated. These events lead to addition or loss of charges on the molecule and therefore may affect separation (Figure 3.2). However, in the pH range of approximately 7–10, the charge-to-size ratio is virtually constant for all single-stranded and double-stranded oligonucleotides, since it depends solely on the number of charges of the phosphate linkages in the backbone (see Figure 3.3).

In addition to its effect on molecule charge, the pH of the medium can have an influence on the stability of the oligonucleotide. RNA can degrade at pH values above 8. However, pH-induced RNA degradation is a slow process and usually does not occur in the time required for sample preparation or chromatographic separation. Several denaturing AEX methods have been described that use mobile phases at pH 11 where no degradation of the analytes has been observed throughout the chromatographic run [3, 4]. On the lower end of the pH scale, depurination becomes likely, especially for DNA at pH of 6 and below. Since depurination leads to strand scission,

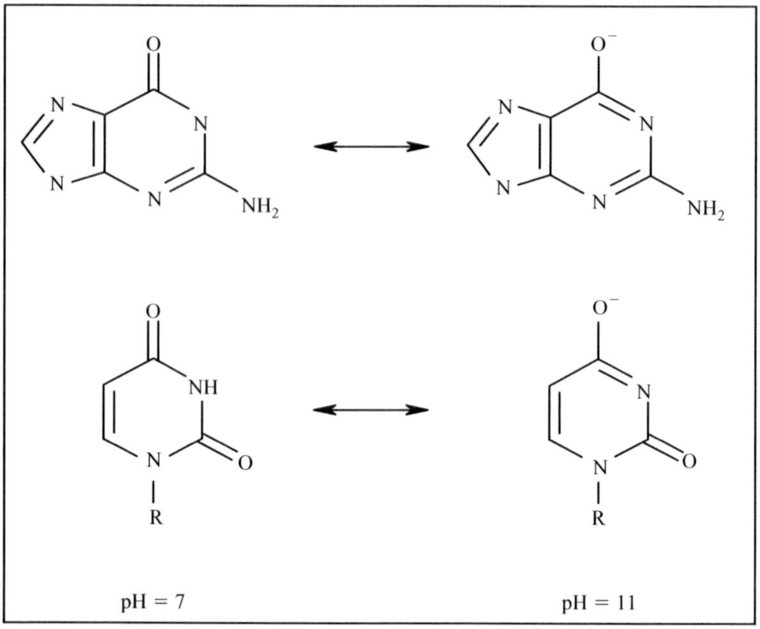

Figure 3.2 Change of charge state of nuclear bases guanine (above) and thymine (below) depending on pH environment.

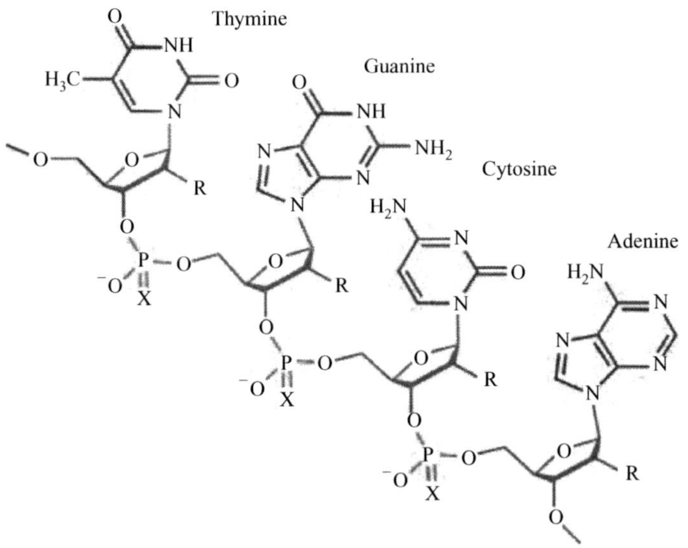

Figure 3.3 Chemical structure of DNA.

pH values below 6.5 are not recommended for the analysis of DNA or related oligonucleotides.

3.2.1.2 Chemical Modifications

Chemically modifying naturally occurring DNA or RNA can have a significant impact on the chromatographic properties of oligonucleotides. Chemical modifications are essential for improving the performance of synthetic nucleotides in therapeutic applications [5, 6]. For example, improvements in the pharmacological properties of these molecules are mainly due to increasing resistance to serum nucleases and/or affinity to the target molecule [7]. Virtually all therapeutic oligonucleotides in development contain some non-natural nucleotides [7]. Moreover, chemical modifications have enabled the rational design of antisense and antigen therapies [8, 9] facilitated the selection and development of aptamers [10], as well as improved the activity of small interfering duplex RNA (siRNA) as research tools and therapeutics. Modifications of the sugar phosphate backbone, in particular, are very common in synthetic oligonucleotides [7]. One of the most common chemical sugar modifications is the replacement of the 2′OH of the ribose moiety by a different chemical group, such as 2′-O-methyl, 2′-deoxy-2′-fluoro or 2′-deoxy. Another important effect of chemical modification is the stabilisation of secondary structures or Watson–Crick helixes. While phosphorothioate modifications typically reduce melting temperature (T_m) of an oligonucleotide duplex ($\Delta T_m = -0.7$°C per introduced modification), 2′MOE, and 2′F increase the T_m ($\Delta T_m = +1.4$°C and +2.7°C per introduced modification, respectively) [11] (see Chapter 6 for more details). The strongest increase in T_m has been reported for sterically constrained sugar moieties like locked-nucleic acids

(LNAs) or ethylene-bridged nucleic acids ($\Delta T_m = +1-8$°C against DNA and $\Delta T_m = +2-10$°C against RNA per introduced modification) [12]. Stability of the duplex in solution can have a significant impact on the separation strategy, for example if complete dissociation of a double-stranded structure is desired (for an example see Section 3.5.2.1).

Phosphorothioate oligonucleotides differ from unmodified oligonucleotides by at least one substitution of non-bridging oxygen (PO) to sulfur (PS) along the phosphate backbone. Phosphorothioate oligonucleotides have been shown to be very stable in media, cells and cell extracts, serum various tissues, serum, urine, and in the presence of most nucleases [13]. Oligonucleotides containing a high percentage of PS linkages show increased protein binding [14]. Presence of PS linkages in an oligonucleotide has a direct impact on chromatographic separation. The lipophilic character of the sulfur atom results in a higher binding affinity of PS oligonucleotides to many stationary phases. Compared to PO, the sulfur analogues are more strongly retained, especially on the AEX resins, resulting in very good separations of fully and partially thioated oligonucleotides (for an example see Section 3.4.2.5).

In addition to changing the lipophilic character of the oligonucleotide, the substitution of the oxygen results in the formation of a chiral centre at the phosphate linkage. For an oligonucleotide with n PS linkages, there can be $2n$ stereoisomers. Every stereoisomer is a distinct chemical entity, which may show small differences in retention times compared to other phosphorothioate stereoisomers. This results in peak broadening during chromatographic separation, which may complicate analysis [15]. Whereas baseline resolution of phosphodiester oligonucleotides (10- to 30-mer) is usually easily accomplished, phosphorothioate oligonucleotides may be more difficult to separate.

Taken together, the retention time of the single strands as well as the double strand in chromatographic separations may be influenced not only by length and base sequence, but also by the modification patterns of the individual strands. Chromatographic separations of molecules of the same length and base sequence, but with different modification patterns, may still require adjustment of the chromatographic conditions. The chromatographer is advised to review carefully not only the sequence, but also the modification pattern of an oligonucleotide, when planning a separation strategy.

3.2.2 Detection Techniques

Depending on the application and the properties of the analyte, different detectors or a combination of detectors can be used. The most widely used mode of chromatographic detection for nucleic acids is UV absorption. Depending on the type of the employed detector, UV detection can be performed at more than one wavelength simultaneously and/or can be combined with other detection techniques. Conductivity detectors are commonly used in combination with other detectors, especially for gel filtration or SEC, to monitor the separation efficiency of nucleic acids from salt or other ionic buffer components.

3.2.2.1 UV Detection

UV detection is usually the detection method of choice for nucleic acids. This is because of the specific absorption characteristics of the heterocyclic bases of the nucleic acids. The purine and pyrimidine bases of the oligonucleotides have a strong UV absorption ($\varepsilon \sim 2 \times 10^5$/M per cm for a 20-mer oligonucleotide) between 255 and 265 nm. The absorption maximum depends on the specific sequence of the nucleic acid, but for reasons of simplicity, UV detection is often performed at 260 nm.

The basis for concentration determination by UV measurement is the linear relationship of analyte concentration to the extent of absorption. This relationship is described by Beer's law [16] in equation (3.1).

$$A = \varepsilon * l * c \tag{3.1}$$

where A is absorbance; ε is the molar absorptivity or extinction coefficient; l is the path length of the sample (i.e. the path length of the cuvette in which the sample is contained); and c is the concentration of the compound in solution, expressed in mol/L.

When a sample of nucleic acids is analysed, it is important to establish whether other components present in the sample are absorbing UV light and thus may interfere with the concentration determination of the main analyte. Other biopolymers like proteins and peptides also absorb at 260 nm owing mainly to the aromatic amino acid residues. However, the absorption maximum of the peptide bond lies at longer wavelength (280 nm). Comparing the absorbance of certain eluent peaks at two or more wavelengths can therefore give important supplementary information on the purity of a sample or the identity of the individual peaks in a chromatographic separation. Hence, using a wavelength detector that is capable of measuring at more than one wavelength (e.g. diode array detector) can be beneficial, especially when the sample contains components other than the nucleic acid.

3.2.2.2 Hyper/Hypochromicity

An important feature of the absorption characteristics of nucleic acids is an effect called hypochromicity. As described by Beer's law (above), the concentration is related to the extent of UV absorption, and the calculation for specific analytes depends on the extinction coefficient, which is compound-specific. In the case of oligonucleotides, the extinction coefficient depends on base sequence, although the variations occur in a relatively limited range. A theoretical model for the calculation of the extinction coefficient of specific base sequences (so-called 'nearest-neighbour model') is commonly used to determine extinction coefficients (ε) of single-stranded oligonucleotides [17], in particular when limited amounts of dried material are available for experimental determination of ε. However, the nearest-neighbour model for calculation of ε applies only to single-stranded, unstructured oligonucleotides. The formation of secondary structures, namely Watson–Crick double helices, significantly impacts the absorption characteristics of the molecule, since the UV absorption of a nucleic acid solution decreases when two complementary single strands form a duplex. This decrease of absorbance (optical density) is called hypochromicity. When the opposite occurs and the absorbance is increased, the phenomenon is called hyperchromicity. A consequence of this effect is that UV absorption of hybridised double strands can be much lower than the absorption of single strands of the same concentration, and vice versa. Hence, in practice, the extinction coefficient of a double strand is typically not the sum of the two ε of both complementary single strands, but may be significantly lower (for an example see Figure 3.4 and Table 3.2). Whereas for single strands, a good approximation of ε is possible using nearest-neighbour calculations, this is not possible in the case of double-stranded oligonucleotides. Extinction coefficients of double strands are usually determined experimentally by accurately weighing solid double-stranded oligonucleotide, correcting for water and salt, dissolving in aqueous solution and measuring UV absorption of the resulting solution.

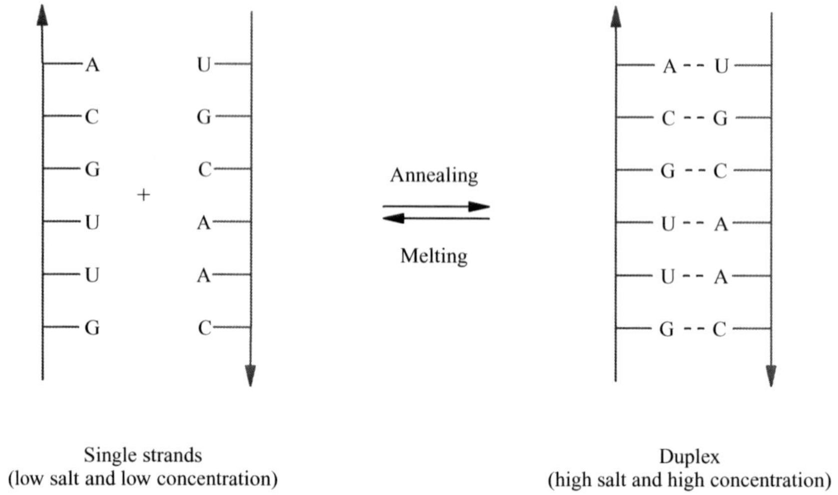

Figure 3.4 Dissociation and association of two complementary single strands. Low salt concentrations as well as low strand concentrations favour dissociation of the duplex. High salt concentrations and high strand concentrations favour the formation of the duplex.

Table 3.2 Extinction coefficients of a siRNA molecule determined experimentally by UV-spectroscopy in different media and different laboratories. The change in UV absorbance at 260 nm could be greater than 30% when comparing measurements in water and in PBS.

	Extinction coefficient (L/mol·per cm) determined in water (CRO)	**Extinction coefficient (L/mol·per cm) determined in water (client)**	**Extinction coefficient (L/mol·per cm) determined in 1 × PBS (client)**
siRNA-Luc	389 000 132%	365 000 124%	294 900 100%

For the estimation of the extent of duplex formation, it has to be recognised that stability of the double strand depends not only on the base sequence, but is also strongly dependent on the medium (i.e. cation concentration) and the concentration of the strands [18–21]. Generally, two perfectly complementary nucleic acid strands bind to each other readily in aqueous solution. In the case of short nucleic acids such as siRNA, hybridisation proceeds very rapidly. Stability of the duplex in solution is determined by the extent of non-covalent, sequence-specific interactions between the two complementary strands. Cations in the solution increase duplex stability by partially neutralising the negative charges at the phosphate groups of the backbone. Duplex stability also increases with strand concentration, which can be attributed to the second-order kinetics of the hybridisation reaction. Duplex formation at higher concentrations is favoured more strongly in solutions where the duplex is not further stabilised by the presence of salt. This has important implications for the measurement of duplex concentrations by UV absorption. In salt-free solutions at concentrations typically used for UV spectroscopic measurements, a given duplex can be partially dissociated. Owing to the different extinction coefficients of double strand and single strands, unaccounted dissociation can result in significant variations in the UV read-out. This has to be taken into account when determining siRNA concentration in water. For example, a 4 μM siRNA sample may show a T_m of 32°C in water but a T_m of 81°C in 1 × phosphate buffered saline (PBS). The same siRNA can display a T_m of 67°C in water, when concentration is 500 μM (see Table 3.3), demonstrating that stable duplex formation can be achieved at ambient temperature in solutions of siRNA that do not contain salt, provided the respective siRNA concentration is sufficiently high [22].

Table 3.3 T_m of two siRNAs of different sequence and modification pattern (siRNA-1 and siRNA-2). T_m of 4 μM solutions in water or 150 mM sodium chloride were measured using thermal UV-spectrometry at a wavelength of 260 nm. T_m values of all other solutions were determined using differential scanning calorimetry. (Source: Modified from *Analyt. Biochem.*, 414, 1, Seiffert, S., Debelak, H., Hadwiger, P., Jahn-Hofmann, K., Röhl, I., Vornlocher, H.P., Noll, B., Characterization of side reactions during the annealing of small interfering RNAs, 47–57, Copyright (2011), with permission from Elsevier.)

T_m (°C)	**siRNA-1**		**siRNA-2**	
	Water	**NaCl**	**Water**	**NaCl**
4 μM	31.8	80.6	40.1	84.1
20 μM	43.1	82.5	50.7	85.5
100 μM	52.2	84.2	59.3	87.7
500 μM	67.1	86.9	75.3	90.6

3.3 General Principles of Chromatographic Separation

3.3.1 Basic Calculations

3.3.1.1 Resolution

The result of a chromatographic separation is often expressed as the resolution (R_s) between the peaks of interest. Resolution is defined as the distance between peak maxima compared with the average width of the two peaks. Resolution can be calculated from chromatograms using equation (3.2).

$$R_s = 2(V_{R2} - V_{R1})/(w_{b1} + w_{b2}) \tag{3.2}$$

where V_{R1} is the elution volume for peak 1, V_{R2} is the elution volume for peak 2, w_{b1} is the peak width of peak 1 and w_{b2} is the peak width of peak 2. Depending on calculation method, peak width can be taken at the base or the half height of the peaks. For baseline resolution R_s has to be greater than 1.5. Note that a single, well-resolved peak is not necessarily a pure substance, but may represent a series of components which could not be separated under the chosen elution conditions.

The resolution achievable in a system is proportional to the product of the selectivity, the efficiency and the capacity of the system, the three most important parameters to control column chromatography. The relation of these parameters can be described by equation (3.3).

$$R_s = \text{selectivity} * \text{efficiency} * \text{capacity} = 14 * (\alpha - 1)/\alpha * (\sqrt{N}) * k/(1 + k) \tag{3.3}$$

where α is the selectivity, N is the column efficiency or number of theoretical plates per meter chromatography bed and k is the capacity factor. The equation implies that to obtain high resolution, the three terms selectivity, efficiency and capacity must be maximised. However, an increase in N (efficiency or number of theoretical plates) by lengthening the column leads to an increase in retention time and increased band broadening – which may not be desirable. Instead, to increase the number of plates, the height equivalent to a theoretical plate can be reduced by reducing the size of the stationary phase particles.

3.3.1.2 Selectivity

The selectivity (α) defines the ability of the system to separate peaks. The selectivity factor can be calculated using the distance between two peaks in a chromatogram (equation (3.4))

$$\alpha = k_2/k_1 = (V_{R2} - V_0)/(V_{R1} - V_0) = V_{R2}/V_{R1} \tag{3.4}$$

where V_0 is the void volume, V_{R1} is the elution volume for peak 1 and V_{R2} is the elution volume for peak 2. Given that peak 1 elutes earlier than peak 2, the selectivity factor is always greater than 1. Good selectivity is a more important factor than high efficiency in determining resolution and depends not only on the nature and number of the functional groups on the matrix (composition of stationary phase), but also on the experimental conditions, such as mobile phase composition (e.g. pH, ionic strength), column temperature and elution conditions.

3.3.1.3 Capacity Factor

The capacity or retention factor k is a measure of the retention of a component and should not be confused with loading capacity (mg sample/mL) or ionic capacity (mmol/mL). The capacity factor can be calculated for each individual peak of a chromatogram using equation (3.5)

$$k = V_{R1} - V_t / V_t \tag{3.5}$$

where V_{R1} is the elution volume for peak 1 and V_t is the total (elution) volume. When an analyte's retention factor is less than 1, elution is so fast that accurate determination of the retention time is very difficult. High retention factors (greater than 20) mean that elution takes a very long time. Ideally, the retention factor for an analyte should be between 1 and 5. By controlling the capacity factor, k, separations can be greatly improved. In liquid chromatography this is usually achieved by modifying the mobile phase.

3.3.1.4 Column Efficiency (Plate Number)

Column efficiency is the ability to elute narrow, symmetrical peaks from a packed bed and relates to the zone broadening which occurs when an analyte moves through the stationary phase. Column efficiency is frequently expressed as the number of theoretical plates per metre of the chromatography bed (N) or as H (height equivalent to a theoretical plate (HETP)), which is the bed length (L) divided by the plate number (equation (3.6))

$$\text{HETP} = L/N \tag{3.6}$$

where L is the length of the column (height of packed bed) and N is the plate number. The plate model postulates that the chromatographic column contains a large number of separate layers, called theoretical plates. Separate equilibrations of the sample components between the stationary and mobile phase occur in these 'plates'. During chromatography, the analytes move down the column by transfer of equilibrated mobile phase from one plate to the next. However, it is important to remember that the plates do not really exist; they are simply a theoretical model that helps to illustrate the processes at work and serve as a measure of column efficiency.

Column efficiency can be calculated from peak width and peak elution volume using equation (3.7)

$$N = 5{\cdot}54 \times (V_R/w_h)^2 \tag{3.7}$$

where V_R is the volume eluted from the start of sample application to the peak maximum and w_h is the width of the recorded peak at half of the peak height. Measurements of V_R and w_h can be made in distance (mm) or volume (ml) but both parameters must be expressed in the same unit. The equation also implies that columns behave as if they have different numbers of plates for different solutes in a mixture.

To obtain optimal separations, band broadening during separations must be limited. Narrow, symmetrical chromatographic peaks result in the highest resolutions and thus the best separations. One of the main causes of zone broadening is longitudinal diffusion of the analyte molecules. Zone broadening can be minimised if the distances available for diffusion are minimised. A homogeneously packed column will contribute significantly to resolution, since variations in diffusion distances are limited. Columns that are packed unevenly, too tightly, too loosely or that contain air bubbles will lead to channelling (uneven passage of buffer through the column), zone broadening and hence loss of resolution.

3.3.1.5 Van Deemter Equation

The Van Deemter equation is a hyperbolic function that correlates resolution and flow rate [23]. It allows for the determination of the minimal theoretical plate height HETP or column efficiency (which results in optimal resolution) and the optimal mobile phase velocity of a chromatographic column. In its graphical description, the Van Deemter equation is termed the Van Deemter plot and is generated by plotting plate height against average linear velocity of mobile phase (Figure 3.5). From its graphical plot it is easy to see that the Van Deemter equation predicts that there is an optimum velocity at which there will be a maximum efficiency (HETP). The factors influencing efficiency are eddy diffusion, longitudinal diffusion and a mass transfer term. The relative importance of these factors varies with mobile phase velocity. Particle size and morphology contribute to HETP, along with a variety of other factors (equation (3.8))

$$\text{HETP} = A + B/\mu + C\mu \tag{3.8}$$

where μ is the average velocity of the mobile phase. A, B and C are the three factors which contribute to band broadening

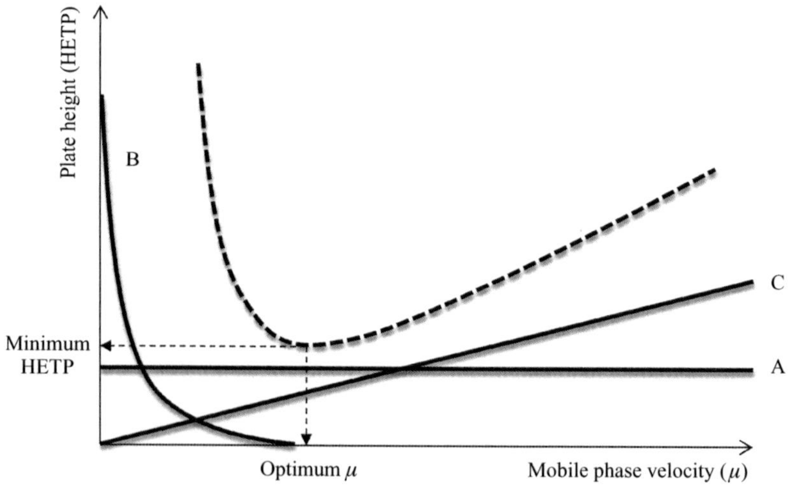

Figure 3.5 Van Deemter plot and equation. Plate height (HETP) is plotted against mobile phase velocity (μ). A: eddy diffusion; B: longitudinal diffusion; C: resistance to mass transfer.

$$\text{Eddy diffusion,} \quad A = 2\lambda d_P \tag{3.09}$$

The mobile phase moves through the column which is packed with stationary phase. Solute molecules will take different paths through the stationary phase at random. This will cause broadening of the solute band, because different paths are of different lengths (equation (3.09))

$$\text{Longitudinal diffusion,} \quad B = 2\Psi D_M \tag{3.10}$$

The concentration of analyte is less at the edges of the band than at the centre. Analyte diffuses out from the centre to the edges. This causes band broadening. If the velocity of the mobile phase is high then the analyte spends less time on the column, which decreases the effects of longitudinal diffusion (equation (3.10))

$$\text{Resistance to mass transfer,} \quad C = K_1^*(d_P)^2/D_M + K_2^*(d_f)^2/D_S \tag{3.11}$$

The analyte takes a certain amount of time to equilibrate between the stationary and mobile phase. If the velocity of the mobile phase is high, and the analyte has a strong affinity for the stationary phase, then the analyte in the mobile phase will move ahead of the analyte in the stationary phase. The band of analyte is broadened. The higher the velocity of mobile phase, the worse the broadening becomes (equation 3.11).

λ is a constant, regarding particle form and homogeneity of the packing, d_P represents the particle diameter, D_M is the diffusion coefficient of the mobile phase, D_S is the diffusion coefficient of the stationary phase, Ψ is the correction factor, describing the space between particles, d_f defines the thickness of the liquid film and K_1 and K_2 are correction factors, regarding the geometry of the column and column capacity (for equations (3.9) to (3.11)).

3.3.2 Chromatographic Stationary Phases

3.3.2.1 Granular Resins

For granular resins, particle size is a significant factor in resolution. In general, the smallest particles will produce the narrowest peaks under the correct elution conditions and in a well-packed column. This is mainly due to the lower diffusion distance in smaller resin particles (see Section 3.3.1.4 on column efficiency). The size of the particle determines the plate number of the column (see Section 3.3.1.5 on the Van Deemter equation: d_P = particle diameter). A reduction in diameter by 50% results approximately in a doubling of the plate count. Therefore, many manufacturers of globular resins are aiming for smaller particle sizes. Currently, many columns are available with particle sizes smaller than 2 μm. Although smaller particles offer better resolution, they also generate higher back pressures which may affect operating flow rate and/or put extra demand on equipment.

In general, it is beneficial to use monodisperse resins for chromatographic separations. Size distribution of these beads is very small, making for identical properties of each bead (see Section 3.3.1.5 on the Van Deemter equation: λ is a constant describing particle form and homogeneity of the packing and Ψ is a correction factor describing the space between particles). In the case of fully porous beads, eluents and analytes are expected to diffuse into and out of the resin beads. The speed of diffusion, and therefore diffusion-induced band broadening, depends on pore size and structure. The pore size of particles should be large enough to allow unhindered diffusion of the analyte into the particle, but not too large to avoid loss of capacity. In size-exclusion chromatography the diffusion of the analytes in the pores becomes the primary mode of column selectivity. By choosing the pore size range, the separation efficiency on different molecule sizes can be determined. Because diffusion rates are inversely proportional to relative molecular mass (M_r), porous bead phases may not provide similar resolution for large and small oligonucleotide molecules. Note that pore size depends not only on base matrix materials but also on the medium/mobile phase which is used in chromatography. Acrylic-based materials tend to increase pore size as a function of salt concentration, whereas silica-based materials seem to decrease in size [24].

3.3.2.2 Monolithic Columns/Hybrid Phases

Despite the many advantages, high-performance liquid chromatography (HPLC) columns packed with particulate, porous stationary phases have limitations, such as the relatively large void volume between the packed beads and the slow diffusional mass transfer of solutes into and out of the stagnant mobile phase present in the pores of the separation medium [25]. One solution to these problems is to use monolithic chromatographic beds. Monolithic columns are not made by packings of single beads, but consist of a single, solid, porous, polymeric structure with interconnecting pores and channels [26]. Thus interstitial void volumes are avoided and all mobile phase is forced to pass through the pores of the stationary phase [27]. Pore structure and polymer surface properties can be independently controlled during polymer synthesis. Thus, monolithic polymers with large pores and greater surface areas as pellicular phases are possible [28, 29]; however, they are limited by having reduced capacities when compared to those of fully porous bead-based resins.

A hybrid phase monolith has recently been introduced for AEX (DNASwift SAX-1S Data Sheet, Dionex/Thermo Scientific) that combines the advantages of high capacity and resolution. This new type of phase has a monolithic matrix that is functionalised with a cation exchange (anionic) surface, which in turn is coated with aminated AEX nanobeads. The porosity of the monolith is tailored to accommodate relatively large nanobeads, which are completely accessible for the mobile phase, resulting in capacities approaching those of fully porous beads. Owing to the faster mass transfer in the monolith, resolution of the hybrid phase monolith is improved over porous bead resins [30].

3.4 Ion Exchange Chromatography

3.4.1 Separation Principles

Ion exchange chromatography separates molecules on the basis of differences in their net surface charge. During chromatography, charged analyte molecules adsorb reversibly to the ion exchange groups of opposite charge on the stationary phase. Separation is obtained owing to differences in relative charge densities and distribution of charges on the surfaces of the analytes, which have different degrees of interaction with the ion exchange resin [31]. Positively charged chemical groups on the stationary phase define an AEX, while negatively charged groups define a cation exchanger. Ion exchangers are traditionally further split up into two categories: strong and weak ion exchangers. The terms strong and weak refer to the extent of variation of ionisation with pH and not the strength of the binding. Strong ion exchangers are completely ionised over a wide pH range, whereas with weak ion exchangers, the degree of dissociation and thus exchange capacity varies much more markedly with pH. Strong AEX ligands are mostly quaternary aminomethyl- or quarternary aminoethyl- (Q or QAE) types, with pH working ranges of ~3–13, and weak AEX ligands are often diethyl aminoethyl- (D or DEAE) types with a pH working range of 3–9. For the analysis of nucleic acids, which are anionic in nature, strong AEXs are typically used. Separations are more controllable, since charge characteristics of the stationary phases do not change with pH. In addition, sample loading capacity does not decrease at high or low pH values owing to loss of charge from the ion exchangers, as would be the case for weak exchangers.

AEX is capable of separating analytes with very minor differences in charge. For double- as well as single-stranded oligonucleotides, total charge increases with size. Hence, separation of oligonucleotides occurs according to length. Over the past 25 years, AEX has been used in denaturing and non-denaturing modes and in analytical and preparative applications for nucleic acids (reviewed in reference [32]). Although the primary mode of AEX separation relies on electrostatic interactions between the charged phosphate ion of the oligonucleotide and the functional group of the matrix AEX stationary phase, secondary interactions, for example non-polar interactions due to van der Waals' forces with the matrix surface of the column and the hydrophobic bases of the nucleotides, can have a strong impact on the elution properties of analytes on an AEX column.

3.4.2 Denaturing AEX

Denaturing applications are most common in analytical applications, where high resolutions are the primary objective. In denaturing AEX, secondary structures are suppressed, and separations are primarily based on the number of charges of the molecule (e.g. length) [30, 33–36]. High temperatures, high pH and the possible inclusion of chaotropic agents into the mobile phases can all serve as means to resolve highly stable secondary structures. For more details on oligonucleotide impurities, please refer to Chapter 2.

3.4.2.1 Impact of Mobile Phase pH on Separation

A key factor in AEX chromatography is pH, since it may have a significant impact on analyte and resin stability. In addition, pH may have an impact on the retention time of oligonucleotides during chromatographic separation, owing to the associated changes in the charge state of analytes and resins (Figure 3.6). Owing to the nature of the hydrogen bonds between the complementary strands that stabilise the Watson–Crick duplex, a double helix is not stable above a pH of 10.5. High pH values are therefore

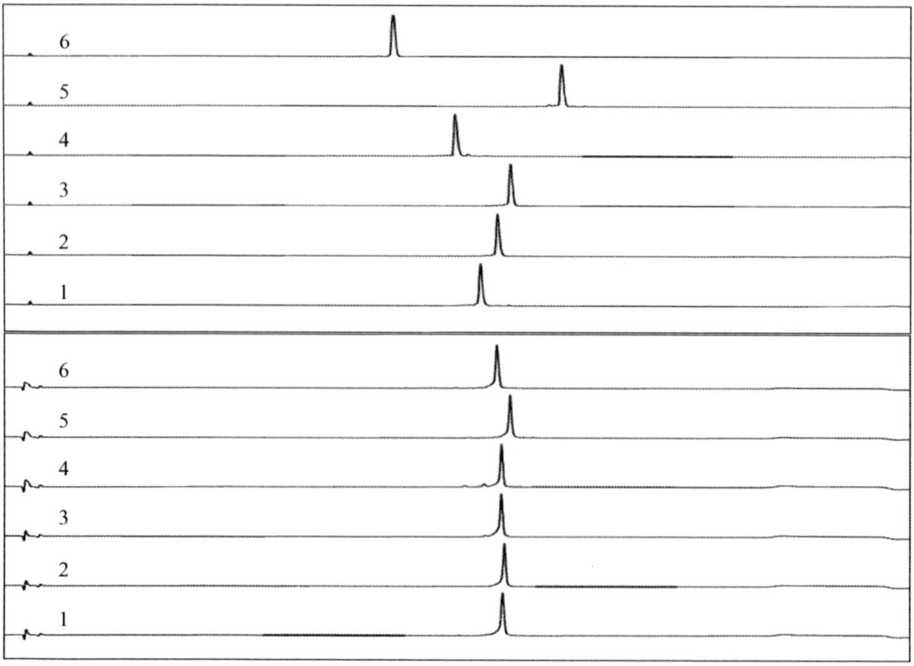

Figure 3.6 Impact of pH on retention time. Retention time in AEX-HPLC shifts in high pH and neutral pH applications, depending on the G/U content and base position within the sequence. Chromatography was performed using a DNA-Pac200 column (2.1 × 100 mm) (Dionex) at a flow rate of 1 mL/min. For high pH (upper panel) mobile phase A was {20 mM sodium phosphate (pH 11.0), 10% acetonitrile (ACN)} and mobile phase B was {20 mM Na_3PO_4, 1.0 M NaBr (pH 11.0), 10% ACN}. Column temperature was 30°C. A gradient was run from 25 to 62% B in 12 min. For neutral pH (lower panel) mobile phase A was {10 mM tris/HCl (pH 8), 10% ACN} and mobile phase B was {10 mM tris/HCl (pH 8), 1.0 M NaBr}. Column temperature was 75°C. A gradient was run from 30 to 60% B in 12 min. Traces 1 to 6 represent RNA-1 to RNA-6 respectively (see also Table 3.4).

Table 3.4 Base composition of RNA single strands analysed in Figure 3.6. Number of guanines and uridines in the sequence of six different RNA strands (RNA-1 to RNA-6) is indicated.

RNA strand	Length	Number of guanines	Number of uridines
RNA-1	21mer	5	5
RNA-2	21mer	5	6
RNA-3	21mer	5	7
RNA-4	21mer	6	2
RNA-5	21mer	7	7
RNA-6	21mer	3	2

often used in chromatographic separations to disrupt otherwise stable secondary structures, especially in RNA or modified DNA. Hence, several denaturing separations of oligonucleotides have been reported that utilise high pH values to separate strands or suppress secondary structures [3, 4].

3.4.2.2 Impact of Column Temperature on Separation

Temperature of the column can be used to improve resolution of the analytes. In AEX, polyanions tend to exhibit increased retention and mass transfer rates with rising temperatures, resulting in improved resolution (Table 3.5 and Figure 3.7). AEX columns have been used with temperatures up to 95°C to control hydrogen bonding or maximise resolution. Many AEX resins are very stable towards conditions involving high temperature and pH, which makes AEX a very attractive technique for denaturing separations. However, for most column materials it has to be noted that high temperatures usually as well as high pH and inclusion of chaotropic agents may limit column longevity.

3.4.2.3 Separation of Single-Strand Impurities

AEX has been utilised to separate major impurities from oligonucleotide synthesis from the full-length product. Among the expected impurities resulting from DNA and RNA synthesis are

- shortmer sequences (N-x; e.g. N-1 or N-2) [37–39] (Figure 3.8)
- longmer sequences (N+x; e.g. N+G or N+A) [40] (Figure 3.9)

Table 3.5 Peak resolution values obtained by AEX chromatography of four different siRNAs at 40°C, 50°C and 60°C. A SuperQ 5PW column was used at a flow rate of 1 mL/min. Mobile phase A was {20 mM sodium phosphate (pH 11.0), 10% ACN} and mobile phase B was {20 mM sodium phosphate, 1.0 M NaBr (pH 11.0), 10% ACN}. A gradient was run from 25 to 62% B in 12 min. RRT denotes relative retention time. For base composition of siRNA-I to siRNA-IV see Table 3.6.

	Column temperature	**RRT PS-strand**	**RRT GS-strand**
siRNA-I	40°C	1.085	1.172
	50°C	1.103	1.198
	60°C	1.124	1.221
siRNA-II	40°C	1.216	1.147
	50°C	1.238	1.176
	60°C	1.238	1.201
siRNA-III	40°C	1.128	1.145
	50°C	1.131	1.160
	60°C	1.144	1.187
siRNA-IV	40°C	1.133	0.968
	50°C	1.126	No
	60°C	1.124	No

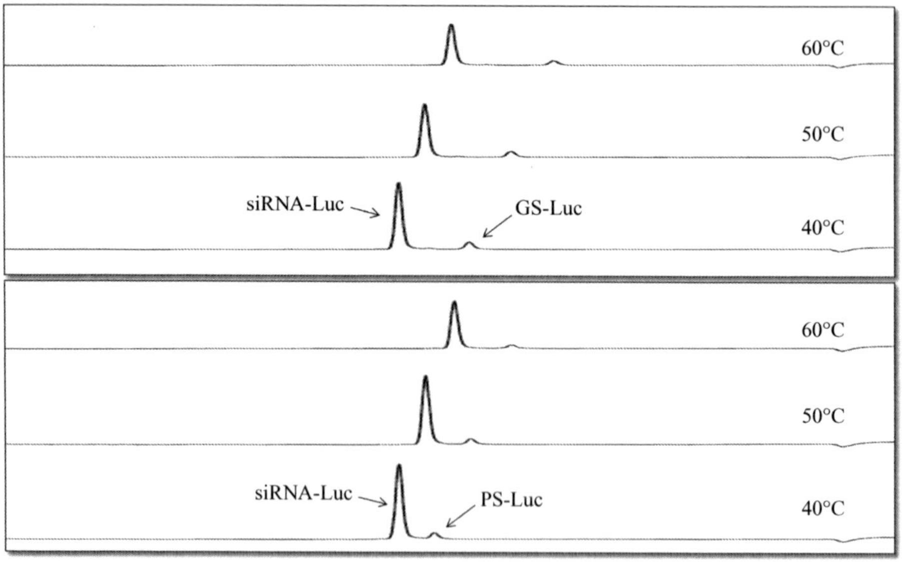

Figure 3.7 Temperature dependency of chromatographic separations of the siRNA-Luc duplex (siRNA-I) and its passenger strands (PS-Luc, lower panels) or guide strands (GS-Luc, upper panels). A SuperQ 5PW column was used at a flow rate of 1 mL/min. Mobile phase A: 10 mM tris/HCl pH8, Mobile phase B: 10 mM tris/HCl pH8, 1 M NaBr. A gradient was run from 20 to 40% B in 12 min.

- sequences containing phosphodieser linkages (PO) instead of phosphorothioate linkages [41] (Figure 3.8).

3.4.2.4 Separation of Isomers

Isomerisation is a common side reaction in solutions of RNA. Isomers consisting of the 2′,5′ configuration (rather than the preferred 3′5′ configuration) are generated when ribonucleic acids are subjected to high temperature and/or basic pH in aqueous solution [42, 43]. AEX lends itself well to separation of 2′,5′-isomers from standard 3′,5′-RNA oligomers. Good peak separations have been reported with several applications [4, 22], suggesting that the steric changes induced by the inclusion of 1 or more 2′,5′-linkages in an oligomer of 3′,5′-RNA, significantly alter the interactions of the charged backbone with the AEX matrix (Figures 3.10 and 3.11).

3.4.2.5 Separation of Phosphorothioates (PS) and Phosphodiesters (PO)

The lipophilic character of the sulfur atom results in a higher binding affinity of oligonucleotides containing PS linkages to many stationary phases (compared to PO). Typically, PS-oligonucleotides are retained more strongly on AEX resins, resulting in very good separations of fully and incompletely thioated ON (Figure 3.12).

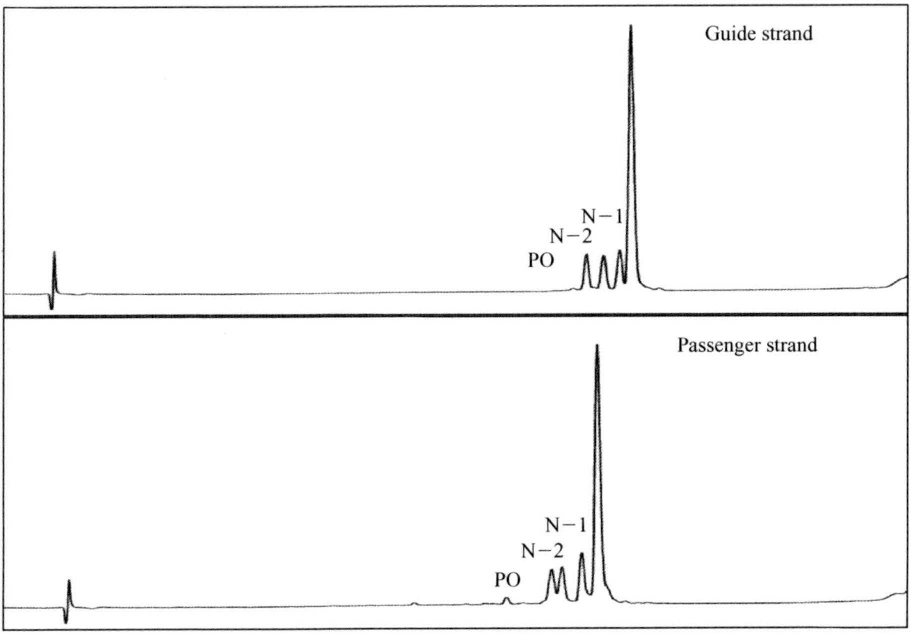

Figure 3.8 Chromatographic separation of three of the major shortmer impurities from guide and passenger strand. High pH was used to avoid secondary structures. A Dionex DNA-Pac PA200 column was used at a flow rate of 1 mL/min. Column temperature was 60°C. Mobile phase A was {20 mM sodium phosphate (pH 11.0), 10% ACN}; mobile phase B was {20 mM sodium phosphate, 1.0 M sodium bromide (pH 11.0), 10% ACN}. A gradient was run from 20 to 40% B in 12 min.

Table 3.6 Base composition of the siRNA strands of siRNA-I to siRNA-IV. Upper case letters indicate 2′-OH RNA nucleotides, lower case letters indicate 2′-O-methyl nucleotides, TT indicates two 2′-deoxy-thymidine nucleotides connected via a phosphorothioate linkage. (Source: Modified from *Nucleic Acid Technol.*, 21, Noll, B., Seiffert, S., Hertel, F., Debelak, H., Hadwiger, P., Vornlocher, H.P., Röhl, I., Purification of small interfering RNA using non-denaturing anion-exchange chromatography, 383–393, Copyright 2011.)

ID	**Guide strand**	**Passenger strand**
siRNA-I	$U_2u_2C_3c_1G_5A_6TT$	$u_6c_5G_4A_4TT$
siRNA-II	$U_7u_1C_3c_1G_5A_2TT$	$u_2c_5G_4A_8TT$
siRNA-III	$U_6C_4c_1G_6A_2TT$	$u_2c_6G_5A_6TT$
siRNA-IV	$U_2u_2C_3c_1G_5A_6TT$	$U_6C_5G_4A_4TT$

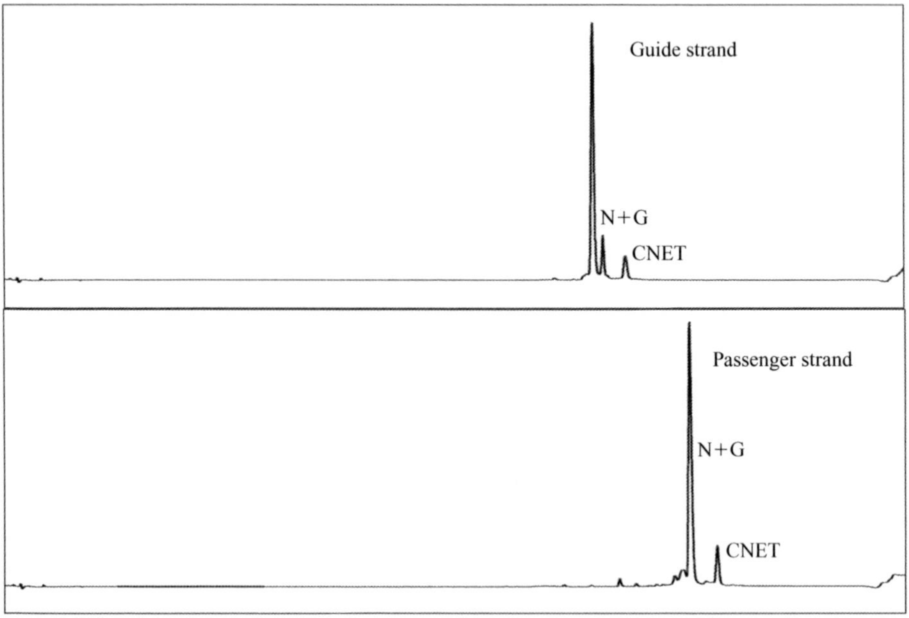

Figure 3.9 Chromatographic separation of two major longmer impurities from guide and passenger strand. High pH was used to avoid secondary structures. A Dionex DNA-Pac PA200 column was used at a flow rate of 1 mL/min. Chromatographic conditions as in Figure 3.8.

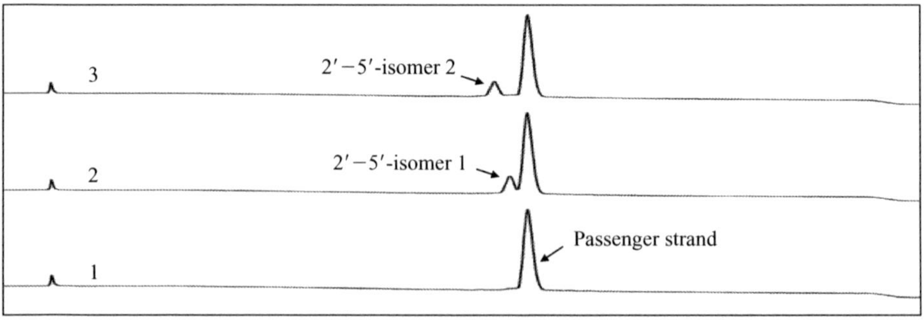

Figure 3.10 Chromatographic separation of two 2′,5′ impurities from passenger strand. High pH was used to avoid secondary structures. A Dionex DNA-Pac PA200 column was used at a flow rate of 1 mL/min. Chromatographic conditions as in Figure 3.8.

3.4.3 Non-denaturing AEX

3.4.3.1 Separation of Duplex and Single Strands

Non-denaturing applications are less common, and are often employed to determine the strand ratio of passenger and guide strand after annealing. Several non-denaturing

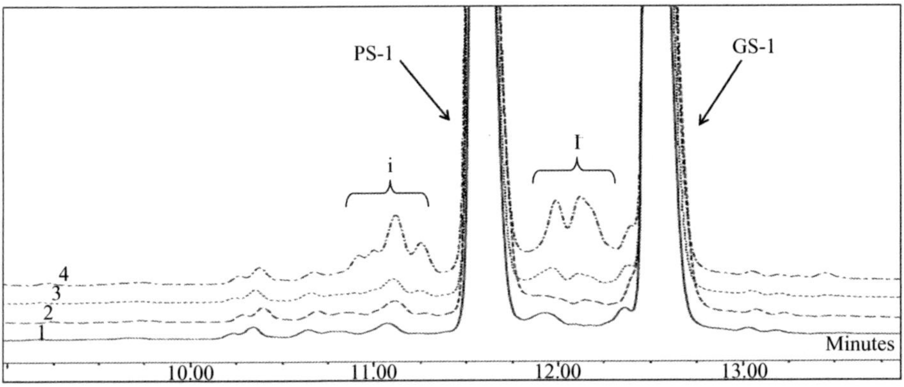

Figure 3.11 Denaturing AEX-HPLC of siRNA-Luc after incubations at room temperature (trace 1), 60°C (trace 2), 70°C (trace 3) and 85°C (trace 4). All incubations were carried out for 4 h using 100μM siRNA solutions in water. Full-length guide strand (GS-1) and passenger strand (PS-1) are indicated by arrows. Impurity peaks are marked by upper case (guide strand) and lower case (passenger strands) letters. 'I' indicates 2′,5′-isomers of GS-1; 'i' indicates 2′,5′-isomers of PS-1. A Dionex DNA-Pac PA200 column was used at a flow rate of 1 mL/min. Chromatographic conditions as in Figure 3.8. (Source: Modified from *Analyt. Biochem.*, 414, 1, Seiffert, S., Debelak, H., Hadwiger, P., Jahn-Hofmann, K., Röhl, I., Vornlocher, H.P., Noll, B., Characterization of side reactions during the annealing of small interfering RNAs, 47–57, Copyright (2011), with permission from Elsevier.)

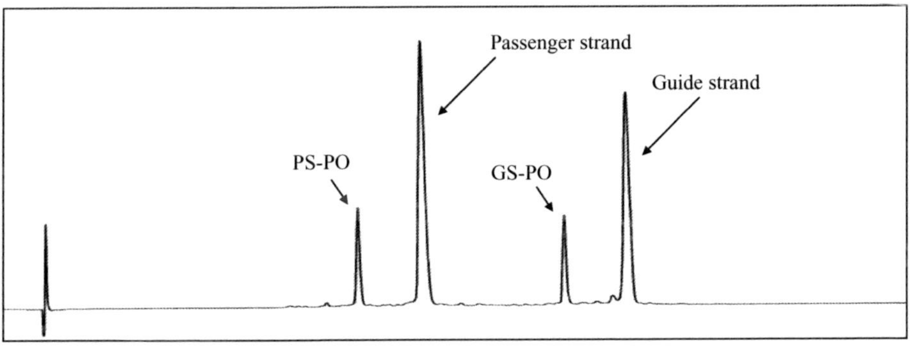

Figure 3.12 Denaturing separation of passenger and guide strand and the respective PO impurities by AEX chromatography at pH 11. A Dionex DNA-Pac PA200 column was used at a flow rate of 1 mL/min. Chromatographic conditions as in Figure 3.8.

applications have been reported for AEX HPLC [4, 44, 45]. However, separation of structured oligonucleotides containing hairpin-forming sequences, hybridised duplex regions or G-tetrad forming sequences can be challenging under non-denaturing conditions.

In Figure 3.13, the chromatographic separation of siRNAs containing duplex and single strand using non-denaturing AEX chromatography is shown. The same mobile phase was used, containing phosphate buffer at pH 6.5 and sodium bromide as eluting agent, but two different AEX resins were employed. One application used MiniQ columns with a bead size of 5 μm (Figure 3.13, left panel), the second application employed SuperQ-5PW columns with bead sizes of 10 μm (Figure 3.13, right panel). Both resins utilised the same functional group, a quaternary amine, but differed in matrix chemistry. Source Q material consisted of polystyrene/divinyl benzene beads. The SuperQ-5PW beads were made from poly-(methyl-methacrylate). In the analyses, a remarkable difference in the elution profiles of the two columns was observed. On the MiniQ column, single strands eluted earlier than the duplex, whereas on the SuperQ-5PW column this elution order was reversed (Figure 3.13, right panel). This difference was attributed to non-electrostatic interactions of duplex and single strands with the resin matrix, namely secondary mode interactions (such as hydrophobic van der Waals' interactions) and was observed for five siRNAs containing different base compositions and modification patterns [46]. Of the three column temperatures evaluated (40°C, 50°C and 60°C), the highest temperature gave the best separations on both columns, which is in accordance with the reported performance of AEX columns for polyanions [30].

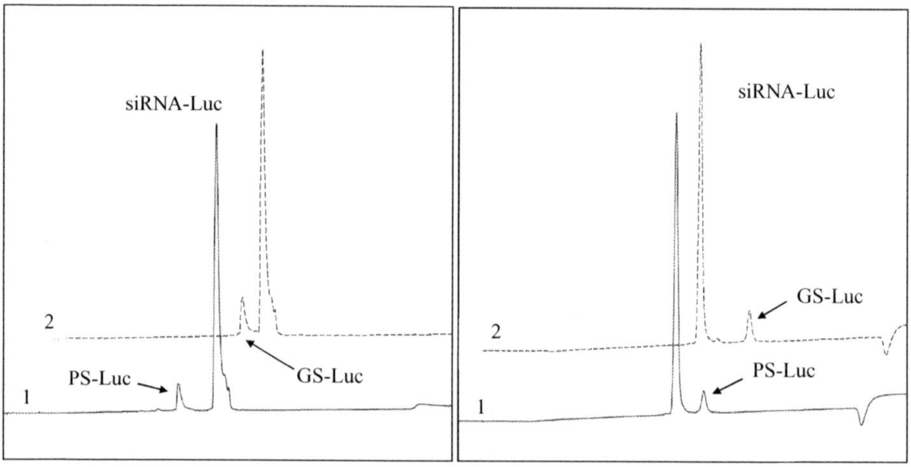

Figure 3.13 Impact of column resin on oligonucleotide separation. Chromatograms of siRNA-Luc spiked with PS-Luc (trace 1) and GS-Luc (trace 2). Separations were performed using MiniQ 4.6 × 50 mm column (left panel) at a flow rate of 1 mL/min and a TSKgel SuperQ-PW5 7.5 × 75 mm column (right panel) at a flow rate of 2 mL/min. Mobile phases contained 20 mM sodium phosphate at pH 6.5. Mobile phase B contained 1M sodium bromide. The gradient was run from 20 to 60% of mobile phase B in 25 min. Column temperature was 60°C. (Source: Modified from *Nucleic Acid Technol.*, 21, Noll, B., Seiffert, S., Hertel, F., Debelak, H., Hadwiger, P., Vornlocher, H.P., Röhl, I., Purification of small interfering RNA using non-denaturing anion-exchange chromatography, 383–393, Copyright 2011.)

3.4.3.2 Impact of Eluting Counterion on Separation

The use of different counterions can have a profound effect on oligonucleotide separations using AEX. Chloride is by far the most commonly used counterion. Changing this eluting anion to bromide, iodide or thiocyanate promotes earlier elution of oligonucleotides under the same gradient conditions [3, 34, 47]. This is due to increasing eluting power of the different anions. The Hofmeisters eluotropic series is a list of solvents ranked according to their eluting power [48]. For ions most common in AEX the Hofmeister series is

$$\text{Cl} < \text{Br} < \text{I} < \text{perchlorate} < \text{SCN}$$

As a consequence, retention times of the analytes containing PO or PS analogues increase in the sequence: SCN > perchlorate > I > Br > Cl. The driving force of this effect are the chaotropic properties of the salts, which render oligonucleotide hydration in solution more disordered, creating a more ‘lipophilic’ environment for the solvatised analyte, thus improving peak shape and reducing the influence of base sequence/composition on separation. Perchlorate, for example, tends to improve peak shape of lipophilic conjugates and is often used to suppress secondary structures and hydrophobic interactions with the resin. The change of eluting anion can lead to different selectivities in the separations of PO and PS variants (see Figure 3.14). Note that concentration of $NaClO_4$ is 0.5M compared to 1M of NaBr and NaCl. Thiocyanate as an eluting agent can be used to purify oligonucleotides containing a high percentage of P-dithioate linkages in lower salt concentrations and provides better separations than chloride as anion [49]. However, it should be noted that using a buffer

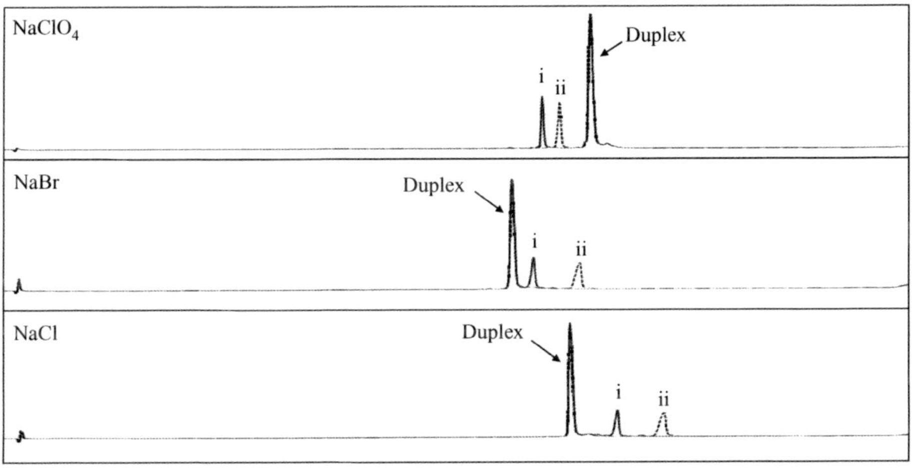

Figure 3.14 Impact of eluting anion on oligonucleotide separation. A Dionex DNA Pac-PA200 column was used at a flow rate of 1 mL/min. Mobile phase A consisted of 25 mM tris/HCl pH 8. Mobile phase B consisted of 25 mM tris/HCl pH 8 and either 1M NaCl, 1M NaBr or 0.5M $NaClO_4$. Column temperature was 40°C. A gradient was run from 20 to 60% B in 12 min. ‘i’ indicates passenger strand, ‘ii’ indicates guide strand.

high in salt concentration may result in a reduced lifetime of the instrument and AEX columns, as well as low resolution of the chromatographic separation, and care has to be taken when operating HPLC systems using buffers containing high concentrations of salt (see also Section 3.4.5).

3.4.3.3 Impact of Solvent in the Mobile Phase

Non-ionic interactions with the chromatographic resin can severely compromise separation properties of oligonucleotides containing strongly hydrophobic moieties such as fluorophores, quenchers or lipids. Addition of solvent to the mobile phases of AEX can help to suppress these undesired hydrophobic interactions between analyte and resin. Addition of ACN (acetonitrile) or methanol to the mobile phases can significantly improve peak shape. Solvent concentrations of 2–30% are used, helped by the fact that most AEX resins are stable towards these concentrations of solvents (Figure 3.15). However, the recommendations of the manufacturers regarding resin stability should be followed when adding solvents to the mobile phases.

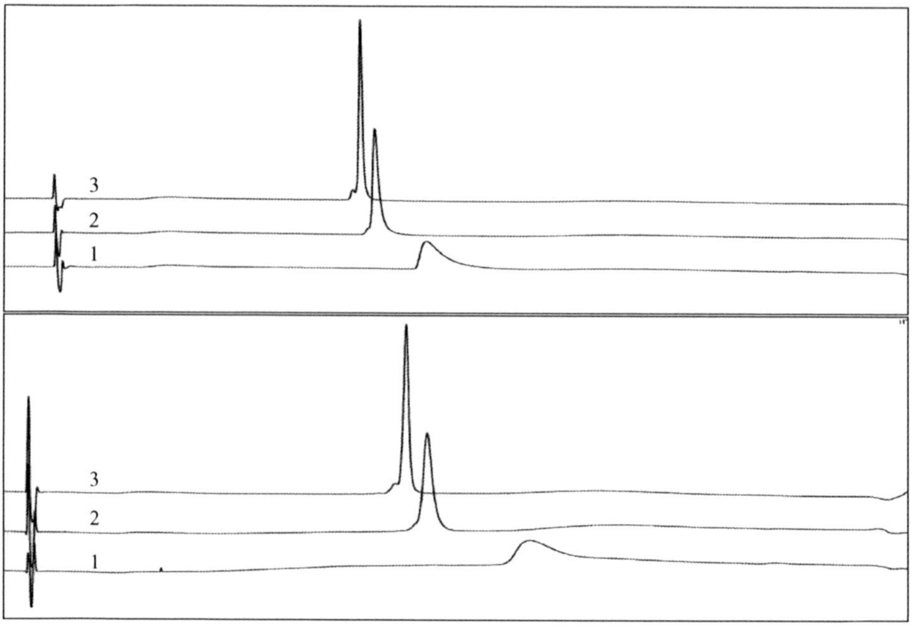

Figure 3.15 Impact of solvent on separation of lipophilic oligonucleotide conjugates. Mobile phase A consisted of 25 mM tris/HCl pH 8. Mobile phase B consisted of 25 mM tris/HCl (pH8) and 1M NaBr. Column temperature was 40°C. A gradient was run from 20 to 60% B in 12 min. Trace 1: mobile phases contained 10% ACN; trace 2: mobile phases contained 20% ACN; trace 3: mobile phases contained 30% ACN. A DNAPac-200 4 × 200 mm column (upper panel) or a PL-AX column (lower panel) was used at a flow rate of 1 mL/min.

3.4.4 Preparative AEX

Preparative purifications of oligonucleotides after synthesis are typically performed using AEX chromatography. Compared to most RP-resins, AEX-resins offer higher loading capacity, and cost of buffer chemicals and separation matrices are considerably lower. The respective methods are well established, scalable and offer high resolution [32]. In addition, buffer constituents can be removed easily, whereas the removal of the IP reagents may prove difficult after IP-RP chromatography. As a result, AEX is commonly used in the preparative purification of antisense, aptameres and siRNA oligonucleotides.

In the manufacture of duplex oligonucleotides, such as siRNA, chromatographic purification is typically performed on the single strands, followed by annealing of the purified strands. However, a direct comparison of two manufacturing methods using either conventional single-strand purification, or duplex purification performed after annealing of the crude single strands, showed improved purity and similar yield of the final siRNA duplex for the latter case [46] (Table 3.7). The manufacture of siRNAs using duplex purification was simpler and faster, and avoided heating of the strands after chromatographic purification. Elevated temperatures during strand annealing can lead to the generation of impurities, which in the case of single-strand purification are not removed from the final product.

3.4.5 Precautions When Using High-Salt Buffers

The previous sections have shown the versatility of AEX as a chromatographic technique for analysis and purification of oligonucleotides. Although the compatibility with organic solvents like methanol or ACN is often limited, the variability of the constituents of the mobile phase within a method makes AEX adaptable to a broad variety of analytes, without the need for major changes in chromatographic conditions. AEX chromatography can typically be operated in a wide pH range, allowing for separations at pH values of 11 or above, which prevent formation of hydrogen bonds between the chemical groups of the analytes, and thus preventing all related secondary structures (see also Section 3.2.1.1). In addition, the inclusion of different eluting agents in the buffers can exert chaotropic effects on the sample constituents and confer different selectivities on the analytes (see also Section 3.4.3.2). However, increasing the concentration of the eluting salts in the elution method can affect long-term operation of the employed instrument. Many salts can exert corrosive effects on instrument surfaces; crystals of buffer salts can form on connections, fittings, piston rods and pump seals and cause abrasion and friction on moving parts. During pump operation small amounts of buffer can leak out of the pump chambers through the piston fittings and, if not rinsed continuously, cause formation of salt crystals on the piston rods. These crystals can lead to abrasion of the pump seal materials during operation. Figure 3.16 depicts parts of HPLC system pumps, and valves employed in AEX chromatography after use in high-salt applications.

Because of the possible detrimental effects of buffer leakage, the use of high-salt buffers in AEX chromatography requires special procedures and precautions to

Table 3.7 Non-denaturing duplex analysis of crude and purified batches of a siRNA duplex (siRNA-Luc). Batch A and batch B were purified by duplex-AEX, batch C was purified by single-strand AEX (batch C). Total duplex content was determined by SEC; optimal duplex content was determined by non-denaturing IP-RP. (Source: Modified from *Nucleic Acid Technol.*, 21, Noll, B., Seiffert, S., Hertel, F., Debelak, H., Hadwiger, P., Vornlocher, H.P., Röhl, I., Purification of small interfering RNA using non-denaturing anion-exchange chromatography, 383–393, Copyright 2011.)

ID	Description	Total duplex by SEC	Optimum duplex by n-d IP-RP	siRNA-Luc denat IP-RP	GS-Luc denat IP-RP	PS-Luc denat IP-RP
Crude siRNA-Luc	Unpurified duplex	90.8%	70.1%	77.8%	39.2%	38.6%
Crude siRNA-Luc (annealed)	Unpurified duplex	90.4%	73.5%	76.9%	38.6%	38.3%
siRNA-Luc (Batch C)	From purified single strands	97.6%	86.1%	90.3%	46.2%	44.1%
siRNA-Luc (Batch B)	Purified by SuperQ 5PW	99.3%	90.2%	91.5%	46.2%	45.3%
siRNA-Luc (Batch A)	Purified by Source 15Q	99.2%	90.5%	92.4%	46.9%	45.5%

Figure 3.16 Corroded pistons (left panel). Salt crystals are visible at the piston base (left piston) and piston rods (both pistons). Corroded injection valve (right panel): metal surfaces show abrasion and salt build-up.

prevent damage to the HPLC system. To circumvent these issues, precautions such as use of circulating fluid in the chambers of the moving piston rods, regular cleaning of tubing fittings and exchange of valve seals are recommended. HPLC systems should therefore be equipped with a continuous pump wash option. Pump, tubing and valve washes should be performed regularly and wash solutions should be changed routinely to avoid accumulation of salt in the wash solution.

3.5 Reverse Phase-Chromatography

3.5.1 Separation Principles

Oligonucleotides are analytes with significant ionic character and have a low binding affinity to lipophilic stationary phases. Hence, RP-HPLC can only be applied using mobile phases containing ion-pairing (IP) reagents. Using IP-RP chromatography, separations of very high resolution can be achieved for a broad variety of oligonucleotides. One important advantage of IP-RP chromatography over AEX chromatography is that this technique can be readily combined with MS detection. IP-RP chromatography with electrospray ionisation–mass spectrometry (ESI-MS) has been used extensively to study single- and double-stranded oligonucleotides and a wide range of IP reagents have been used for the interfacing of IP-RP-HPLC to ESI-MS (reviewed in reference [50]). IP-RP can be considered the method of choice, when chromatographic separation in combination with MS detection is desired.

3.5.1.1 Mobile Phases

Selectivity in IP-RP is determined by the composition of the mobile phase. Mobile phases usually consist of a specific IP reagent, the respective counter ion and the organic eluent. IP reagents are large ionic molecules which have a charge opposite to the analyte of interest, as well as a hydrophobic region (usually an alkyl chain) to interact with the stationary phase. Ion pairs are formed between the charged analyte and the oppositely charged IP reagent, which is equally distributed between the mobile and the stationary phases. Depending on the apparent charge of the analyte, this results in different retention on the stationary phase of the column.

Numerous IP-RP methods have been reported for the physico-chemical characterisation of single-stranded oligonucleotides, as well as the hybridised duplex. IP-RP can be applied in denaturing [39, 51–54] and non-denaturing modes [55–59]. Buffer systems that have been successfully used for separations of nucleic acids include triethylammonium bicarbonate (TEAB) [56], butyldimethylammonium bicarbonate (BDMAB)/butyldimethylammonium acetate [60, 61] and cyclohexyldimethylammonium acetate [62]. Buffers containing triethylammonium acetate (TEAA), triethylammonium (TEA)/1,1,1,3,3,3,-hexafluoro-2-propanol (HFIP) [51, 63] and hexylammonium acetate (HAA) [64] in particular have been reported as giving excellent separation for single- and double-stranded oligonucleotides. The use of HFIP in IP buffers is particularly attractive because it is compatible with ESI-MS and often provides excellent separation [51, 52]. HFIP lends itself better to ESI-MS, primarily because it is more volatile than acetate and readily evaporates from the droplets produced during the electrospray process [50]. Because of its unique IP properties, high separation efficiency and high mass sensitivity, TEA/HFIP is now used as widely as TEAA in the analysis of oligonucleotides.

Another important feature of HFIP is that it reduces the impact of oligonucleotide hydrophobicity upon retention [65]. Usually retention times of single strands in IP-RP are sequence dependent [39, 55, 64, 66]. However, in a buffer system containing HFIP as the counter ion of the IP reagent, the hydrophobicity of the single strands has a lower influence on peak separation than in systems containing acetate, where elution order of oligonucleotides have been reported to show dramatic changes with base sequence and hydrophobicity [65].

3.5.1.2 Stationary Phases

Both silica particles and rigid organic polymers of differing particle and pore sizes are used in RP-HPLC and IP-RP-HPLC. The hydrophobic characteristics of these base materials are usually obtained by bonding C4, C8 or C18 carbon chains to the resin particles. Although polymer-based materials are typically more pH tolerant, functionalised silica-based resins have been used more widely. Silica-based resins have a number of benefits. They withstand greater back pressures, and are compatible with a wider variety of different organic solvents. However, RP columns have their limitations, such as moderate loading capacity and a limited pH range. Although silica-based resins are stable at pH as low as pH 2, pH values above pH 8 can lead to partial hydrolysis of the bonded phase and thus degradation of the column. The limited pH stability of silica-based phases is a major drawback for separations of oligonucleotides by IP-RP, since most of the reported methods (see Table 3.8) use high pH mobile phases. In the pH range of approximately 7–10, the charge-to-size ratio is virtually constant for all single-stranded and double-stranded oligonucleotides, as it depends

Table 3.8 Mobile phases commonly used in IP-RP HPLC of oligonucleotides.

Buffer	IP reagent	MS sensitivity	Remarks	Literature
TEAA	Triethylammonium acetate	Low	Peak broadening by diastereomeres, RT variation with sequence	[67]
TEAB	Triethylammonium bicarbonate	High	RT variation with sequence (single strand)	[56]
BDMAB	Butyldimethylammonium bicarbonate	High	RT variation with sequence (single strand)	[60, 61]
HAA	Hexylammonium acetate	Low	RT variation with sequence (single strand)	[63]
TEA/ HFIP	Triethylamine/ hexafluoroisopropanol	High	Low peak broadening by PS diastereomeres, little RT variation with sequence	[51, 53]
HA/ HFIP	Hexylamine/ hexafluoroisopropanol	High	Low peak broadening by PS diastereomeres, little RT variation with sequence	[68]

solely on the number of charges of the phosphate linkages in the backbone. In addition, higher pH values can lead to higher charge states and greater signal intensity in ESI-MS applications [66]. Stability at different pH values is therefore a prerequisite to achieve good separations for oligonucleotides of different sequences. To increase pH stability of silica-based RP resins, manufacturers have introduced surface modifications to the silica base. Modifications include polar functional groups, sterically hindered silanes and bidentates. In addition, resins made from hybrid materials which combine organic (polymeric) and inorganic (silica) matrix chemistry, have significantly increased the stability of silica-based stationary phases toward very low and high pH. As a result, resins are available that can be used in pH ranges of 2–12. For most column materials it has to be noted that high temperatures, as well as high pH and inclusion of chaotropic agents may limit column longevity. Therefore, it is recommended to check the pH stability of the column towards the intended mobile phase prior to chromatography, to make sure that no unwanted degradation occurs.

3.5.2 Denaturing IP-RP-HPLC

3.5.2.1 Impact of Column Temperature on Separation

Denaturing chromatography involves separation conditions that allow for the disruption of all possible secondary structure of a given oligonucleotide. Denaturing conditions enable separations dependent only on length. This facilitates chromatographic separation and usually improves resolution, helping to determine oligonucleotide purity. Secondary structures of oligonucleotides are primarily caused by the formation of hydrogen bonds between the chemical groups of the analytes, as is the case for the pairing bases in Watson–Crick helices. Many IP reagents, organic additives and some counter ions can exert a destabilising effect on these hydrogen bonds. However, the ambient temperature buffer conditions used in regular IP-RP-HPLC are often not sufficient to fully resolve secondary structures in oligonucleotides containing complementary base sequences. Therefore, elevated column temperatures are required to achieve the conditions necessary for full denaturing of the oligonucleotide. For example, high column temperature in combination with high amounts of the neutralising acid HFIP can assist the disruption of oligonucleotide secondary structures. Figure 3.17 depicts the effect of temperature on the separation of two siRNAs with different T_m in a hexylamine (HA)/HFIP buffer system. Depending on duplex stability, higher column temperatures have to be used for complete resolution of the secondary structures in the duplexes.

3.5.2.2 Separation of Single-Strand Impurities

High-resolution separations of most impurities of single- and double-stranded oligonucleotides can be achieved using denaturing conditions. Chapter 2 provides a comprehensive overview of the types of oligonucleotide-related impurities. Figure 3.18 depicts separations of typical shortmer impurities N-1 and N-2 as well as PO impurities from guide and passenger strands of a typical siRNA.

Good separations in IP-RP-HPLC can typically be achieved for analytes that show differences in lipophilicity. Figure 3.19 depicts separations of guide and passenger strands of their respective N+G and CNET impurities. Lipophilic impurities such as CNET bind more strongly to the RP column as the full-length oligonucleotide and the N+G longmer.

High-resolution IP-RP-HPLC using denaturing conditions is regularly employed in purity analysis of oligonucleotides and allows for the separation of shortmer and longmer, as well as impurities containing lipophilic adducts. Figure 3.20 depicts overlaid chromatograms of IP-RP separations of a RNA single strand before and after chromatographic purification.

3.5.2.3 Separation of Degradation Products

Denaturing applications of IP-RP-HPLC also offer high-resolution purity analysis of oligonucleotides containing chemical modifications. Figure 3.21 depicts chromatograms of siRNA duplexes containing different ribose modifications that have been subjected to elevated temperatures leading to strand degradation. The panel on the left depicts chromatograms of one siRNA containing only 2′OMe modifications, whereas the right panel depicts chromatograms of a siRNA containing additional 2′-fluoro modifications. The degradation products generated by the exposure to the elevated temperatures are 2′,5′- isomers for regular 3′,5′-oligonucleotides and arabinosyl-nucleotides, as well as O-2,2′-anhydro-nucleotides for the 2′-fluoro-oligonucleotides. These degradation products were separated and quantified using denaturing IP-RP-HPLC.

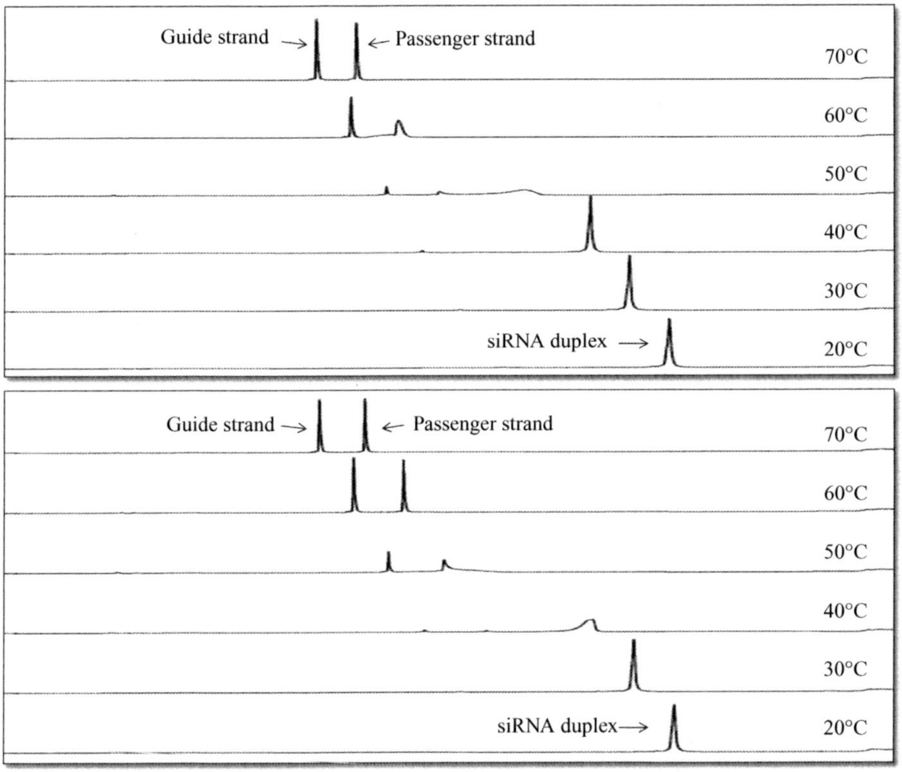

Figure 3.17 Impact of temperature on separation of duplex oligonucleotides. Two siRNA of identical sequence but different 2′-modification pattern were analysed by IP-RP. siRNA-1 (lower panel) contained 2′-OH only, siRNA-2 (upper panel) contained 11 2′F modifications in the passenger strand. siRNA-1 was completely denatured at temperatures of 60°C, siRNA-2 was completely denatured at temperatures of 70°C. Chromatographic conditions: samples were applied to an Ultimate 3000 RS series HPLC system (Dionex, Sunnyvale, CA, USA) using UV detection at 260 nm. Oligonucleotides were separated by ACN-gradient elution on an Acquity UPLC OST C18 2.1 × 100 mm column with a 1.7 μm bead size (Waters, Milford, MA, USA). Mobile phase A consisted of 100 mM HFIP, 16.3 mM TEA and 1% methanol. Mobile phase B consisted of 100 mM HFIP, 16.3 mM TEA and 95% methanol. The gradient was run from 5 to 35% of mobile phase B in 25 min at a flow rate of 250 μl/min. Injection volume was 2 μL of a 50 μM RNA solution in 1 × PBS. Column temperature is indicated at the respective traces.

3.5.3 Non-denaturing IP-RP-HPLC

3.5.3.1 Impact of Mobile Phase Composition

Non-denaturing techniques are typically used to determine secondary structure features of oligonucleotides. One example would be the verification of the formation of a correct double-stranded α-helix structure of a siRNA duplex. Another example

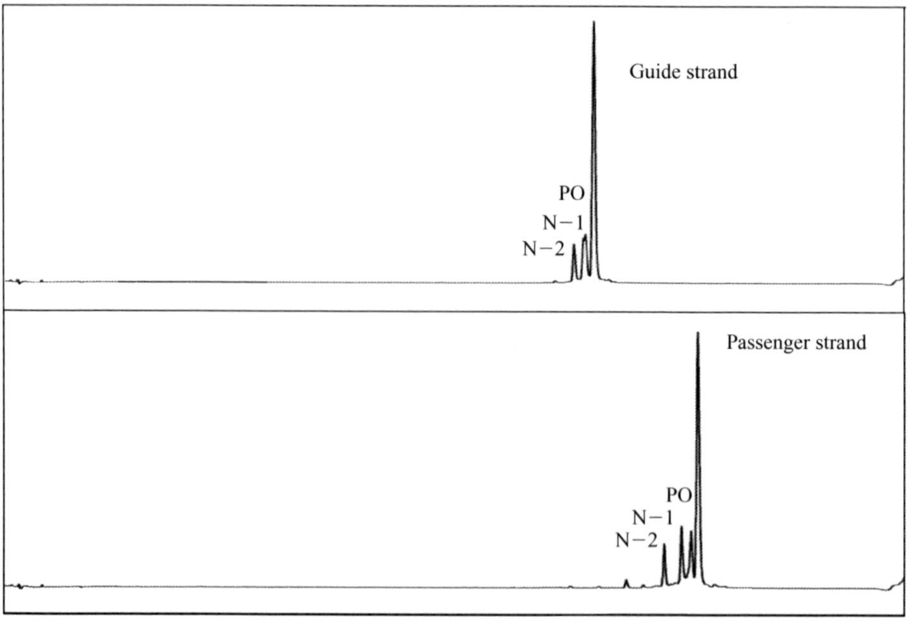

Figure 3.18 Separation of PO and N-1 and N-2 impurities. Ultimate 3000 RS series HPLC system (Dionex, Sunnyvale, CA, USA) using UV detection at 260 nm wavelength. Oligonucleotides were separated by gradient elution on a XBridge C18 OST 2.1 × 50 mm column with a 2.5 μm particle size (Waters, Milford, MA, USA). Mobile phase A consisted of 100 mM HFIP, 16.3 mM TEA and 1% methanol. Mobile phase B consisted of 100 mM HFIP, 16.3 mM TEA and 95% methanol. Column temperature was 65°C. A gradient was run from 1 to 18% of mobile phase B in 30 min at a flow rate of 250 μL/min. Injection volume was 20 μL of a 5 μM RNA solution in water.

Figure 3.19 Separation of N+G and CNET impurities. Samples were separated by methanol-gradient elution on a XBridge C18 OST 2.1 × 50 mm column with a 2.5 μm particle size (Waters, Milford, MA, USA). Chromatographic conditions were the same as described in Figure 3.18.

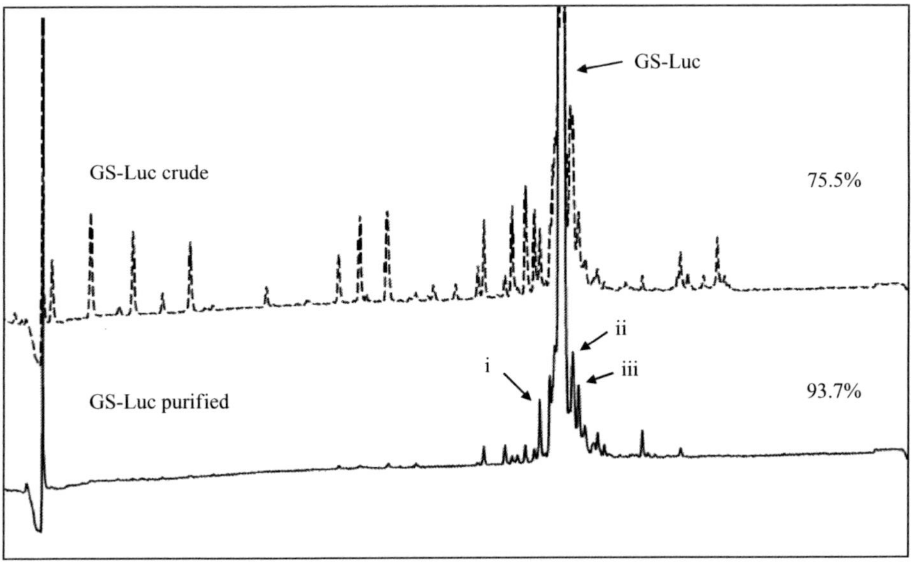

Figure 3.20 Denaturing IP-RP analysis of crude and purified GS-Luc (upper and lower trace, respectively). Peak 'i': 5′(N-2) shortmer. Peak 'ii': N+G longmers and N+U/C longmers. Peak 'iii': N+A longmers and N+U/C longmers. Samples were separated by methanol-gradient elution on an XBridge C18 OST 2.1 × 50 mm column with a 2.5 μm particle size (Waters, Milford, MA, USA). Chromatographic conditions were the same as described in Figure 3.18. (Source: Modified from *Nucleic Acid Technol.*, 21, Noll, B., Seiffert, S., Hertel, F., Debelak, H., Hadwiger, P., Vornlocher, H.P., Röhl, I., Purification of small interfering RNA using non-denaturing anion-exchange chromatography, 383–393, Copyright 2011.)

would be separation of a siRNA duplex from non-hybridised single strands, or other competing structures. Possible secondary structures other than the duplex could be hairpins formed by one or both of the strands as well as triple-helices or multimeric aggregates of one or both of the strands. A more complex task would be the identification of an aptameric structure or a G-tetrad-containing structure. In the analysis of siRNA, non-denaturing IP techniques have yet to achieve the level of peak separation that is commonly achieved using denaturing chromatography of single strands. Chromatographic separation of the optimal duplex (N_{PS}/N_{GS}) from unwanted duplex variants is not easy to achieve because the physico-chemical properties of both species are usually very similar. Insufficient chromatographic resolution is currently the main reason why non-denaturing techniques do not permit detection and quantification of individual duplex impurities. Several IP-RP-HPLC methods have been reported which allow high-resolution separations of siRNA or their metabolites under non-denaturing conditions and are also compatible with ESI-MS detection. The most common buffer systems are either HAA [64] or TEA/HFIP [22, 55]. Improved duplex separation has been reported by McCarthy *et al.* [63] who replaced the weaker ion-pair reagent TEA/acetate with the stronger ion-pair reagent HA/acetate in the chromatographic separation of siRNAs, obtaining higher separation efficiencies.

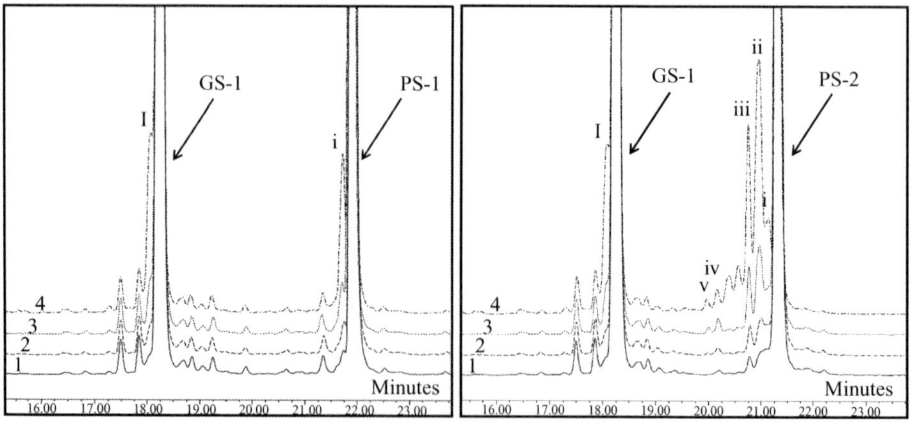

Figure 3.21 Denaturing IP-RP-HPLC of siRNA-1 (left panel) and siRNA-2 (right panel) after incubations at room temperature (trace 1), 60°C (trace 2), 70°C (trace 3) and 85°C (trace 4). All incubations were carried out for 4 h using 100 μM siRNA solutions in water. The full-length guide strand peak (GS-1) and passenger strand peaks (PS-1 or PS-2) are indicated by arrows. Impurity peaks are marked by upper case (guide strand) and lower case (passenger strands) letters. 'I' indicates 2′,5′-isomers of GS-1; 'i' indicates 2′,5′-isomers of PS-1 or PS-2; 'ii' and 'iii' indicate HF-elimination products of PS-2 (M_r(parent)-2 and M_r(parent)-4); 'iv' and 'v' indicate HF-elimination products of PS-2 (M_r(parent)-20). Samples were separated by methanol-gradient elution on a XBridge C18 OST 2.1 × 50 mm column with a 2.5 μm particle size (Waters, Milford, MA, USA). Chromatographic conditions were the same as described in Figure 3.18. (Source: Modified from *Analyt. Biochem.*, 414, 1, Seiffert, S., Debelak, H., Hadwiger, P., Jahn-Hofmann, K., Röhl, I., Vornlocher, H.P., Noll, B., Characterization of side reactions during the annealing of small interfering RNAs, 47–57, Copyright (2011), with permission from Elsevier.)

Another non-denaturing IP-RP system containing HA/HFIP has been described that further improved the separation efficiency of siRNA [68]. A buffer system containing 25 mM HA and 100 mM HFIP was used to analyse samples containing a siRNA and three duplex variants (variant 1 to variant 3, Table 3.9). Peak resolution (R_s) for variant 1 using the HA/HFIP system was improved ($R_s = 1.34$ versus $R_s = 0.82$ using the HAA system), whereas no separation was obtained for the TEA/HFIP system (Figure 3.22). The observed effects were attributed primarily to the increased alkyl chain length and hydrophobicity of the IP reagent HA compared to TEA and its impact on the retention of the oligonucleotides [69]. A comparison of buffer compositions was performed using different concentrations of HA in mobile phase A. Average peak resolution values were calculated and plotted (Figure 3.23). At HA concentrations of 5–10 mM, resolution between the optimal duplex and its variants was poor. Increasing the HA levels to 15 mM resulted in a significant improvement in resolution, with the best separations achieved at a maximum HA of 20 mM ($R_s = 1.46$). At even higher HA levels, peak resolution began to decrease because the sorbent surface may have become saturated, or micelles may have formed

Table 3.9 Passenger and guide strands of siRNA-Luc duplex variants and respective relative retention time (RRT). Base sequences of the respective strands are listed in Table 8. nd = low peak separation, RRT was not determined (for base sequences see Table 3.10).

Duplex ID	Passenger ID	Guide ID	RRT
siRNA-Luc	PS-Luc	GS-Luc	1.000
Variant 1	PS-Luc	Short-G1	0.984
Variant 2	PS-Luc	Short-G2	0.967
Variant 3	Short-P1	GS-Luc	0.950
Variant 4	Short-P2	GS-Luc	0.945
Variant 5	PS-Luc	GS-PO	0.974
Variant 6	PS-PO	GS-Luc	0.984
Variant 7	Long-P1	GS-Luc	0.952
Variant 8	Long-P2	GS-Luc	1.045
Variant 9	Long-P3	GS-Luc	nd
Variant 10	PS-Luc	Long-G1	0.977
Variant 11	PS-Luc	Long-G2	0.985
Variant 12	Iso-P1	GS-Luc	0.975
Variant 13	Iso-P2	GS-Luc	1.025
Variant 14	PS-Luc	Iso-G1	nd
Variant 15	PS-Luc	Iso-G2	nd
Variant 16	PS-Luc	Iso-G3	0.939

in solution, leading to a reduced availability of adsorption sites and hence decreased retention.

In addition to the beneficial effect of HA on duplex separation, the hydrophobic character of HFIP itself may have led to a higher adsorbed concentration of HA on the stationary phase, resulting in improved IP efficiency and, in turn, higher separation selectivity [50–52].

3.5.3.2 Separation of Duplex Variants

Non-denaturing IP-RP-HPLC can be optimised to separate a wide variety of duplex variants using HA and HFIP added to both the aqueous and organic mobile phases. In the example in Figure 3.24, four sets of siRNA duplex variants were analysed using 15 mM HA and 50 mM HFIP (variant 1 to variant 16, Tables 3.9 and 3.10). All duplex variants containing shortmers (variant 1 to variant 4) were separated from optimal duplex (Figure 3.24, upper left panel). A similar level of peak separation was observed for the two duplex variants containing one phosphodiester linkage (PO) in either passenger or guide strand (variant 5 and variant 6; Figure 3.24, upper right panel). Of the five duplex variants containing longmers (variant 7 to variant 11), four were separated from the optimal duplex. One duplex variant (variant 9) eluted with the optimal duplex (Figure 3.24 lower left panel). Three of five duplex variants containing 2′,5′-isomers in either passenger or guide strand were separated from the optimal duplex (variant 12, variant 13 and variant 16; Figure 3.24 lower right panel).

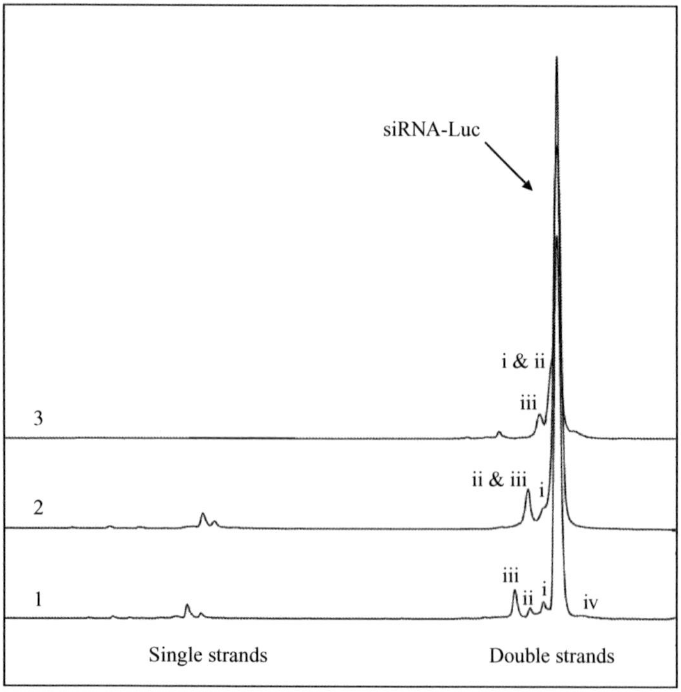

Figure 3.22 Comparison of non-denaturing IP-RP-HPLC methods. UV traces of buffer systems containing 25 mM HA and 100 mM HFIP (trace 1), 25 mM HAA (trace 2) or 16.5 mM TEA and 100 mM HFIP (trace 3) are shown. The analysed samples were siRNA-Luc spiked with 10% short-P1. Non-hybridised single strands eluted earlier than the duplex (single strands). Peak 'i' indicates variant 1; peak 'ii' indicates variant 2, peak 'iii' indicates variant 3 (see Table 3.9) and peak 'iv' indicates late eluters. Samples were applied to an Ultimate 3000 RS series HPLC system (Dionex, Sunnyvale, CA, USA) using UV-detection at 260 nm. Samples were separated by methanol-gradient elution on an Acquity UPLC OST C18 OST 2.1 × 100 mm column with a 1.7 μm particle size (Waters, Milford, MA, USA). Column temperature was 20°C. Injection volume was 2 μL of a 50 μM RNA solution in 1 × PBS. (Source: Modified from *J. Chromatogr. A*, 1218, 33, Noll, B., Seiffert, S., Vornlocher, H.P., Röhl, I., Characterization of small interfering RNA by non-denaturing ion-pair reversed-phase liquid chromatograhy, 5609–5617, Copyright (2011), with permission from Elsevier.)

3.5.4 Strategies for Sample Preparation

Rational approaches on sample preparation of oligonucleotides are often under-represented in the scientific literature. As discussed previously, DNA and RNA molecules are chemically complex and have the propensity to form secondary and tertiary structures. By nature, many oligonucleotides are likely to engage in structure formation when in solution. In solution, the natural tendency of many oligonucleotides is to engage in some degree of structure formation and the configurations adopted may vary from simple base stacking in a 'linear', single-stranded molecule, to bulges in hairpins or partially double-stranded helix, parallel and anti-parallel stranded multi-

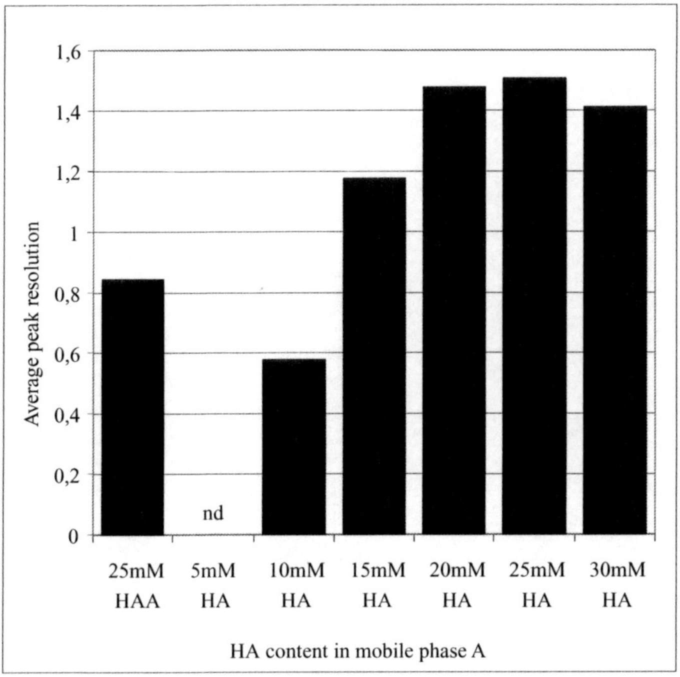

Figure 3.23 Effect of buffer composition on peak resolution. Average peak resolutions of optimal duplex, variant 1, variant 2 and variant 3 (see Table 3.9) peaks in non-denaturing IP-RP-HPLC using the indicated HA concentrations and 100 mM HFIP in mobile phase A. Mobile phase B consisted of 80% ACN. For the 25 mM HAA system, peak resolutions were calculated using optimal duplex, variant 1, variant 2 and variant 4 peaks, since variant 2 and variant 3 peaks were co-eluting. 'nd' indicates that, owing to low peak separation, peak resolution values could not be determined. The gradient was run from 15 to 35% of mobile phase B in 25 min at a flow rate of 250 μL/min. Other chromatographic conditions were as in Figure 3.18. (Source: Modified from *J. Chromatogr. A*, 1218, 33, Noll, B., Seiffert, S., Vornlocher, H.P., Röhl, I., Characterization of small interfering RNA by non-denaturing ion-pair reversed-phase liquid chromatograhy, 5609–5617, Copyright (2011), with permission from Elsevier.)

mers, as well as combinations thereof. Many factors influence the stability of these structures (such as temperature or content of organic solvents or salt) and have been described in more detail in earlier sections. The concentration of the oligonucleotide itself can also have a significant impact on a secondary structure. These factors have to be kept in mind when preparing samples for non-denaturing IP-RP-HPLC. Figure 3.25 depicts an example where non-denaturing analysis of a hybridised siRNA was attempted and the addition of buffer salts to the sample was required to maintain the stability of the duplex in the diluted sample. To ensure completely non-denaturing conditions, the addition of salt or ionic buffers such as 1 × PBS is recommended in the analysis of duplex oligonucleotides.

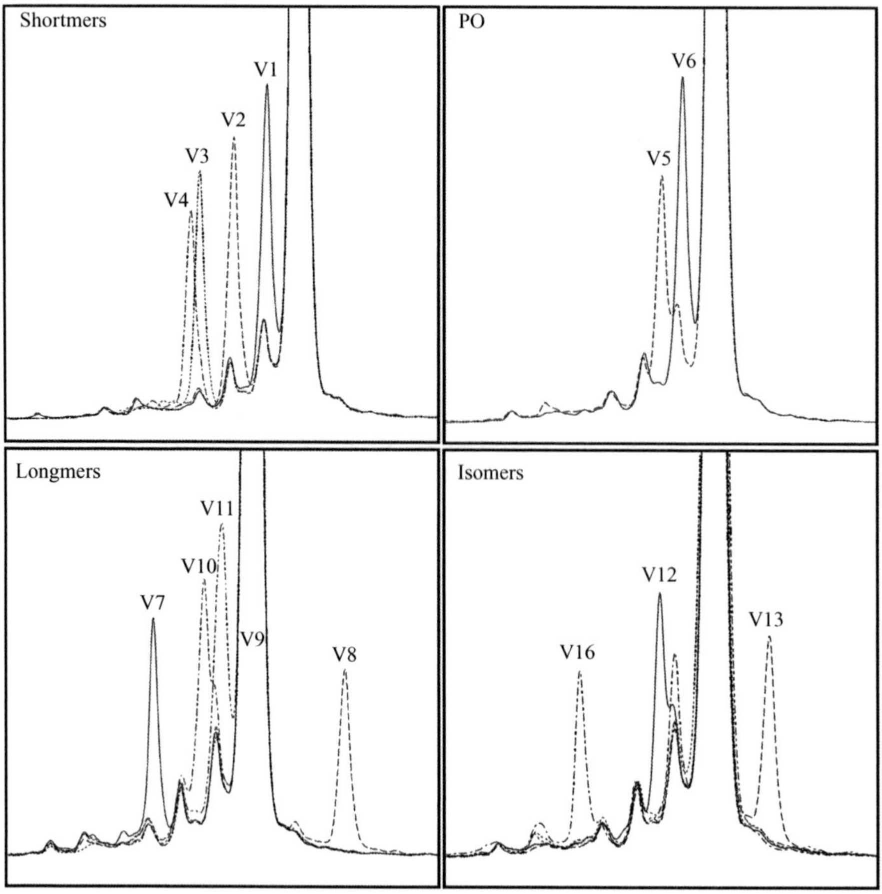

Figure 3.24 Chromatographic separation of duplex variants from optimal duplex. Solid lines indicate unspiked samples of siRNA-Luc; dashed and dotted lines indicate siRNA-Luc samples spiked with duplex impurity markers. Upper left: siRNA-Luc spiked with shortmers. Upper right: siRNA-Luc spiked with PO. Lower left: siRNA-Luc spiked with longmers. Passenger strand variant long-P3 ($M_r = 7123$) was identified under the main peak by ESI-MS (variant 9). Lower right: siRNA-Luc spiked with isomers. Samples were applied to an Ultimate 3000 RS series HPLC system (Dionex, Sunnyvale, CA, USA) using UV detection at 260 nm. Oligonucleotides were separated by ACN-gradient elution on an Acquity UPLC OST C18 2.1 × 100 mm column with a 1.7 μm bead size (Waters, Milford, MA, USA). Mobile phase A consisted of 15 mM HA, 50 mM HFIP and 5% ACN. Mobile phase B consisted of 15 mM HA, 50 mM HFIP and 50% ACN. The gradient was run from 30 to 62 % of mobile phase B in 25 min at a flow rate of 250 μl/min. Column temperature was 20°C. (Source: Modified from *J. Chromatogr. A*, 1218, 33, Noll, B., Seiffert, S., Vornlocher, H.P., Röhl, I., Characterization of small interfering RNA by non-denaturing ion-pair reversed-phase liquid chromatograhy, 5609–5617, Copyright (2011), with permission from Elsevier.)

Table 3.10 Single strands and impurity markers of siRNA-Luc. Upper case letters indicate 2′-OH RNA nucleotides, lower case letters indicate 2′-*O*-methyl nucleotides, TT indicates two 2′-deoxy-thymidine nucleotides connected via a phosphorothioate linkage, G underlined indicates the additional guanine-nucleotides. Italic letters indicate the position of the 2′, 5′-linkage.

Strand ID	Sequence	Orientation	Description	Molecular mass (calculated)
PS-Luc	cuuAcGcuGAGuAcuucGATT	Passenger	Full-length	6778.5
PS-PO	cuuAcGcuGAGuAcuucGATT	Passenger	PO variant	6762.4
Short-P1	uuAcGcuGAGuAcuucGATT	Passenger	5′(n – 1) shortmer	6459.2
Short-P2	uAcGcuGAGuAcuucGATT	Passenger	5′(n – 2) shortmer	6139.0
Long-P1	cuuAcG<u>G</u>cuGAGuAcuucGATT	Passenger	N+G (Pos 1)	7123.7
Long-P2	cuuAcGcuG<u>G</u>AGuAcuucGATT	Passenger	N+G (Pos 2)	7123.7
Long-P3	cuuAcGcuGAGuAcuucG<u>G</u>ATT	Passenger	N+G (Pos 4)	7123.7
Iso-P1	cuu*A*cGcuGAGuAcuucGATT	Passenger	2′,5′-Isomer	6778.5
Iso-P2	cuuAcGcu*G*AGuAcuucGATT	Passenger	2′,5′-Isomer	6778.5
GS-Luc	UCGAAGuACUcAGCGuAAGTT	Guide	Full-length	6752.4
GS-PO	UCGAAGuACUcAGCGuAAGTT	Guide	PO variant	6736.3
Short-G1	CGAAGuACUcAGCGuAAGTT	Guide	5′(n – 1) shortmer	6446.3
Short-G2	GAAGuACUcAGCGuAAGTT	Guide	5′(n – 2) shortmer	6141.0
Long-G1	UCG<u>G</u>AAGuACUcAGCGuAAGTT	Guide	N+G (Pos 1)	7097.6
Long-G2	UCGAAGuACUcAGCGuAAG<u>G</u>TT	Guide	N+G (Pos 5)	7097.6
Iso-G1	UCG*A*AGuACUcAGCGuAAGTT	Guide	2′,5′-Isomer	6752.4
Iso-G2	UCGAAGuAC*U*cAGCGuAAGTT	Guide	2′,5′-Isomer	6752.4
Iso-G3	UCGAAGuACUcAGC*G*uAAGTT	Guide	2′,5′-Isomer	6752.4

3.5.5 Artefacts and Contaminants Observed in IP-RP-HPLC

The high sensitivities that can be achieved in chromatographic methods using UV detection can sometimes lead to the presence of unspecific signal, due to contaminants and other artefacts which absorb light in the employed wavelength range. These contaminants can bind to the RP column and elute within the range of the gradient. They can originate from a number of sources, for example the use of low-purity buffering reagents, wash solutions containing contaminated components, dirty buffer containers and flasks, contaminated pipettes, diluents and other sources. In IP-RP-HPLC, chromatographic buffers usually contain a significant amount of organic solvents and often other potentially corrosive chemicals, which may affect the surfaces of vials and plastics used in sample preparation and storage. Figure 3.26 shows chromatograms obtained with different samples using IP-RP-HPLC. In a number of sample solutions a contaminant peak was repeatedly observed when the injection system was used. Conversely, the peak was not present in any of the solvents and

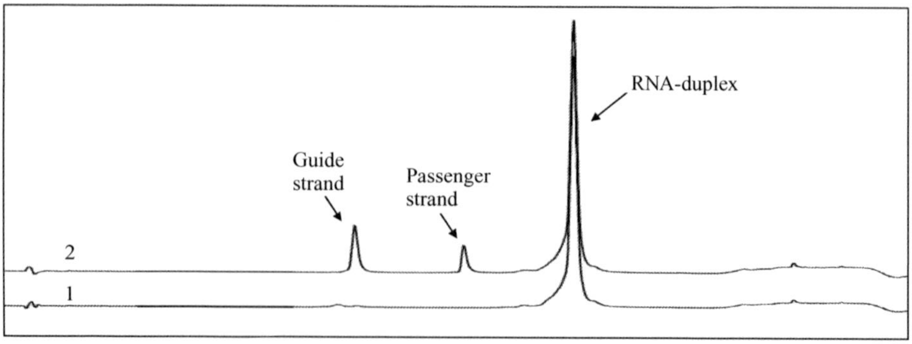

Figure 3.25 Impact of sample preparation on a siRNA in non-denaturing IP-RP analysis. Sample in 1 × PBS (trace 1) and sample in water (trace 2). Dissociation of the duplex is observed in trace 2, single strand and duplex peaks are indicated. Chromatographic conditions were the same as described in Figure 3.24.

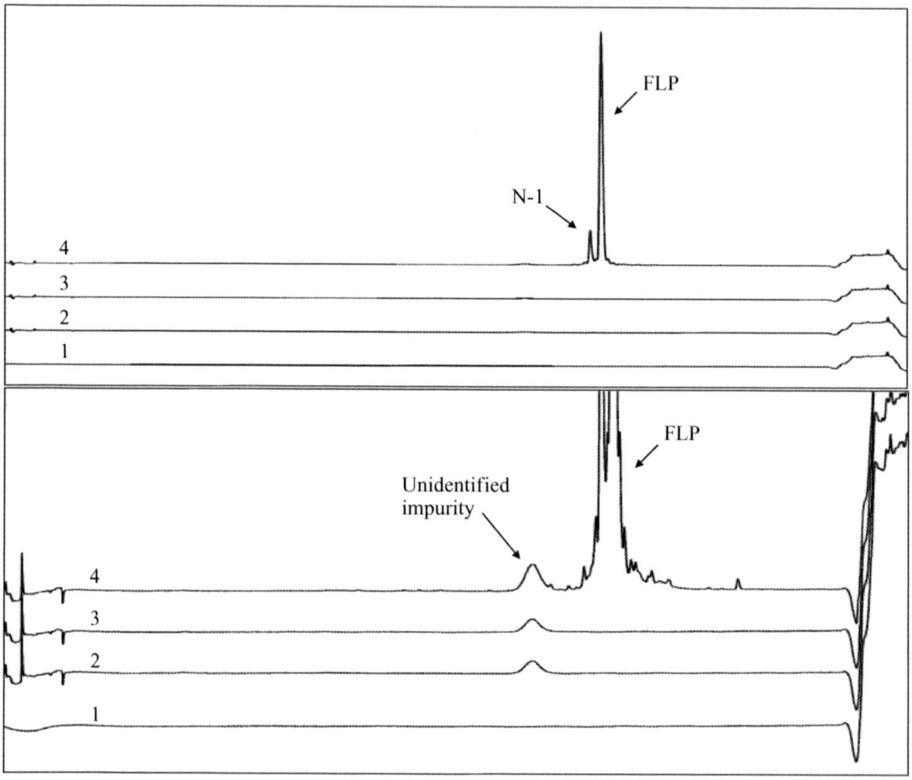

Figure 3.26 IP-RP-HPLC chromatograms of samples containing either water (trace 2), buffer A (trace 3) or an oligonucleotide sample (trace 4). No injection was performed in chromatographic run (trace 1). The upper panel depicts full chromatograms, the lower panel depicts zoomed curves. Chromatographic conditions were the same as described in Figure 3.24 (FLP denotes full-length product).

chemicals used in the analysis; instead it appeared to originate from the vials and caps. When vials and caps made from different materials were examined, this impurity could only be detected when polypropylene (PP) vials or caps containing PTFE membranes were used (Figure 3.27). Other materials such as polymethylpentene (TPX) (for vials) and Teflon/silicon (for membranes) were free of these contaminants.

3.6 Size Exclusion Chromatography

3.6.1 Principles of Separation

SEC is widely used in purification and analysis of DNA, RNA, oligonucleotides, oligonucleotide complexes and conjugates. SEC separates peaks according to the hydrodynamical size of the analytes and its relative rate of diffusion into the pores of the stationary-phase resin. Hence, larger molecules elute earlier than smaller ones. Typically, the elution order in SEC is the reverse of the other commonly used chromatographic techniques and this is a key feature of SEC. The principles governing the mechanisms of separation in SEC have been thoroughly described in various publications [70, 71] and only a brief description will be presented in the following sections.

In SEC, the stationary phase is a heterogeneous phase system, consisting of a liquid phase, usually aqueous, which is contained within the pores of a solid phase. The pores of the solid-phase particles are comparable in size to the hydrodynamic size

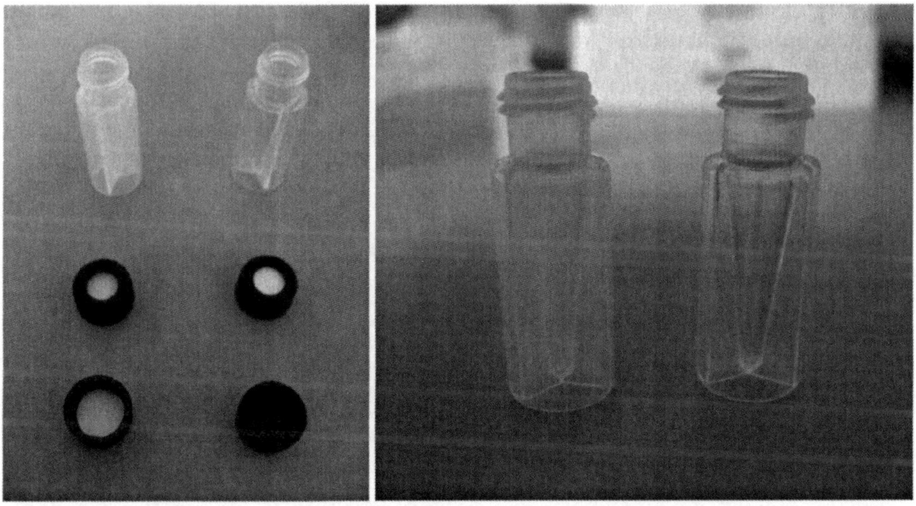

Figure 3.27 Chromatographic vials and respective caps used in the sample preparation for RP-chromatography. Vials made from polypropylene (left, opaque) or polymethylpentene (right, clear). Caps containing Teflon/silicon membranes (left, white inside surface) or polytetrafluorethylene membranes (right, dark (actually blue when seen in colour) inside surface).

of the analytes. During the chromatographic process, only small molecules with the appropriate size can diffuse into the pores of the matrix from the surrounding solution. Molecules that exceed the size of the pores cannot diffuse into the matrix and thus pass through the column with the mobile phase solvent front, also known as the total exclusion or void volume (V_0). In SEC, V_0 is equivalent to the volume of the interstitial space. Molecules with complete access to the pores elute at the total permeation volume (V_t), which is the void volume V_0 plus the pore volume V_i (equation (3.12)).

$$V_t = V_0 + V_i \tag{3.12}$$

Molecules of intermediate sizes, which are able to diffuse into the matrix, are delayed in their passage through the column, eluting between these two volumes, proportional to their sizes (equation (3.13)).

$$V_R = V_0 + K_{sec} V_i \tag{3.13}$$

where K_{sec} is the SEC distribution coefficient and can be defined in equation (3.14)

$$K_{sec} = [S]_i/[S]_0 \tag{3.14}$$

where $[S]_i$ is the solute analyte concentration in the pore volume and $[S]_0$ is the average solute concentration in the void volume. K_{sec} values can range from 0 to 1. $K_{sec} = 0$ is the case when a large molecule is excluded from the pores and elutes with the void volume; when $K_{sec} = 1$ the sample elutes with the permeation volume, which is the case for very small compounds, such as buffer salts.

The most important characteristic of a SEC column is the way the matrix has been packed. The resolution of analytes using SEC is very sensitive to inhomogeneities in the stationary phase. Moreover, ionic strength of the buffer can also have a positive effect in SEC, where at an ionic strength of 0.15 or greater, unwanted ionic interactions between the analytes and the column matrix can be avoided. In contrast to most of the other chromatographic methods, the volume of the column (and the respective V_t) determines the resolution that can be achieved with a SEC column. The length of the column is therefore significant, since it affects both the resolution and the time taken to elute it.

3.6.2 Application of SEC in Sample Preparation

When considering using SEC for sample preparation of oligonucleotides, there are a number of factors that need to be considered. First, the sample loading capacity in SEC is low to moderate with maximum sample load of 0.5–1 mg/mL of packing material, hence SEC columns are often 4–5 times the size of a standard RP or AEX column; second, SEC column size also defines the maximal volume of the sample; third, sample concentration is limited by the viscosity of the sample and this should be low enough not to cause inconsistencies in the hydrodynamic flow of the mobile phase; fourth, the composition, pH and ionic strength of the sample are of lesser significance, as long as they do not affect the sizes or stability of the analyte molecules and are not outside the stability range of the stationary phase.

3.6.3 SEC Performance for Different Analytes

SEC does not have the same high-resolution power of other chromatographic techniques such as AEX or IP-RP-HPLC; however, it does have the resolving power to separate duplex oligonucleotides from monomers and multimers, as well as unstructured monomers from hairpin structures. As a result, SEC is well suited for the non-denaturing analysis of oligonucleotides. This is largely due to the unusually mild experimental conditions that are typically employed in SEC analysis which, in some cases, resemble physiological conditions.

The analysis of intact oligonucleotide duplexes and multiplexes (such as triple helices and tetramers of G-rich oligonucleotides involving Hoogsteen bindings) is often challenging, since most analytical chromatographic techniques (AEX or IP-RP-HPLC) exert some degree of destabilising effects on the non-covalent bindings between the components of the complex. The presence of IP reagents, low amounts of inorganic buffer salts, as well as high amounts of chaotropic salts in the mobile phases, and also high pH and temperature during analysis, can result in dissociation of the complex or at least a significant change in the duplex to monomer or multimer ratio. In SEC, however, conditions like high pH, high temperatures or high concentration of buffer salts are seldom used. Instead, ambient temperature, physiological buffer compositions and pH are common, and make this chromatographic technique an ideal tool to analyse secondary or tertiary structures of nucleic acids such as siRNA, duplex decoy or aptamers. SEC buffers are often the same as those commonly used for storage or formulation of the respective drug products. Hence SEC is the method of choice for analysing oligonucleotide aggregation.

In the case of PEGylated aptamers, where the presence of a polydispersed synthetic polymer (e.g. polyethylene glycol (PEG)) greatly complicates the analysis using other techniques, SEC has greater utility. The same is true for lipophilic conjugates such as oligonucleotides containing a cholesterol moiety, where the significantly altered adsorption characteristics to the functionalised surfaces impede analysis by AEX or IP-RP-HPLC. In SEC, the analysis of these large hydrophobic conjugates can be enhanced when organic solvents are used in the mobile phase (see Section 3.6.9).

3.6.4 Impact of Flow Rate on Resolution

Improvement in resolution in SEC can be achieved by varying the flow rates through the column rather than column temperature. Owing to the separation principle described earlier, the employed flow rate should allow diffusion of analytes into the pores. Generally, the lower the flow rate, the better the resolution, at least for large molecules. This is demonstrated in Figure 3.28, where flow rate was reduced gradually from 750 μL/min to 200 μL/min. Resolution between the peaks for 35mer, 18mer and 9mer oligonucleotides increased correspondingly (Table 3.11).

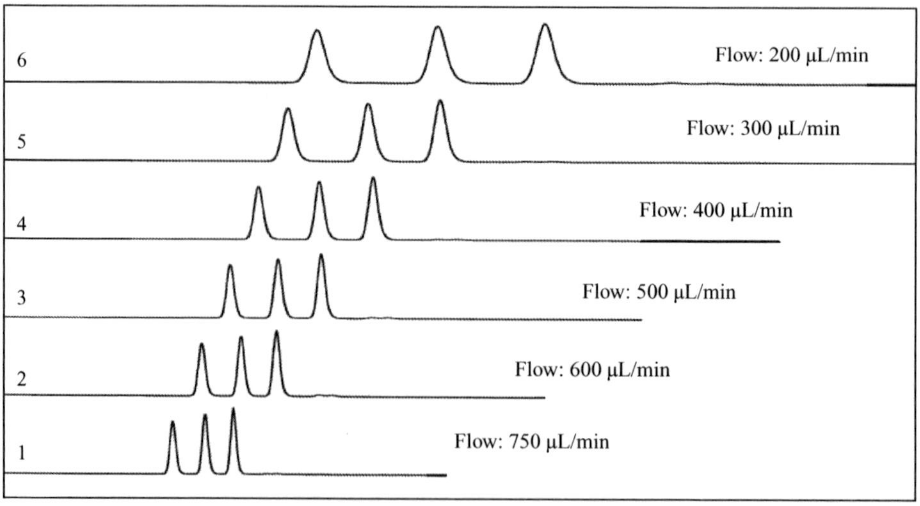

Figure 3.28 Effect of different flow rates on SEC separation. Trace 1: flow = 750 μL/min; trace 2: flow = 600 μL/min; trace 3: flow = 500 μL/min; trace 4: flow = 400 μL/min; trace 5: flow = 300 μL/min; trace 6: flow = 200 μL/min. SEC was carried out on a Dionex Ultimate 3000 series HPLC system (Dionex, Sunnyvale, CA, USA) using UV detection at 260 nm. Oligonucleotides were separated on a Superdex 75 column, 10/300 GL (GE Healthcare, Pollars Wood, UK) at room temperature. Mobile phase consisted of 1 × PBS containing 10% ACN. Injection volume was 5 μL of a 50 μM RNA solution in 1 × PBS.

Table 3.11 Peak resolutions between peaks of 36mer and 18mer oligonucleotides (R_s1) and peaks of 18mer and 9mer oligonucleotides (R_s2) in SEC analyses using flow rates between 750 μL/min and 200 μL/min. Chromatographic conditions were as described in Figure 3.28.

Flow (μL/min)	R_s1	R_s2
750	2.84	2.57
600	2.98	2.69
500	3.20	2.85
400	3.40	3.00
300	3.57	3.12
200	3.84	3.30

3.6.5 Impact of Buffer Salt Concentration on Resolution

Other than flow rate, the concentration of buffer salt has a moderate impact on peak resolution in SEC. When a 20 mM phosphate buffer at pH 7 was supplemented with increasing amounts of sodium chloride, separation time increased, but peak resolution decreased (Table 3.12 and Figure 3.29). The increase in separation time may reflect

Table 3.12 Peak resolutions in SEC. Mobile phases containing increasing concentrations of sodium chloride were used (0 to 500 mM NaCl). R_s1 is the resolution between peaks of 36mer and 18mer oligonucleotides. R_s2 is the resolution between peaks of 18mer and 9mer oligonucleotides. Chromatographic conditions were as described in Figure 3.29.

NaCl (mM)	R_s1	R_s2
0	3.01	3.14
25	3.03	3.01
50	3.01	2.90
150	2.89	2.66
500	2.75	2.44

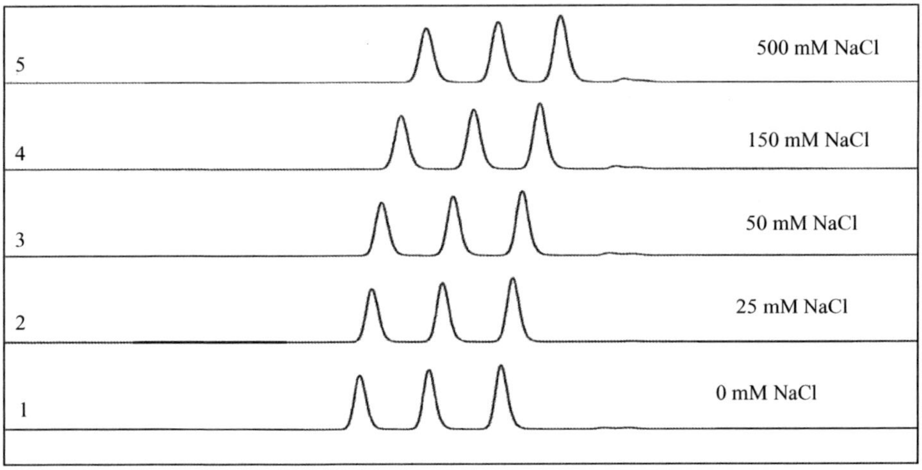

Figure 3.29 Effect of salt concentration on SEC separation. Mobile phase is 20 mM sodium phosphate pH 7 containing (i) 0 mM NaCl (trace 1); (ii) 25 mM NaCl (trace 2); (iii) 50 mM NaCl (trace 3); (iv) 150 mM NaCl (trace 4); (v) 500 mM NaCl (trace 5). SEC was carried out on a Dionex Ultimate 3000 series HPLC system (Dionex, Sunnyvale, CA, USA) using UV detection at 260 nm. Oligonucleotides were separated on a Superdex 75 column, 10/300 GL (GE Healthcare, Pollars Wood, UK) at room temperature. Mobile phase consisted of 20 mM phosphate buffer containing the indicated amount of NaCl and 10% ACN. Injection volume was 5 μL of a 50 μM RNA solution in 1 × PBS. Flow rate was 750 μL/min.

the reduction in hydrodynamic size of the analytes in solutions of higher salt concentration. The reduction in resolution appears to be attributable primarily to an increase in peak width, rather than differences in elution volume. The increase in peak width may be the result of larger inconsistencies in the mobile phase flow over the course of the column, caused by the increased viscosity of the solutions containing higher salt concentrations.

3.6.6 Separation of Duplex and Single Strand

Generally, SEC can provide good separation between duplex siRNA and its non-hybridised single strands. Figure 3.30 shows the separation of samples containing duplex only (trace 3), duplex including a small excess of guide strand (trace 2) and duplex including a small excess of passenger strand (trace 1). When chromatograms are enlarged, different separation efficiencies for the single strands can be identified. Improved separations were observed for the passenger strand from duplex when compared to that of the guide strand. Nevertheless, in both cases, good separation between duplex and single strands was achieved.

Figure 3.31 shows a chromatogram of a crude duplex, generated by annealing of crude guide and passenger strand after oligonucleotide synthesis. Compared to a purified duplex, this mixture contains larger amounts of single-strand impurities such as shortmer and longer, as well as non-optimal duplexes containing one or two impurity strands. In this example, single strands and shortmer sequences were resolved using SEC, whereas non-optimal duplexes, which have only a small difference in hydrodynamic size compared to the optimal duplex, are detected only as a shoulder of the main duplex peak. Aggregates, due to the larger size differences, are visible as distinct peaks, eluting earlier than the main peak.

3.6.7 Strand Titration Using SEC

During the manufacture of a siRNA, the final siRNA duplex is formed by mixing the two purified single strands in near equimolar concentration in aqueous solution,

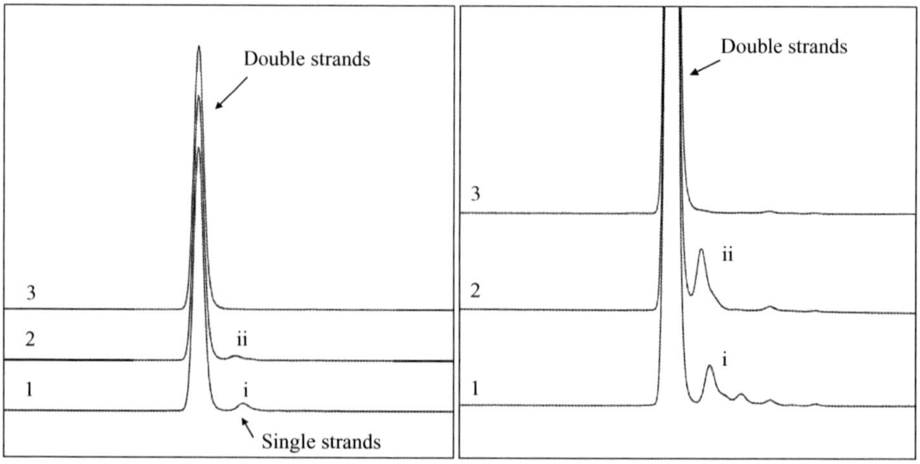

Figure 3.30 SEC of samples containing a small excess of passenger strand (trace 1), a small excess of guide strand (trace 2) or equimolar amounts of guide and passenger strand (trace 3). Peak 'i' indicates non-hybridised passenger strand; peak 'ii' indicates non-hybridised guide strand. Chromatographic conditions were as described in Figure 3.28. Flow rate was 750 μL/min.

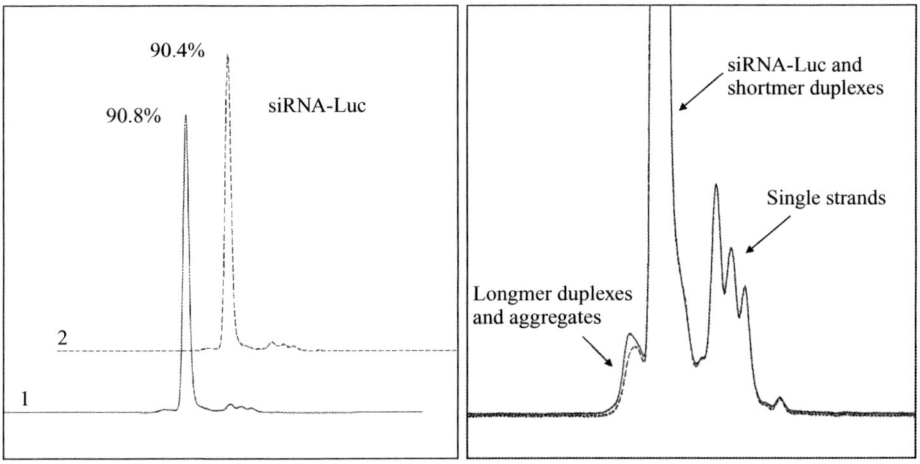

Figure 3.31 Non-denaturing analysis of the crude single-strand mixture before and after annealing. Equimolar amounts of desalted crude GS-Luc and crude PS-Luc were mixed and analysed using SEC. Chromatography was performed before (solid line) and after (dashed line) annealing (85°C, 10 min). Full chromatograms (left panel) and enlarged chromatograms (right panel) are depicted. Chromatographic conditions were as described in Figure 3.28. Flow rate was 750 μL/min. (Source: Modified from *Nucleic Acid Technol.*, 21, Noll, B., Seiffert, S., Hertel, F., Debelak, H., Hadwiger, P., Vornlocher, H.P., Röhl, I., Purification of small interfering RNA using non-denaturing anion-exchange chromatography, 383–393, Copyright 2011.)

typically followed by a heating and cooling phase (i.e. annealing). Strand association is mediated non-covalently *via* base-pairing and stacking interactions. The resulting siRNA typically contains a duplex fraction, including optimal duplex (two full-length single strands), non-optimal duplex variants (containing at least one single-strand impurity and/or mismatched sequences) as well as non-hybridised single strands. In theory, a 1:1 molar mixture of guide and passenger strand should result in the formation of the maximal amount of optimal duplex. In practice, this is not always the case since each strand contains significant amounts of longmer and shortmer impurities, some of which may not form stable duplexes. Thus, the molar equivalents needed to form a maximal amount of duplex depend not only on the full-length purity of the strands, but also on the nature of the impurities present. Hence, determination of optimal strand ratio solely by theoretical calculations based on the extinction coefficients of the full-length strands and UV absorption measurements of the respective solution may prove inaccurate or impractical during the manufacturing process.

In practice, more often than not, accurate strand ratio is adjusted experimentally during the manufacture of duplex oligonucleotides by determining content of a non-hybridised strand. SEC is an excellent tool to determine duplex and single-strand content in a sample under non-denaturing conditions. The optimal strand ratio can be determined either in real time, at different stages of the annealing process, or by preparing analytical samples of different mixing stoichiometries and plotting the results to determine the endpoint of the titration. Figure 3.32 displays a plot of percentage single-strand peak area over percentage molar excess of guide strand (GS-Luc). In this case, the lowest amount of non-hybridised single strand was reached at an 8% excess of passenger strand.

3.6.8 Separation of Encapsulated Liposomal siRNA from Free siRNA

Therapeutic oligonucleotides, in particular unmodified analogues, can be encapsulated into liposomes or nanoparticles for better bioavailability and pharmacokinetic profile [72–74]. Free and encapsulated oligonucleotides can be measured by SEC due to their significant size difference. Delivery systems such as liposomes typically display size ranges from 50 to 500 μm, whereas oligonucleotides show sizes that are more in the 2–10 μm range. Figure 3.33 shows a chromatogram of a liposomal solution containing encapsulated siRNA as well as free siRNA. Note that the recorded UV absorption of the encapsulated siRNA may not be an accurate reflection of the concentration of encapsulated oligonucleotides, since other components of the liposome such as the

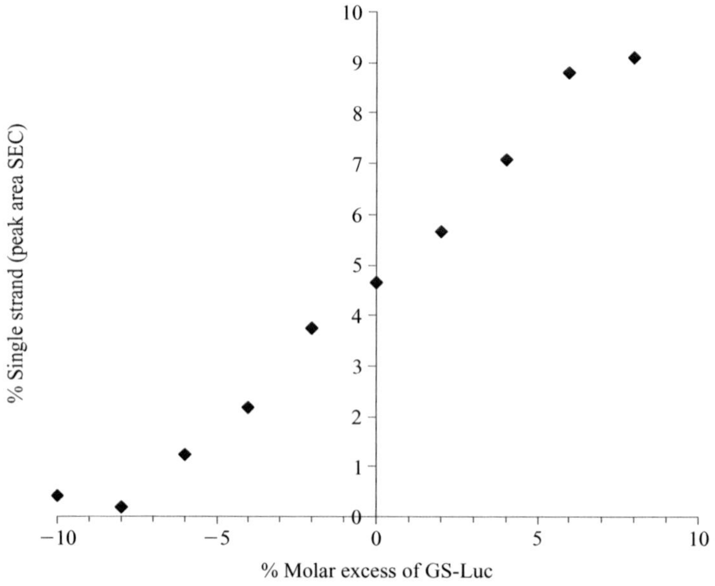

Figure 3.32 Titration of excess single strands of siRNA. Percentages of single-strand peak areas were plotted over the molar excess of guide strand GS-Luc employed in the preparation of the duplex sample. SEC was carried out on a Dionex Ultimate 3000 series HPLC system (Dionex, Sunnyvale, CA, USA) using UV detection at 260 nm. Injection volume was 5 μL of a 50 μM RNA solution in 1 × PBS. Chromatographic conditions were as described in Figure 3.28. Flow rate was 750 μL/min.

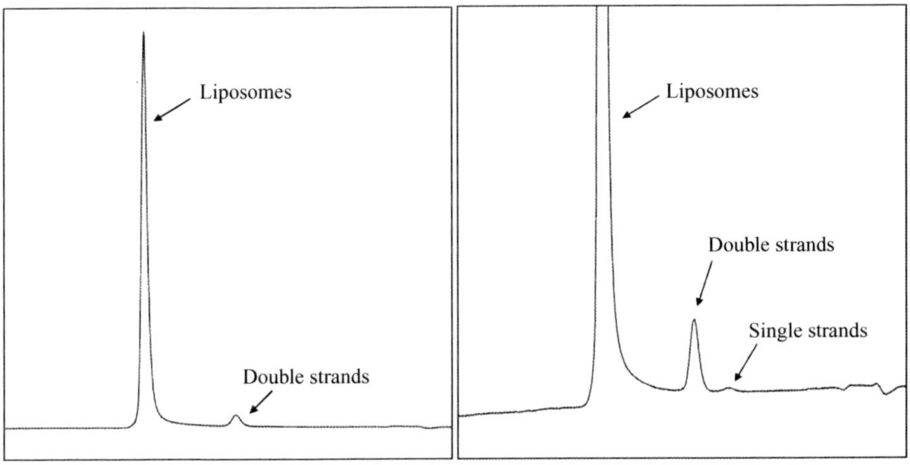

Figure 3.33 Separation of encapsulated siRNA (liposomes) and free siRNA. The full chromatogram (left panel) and an enlarged section of the chromatogram are depicted (right panel). SEC was carried out on a Dionex Ultimate 3000 series HPLC system (Dionex, Sunnyvale, CA, USA) using UV detection at 260 nm. Samples were separated on a Superdex 75 column, 10/300 GL (GE Healthcare, Pollars Wood, UK) at room temperature. Mobile phase consisted of 1 × PBS containing 10% ACN. The flow rate was 750 μL/min.

lipids may absorb at the recorded wavelength. In addition, the liposomal particles may cause light diffraction in the detector cell of the HPLC system, further reducing the recorded signal.

3.6.9 Separation of Lipophilic-Conjugated Oligonucleotides

In some applications, oligonucleotides are chemically conjugated to other chemical entities, such as biotin, galactose, cholesterol or other moieties that can improve receptor binding or bioavailability. Conjugation to lipophilic moieties can significantly complicate chromatographic analysis of oligonucleotides. Interactions of the lipophilic moiety by hydrophobic interactions with the size-exclusion matrix can cause poor resolution of the conjugated analytes, delayed passage through the column or, in the worst case, complete retention of the analyte on the column. Figure 3.34 shows examples of SEC separations where cholesterol-conjugated oligonucleotides did not elute from SEC columns when the usually employed mobile phase (1 × PBS containing 10% ACN) was used. After the mobile phase was gradually supplemented with a solvent of higher lipophilicity such as methanol, hydrophobic interactions of oligonucleotide-conjugates with the size-exclusion matrix could be suppressed. However, high content of organic solvent in the mobile phase can affect the hydrodynamic size of the analytes, as well as the diffusion behaviour during chromatography, and may affect resolution (Figure 3.35).

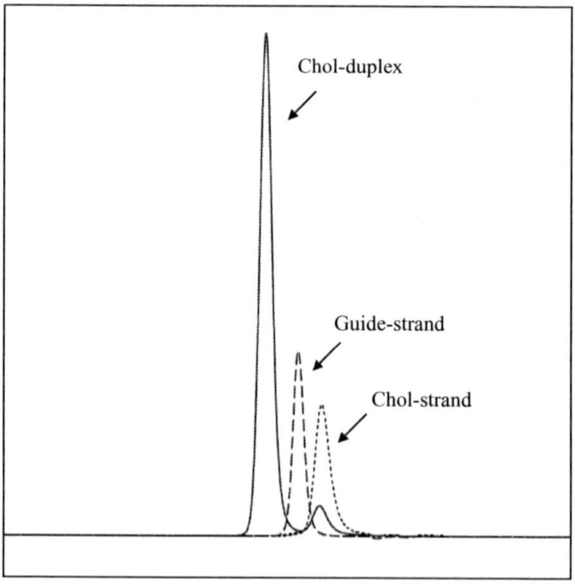

Figure 3.34 Separation of a siRNA containing a strand conjugated to a cholesterol-moiety. Chromatograms depicting separations of the duplex (straight line), guide strand (dashed line) and conjugated passenger strand (dotted line). Chromatographic conditions were as in Figure 3.35. Mobile phase was 1 × PBS containing 30% ACN.

3.6.10 Future Trends

3.6.10.1 UHPLC

The main focus of this chapter has been on the application of IP-RP-HPLC methods for the quantitation and characterisation of oligonucleotide samples. By optimising the method conditions and parameters, a high degree of resolution can be achieved between oligonucleotides and related substances using conventional HPLC systems. However, further increases in separation efficiency can be achieved by reducing particle size and increasing column length. While commercial columns with particle sizes between 3 and 10 μm are common in conventional HPLC, improvements in system stability towards extreme pressures led to the introduction of UPLC. UPLC uses sub-2-micron resin particles in columns of up to 150 mm length and can operate at back pressures of up to 1200 bar, compared to 400 bar in standard HPLC. Owing to very high plate counts, UPLC shows very high chromatographic resolution, higher sensitivity and narrow peaks at significantly faster separations [75]. More information on UPLC can be found in the review by Neue *et al.* [76]. UPLC can be particularly beneficial in separations involving analytes with a large molecular weight, such as oligonucleotides. A four- to ten-fold increase in speed can be obtained compared to conventional HPLC, with less buffer consumed, reducing time and cost of a separation [77].

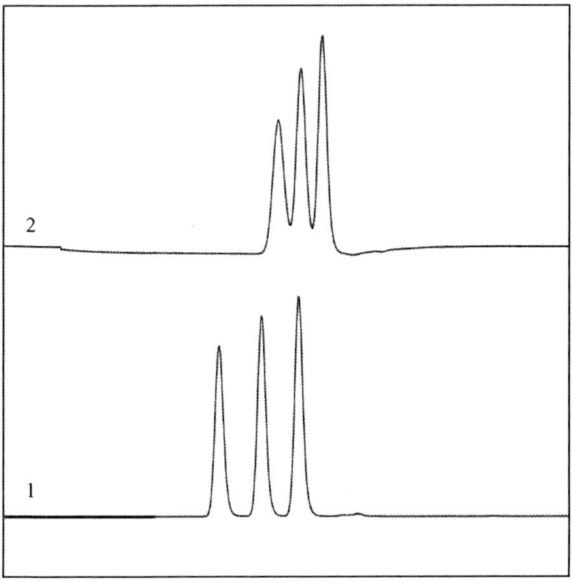

Figure 3.35 Effect of organic solvent content on separation. A sample containing 36mer, 18mer and 9mer polyT oligonucleotides was separated on a Superdex 75 column, 10/300 GL (GE Healthcare, Pollars Wood, UK) at room temperature. Chromatography was carried out on a Dionex Ultimate 3000 series HPLC system (Dionex, Sunnyvale, CA, USA) using UV detection at 260 nm. The flow rate was 750 μL/min. Mobile phase was 1 × PBS containing 10% ACN (trace 1) or 30% ACN (trace 2).

For the separation of oligonucleotides, UPLC is typically used in IP-RP applications. A number of columns suitable for UPLC applications for oligonucleotides are available to date. A comprehensive list is available by Cramer *et al.* [78]. Examples of separations on RP-UHPLC resins are shown in this chapter (Figures 3.22 to 3.26). The development of small-particle AEX chromatography has not progressed as rapidly; however, the first commercial SAX columns suitable for UPLC applications have been introduced recently [79] and significantly improved resolution and speed compared to commonly used AEX columns has been demonstrated. Taken together, owing to its improved resolution and speed, as well as its capability of reducing time and cost of a separation, it is expected that in the future, UPLC application will play an increasingly important role in the analysis of oligonucleotides.

3.6.10.2 LC-MS

Compared with AEX or SEC analysis, IP-RP-HPLC offers not only higher separation efficiency, but also compatibility with MS. HPLC coupled with ESI-MS has increasingly become the method of choice for impurity profiling of oligonucleotides. ESI-MS generates intact, multiply charged gas-phase ions of high-molecular-weight biomolecules [80, 81]. Owing to the multiple charges, the mass-to-charge rates of typical single- and double-stranded oligonucleotides are well within the mass range of most commercial mass spectrometers (relative molecular mass, $M_r < 4000-6000$). ESI-MS can provide accurate mass detection after chromatographic separation and can be used to identify as well as quantify oligonucleotides, related substances, impurities and degradants, even when the analytes are not or are only partially chromatographically resolved. Consequently, HPLC using MS detection is becoming a popular characterisation tool for oligonucleotide analysis.

The main prerequisite for LC-MS compared to regular LC applications is the use of IP buffers that are compatible with ESI-MS detection [50]. To date, a large number of denaturing IP-RP-HPLC methods [39, 50, 51, 54, 63, 68] have been reported for the characterisation of single- and double-stranded oligonucleotides. Among the described IP systems TEA/HFIP and HAA systems were reported as giving the best separation [53, 63, 68]. The use of HFIP as counter ion of the IP agent is particularly attractive because it is compatible with ESI-MS and often provides excellent separation [51, 52]. Another important feature of HFIP appears to be that it reduces the impact of oligonucleotide hydrophobicity upon retention [64] and suppresses diastereomeric resolution [52, 53].

In addition to denaturing LC-MS applications, several IP-RP-HPLC methods have been reported which allow high-resolution separations under non-denaturing conditions. These include ESI-MS compatible methods using

- an HAA buffer system [63]
- a TEA/HFIP buffer system [22, 53, 55]
- an HA/HFIP buffer system [68].

All three buffer systems showed high separation efficiencies for single, as well as duplex oligonucleotides. As in denaturing applications, using HFIP as counter ion allowed for significantly higher mass sensitivity. In the future, a combination of both UPLC and MS detection is likely to gain importance in routine quality control analysis of oligonucleotides.

References

1. Bloomfield, V., Crothers, D., Tinoco, I. Jr (Eds), *Nucleic Acids, Structures, Properties, and Functions,* 2000, University Science Books, Sausalita, CA, USA.
2. Watson, J.D., Crick, F., A structure for deoxyribose nucleic acid, *Nature*, 1953, 171, 4356, 737–738.
3. Thayer, J.R., Barreto, V., Rao, S., Pohl, C., Control of oligonucleotide retention on a pH-stabilized strong anion exchange column, *Analyt. Biochem.*, 2005, 338, 39–47.
4. Thayer, J.R., Rao, S., Puri, N., Burnett, C.A., Young, M., Identification of aberrant 2'-5' RNA linkage isomers by pellicular anion exchange chromatography, *Analyt. Biochem.*, 2007, 361, 132–139.
5. Prakash, T.P., Bhat, B., 2'-Modified oligonucleotides for antisense therapeutics, *Curr. Topics in Med. Chem.*, 2007, 7, 641–649.
6. Appella, D.H., Non-natural nucleic acids for synthetic biology, *Curr. Opin. Chem. Biol.*, 2009, 13, 5–6, 687–696.

7. Faria, M., Ulrich, H., Sugar boost: when ribose modifications improve oligonucleotide performance, *Curr. Opin. Molecular Ther.*, 2008, 10, 2, 168–175.
8. Helene, C., Sequence-selective recognition and cleavage of double-helical DNA, *Curr. Opin. Biotechnol.*, 1993, 4, 1, 29–36.
9. Crooke S.T., Oligonucleotide analogues might be designed to bind to mRNA, *Anticancer Drug Des.*, 1997, 12, 311–313.
10. Ng, E.W., Shima, D.T., Calias, P., Cunningham, E.T., Jr, Guyer, D.R., Adamis, A.P., Pegaptanib, a targeted anti-VEGF aptamer for ocular vascular disease, *Nature Rev. Drug Discovery*, 2006, 5, 2, 123–132.
11. Egli, M., Towards the structure-based design of oligonucleotide therapeutics, *Adv. Enzyme Regul.*, 1998, 38, 181–203.
12. Petersen, M., Wengel, J., LNA: a versatile tool for therapeutics and genomics, *Trends Biotechnol.*, 2003, 21, 2, 74–81.
13. Campbell, J.M., Bacon, T.A., Wickstrom, E., Oligodeoxynucleoside phosphorothioate stability in subcellular extracts, culture media, sera and cerebrospinal fluid, *J. Biochem. Biophys. Methods*, 1990, 20, 3, 259–267.
14. King, D.J., Ventura, D.A., Brasier, A.R., Gorenstein, D.G., Novel combinatorial selection of phosphorothioate oligonucleotide aptamers, *Biochemistry*, 1998, 37, 16489–16493.
15. Thayer, J.R., McCormick, R.M., Avdalovic N., High-resolution nucleic acid separations by high-performance liquid chromatography, *Methods Enzymol.*, 1996, 271, 147–174.
16. Beer, A., Bestimmung der Absorption des rothen Lichts in farbigen Flüssigkeiten, *Annln Phys. Chem.*, 1852, 86, 78–88.
17. Cavaluzzi, M.J., Borer, P.N., Revised UV extinction coefficients for nucleoside-5'-monophosphates and unpaired DNA and RNA, *Nucleic Acids Res.*, 2004, 32, e13.
18. Borer, P.N., Dengler, B., Tinoco, I. Jr, Uhlenbeck, O.C., Stability of ribonucleic acid double-stranded helices, *J. Molecular Biol.*, 1974, 86, 4, 843–853.
19. Owczarzy, R., You, Y., Moreira, B.G., Manthey, J.A., Huang, L., Behlke, M.A., Walder, J.A., Effects of sodium ions on DNA duplex oligomers: improved predictions of melting temperatures, *Biochemistry*, 2004, 43, 12, 3537–3554.
20. Owczarzy, R., Moreira, B.G., You, Y., Behlke, M.A., Walder, J.A., Predicting stability of DNA duplexes in solutions containing magnesium and monovalent cations, *Biochemistry*, 2008, 47, 19, 5336–5353.
21. Tan, Z.J., Chen, S.J., Ion-mediated nucleic acid helix-helix interactions, *Biophys. J.*, 2006, 91, 518–536.
22. Seiffert, S., Debelak, H., Hadwiger, P., Jahn-Hofmann, K., Roehl, I., Vornlocher, H.P., Noll, B., Characterization of side reactions during the annealing of small interfering RNAs, *Analyt. Biochem.*, 2011, 414, 1, 47–57.
23. van Deemter, J.J., Zuiderweg, F.J., Klinkenberg, A., Longitudinal diffusion and resistance to mass transfer as causes of non ideality in chromatography, *Chem. Engng Sci.*, 1956, 5, 271–289.
24. DePhillips, P., Lenhoff, A.M., Pore size distributions of cation-exchange adsorbents determined by inverse size-exclusion chromatography, *J. Chromatogr. A*, 2000, 883, 1–2, 39–54.
25. Unger, K.K., *Packings and Stationary Phases in Chromatographic Techniques*, 1990, Chromatographic Science Series, CRC Press, New York, USA.
26. Svec, F., Frechet, J.M., Continuous rods of macroporous polymer as high-performance-liquid-chromatography separation resin, *Analyt. Chem.*, 1992, 64, 7, 820–822.
27. Petro, M., Svec, F., Fréchet, J.M., Molded continuous poly(styrene-co-divinylbenzene) rod as a separation medium for the very fast separation of polymers. Comparison of the chromatographic properties of the monolithic rod with columns packed with porous and non-porous beads in high-performance liquid chromatography of polystyrenes, *J. Chromatogr. A*, 1996, 752, 1–2, 59–66.
28. Peters, E.C., Petro, M., Svec, F., Fréchet, J.M., Molded rigid polymer monoliths as separation media for capillary electrochromatography. 2. Effect of chromatographic conditions on the separation, *Anayt. Chem.*, 1998, 70, 11, 2296–2302.

29. Svec, F., Huber, C.G., Monolithic materials: Promises, challenges, achievements, *Analyt. Chem.*, 2006, 78, 7, 2101–2107.
30. Thayer, J.R., Flook, K.J., Woodruff, A., Rao, S., Pohl, C.A., New monolith technology for automated anion-exchange purification of nucleic acids, *J. Chromatogr. B*, 2010, 878, 933–941.
31. Staby, A., Jensen, R.H., Bensch, M., Hubbuch, J., Dünweber, D.L., Krarup, J., Nielsen, J., Lund, M., Kidal, S., Hansen, T.B., Jensen, I.C.H., Comparison of chromatographic ion-exchange resins VI. Weak anion-exchange resins, *J. Chromatogr. A*, 2007, 1164, 1–2: 82–94.
32. Warren, W.J., Vella, G., Principles and methods for the analysis and purification of synthetic deoxyribonucleotides by high-performance liquid chromatography, *Molecular Biotechnol.*, 1995, 4, 179–199.
33. Srivatsa, G.S., Klopchin, P., Batt, M., Feldman, M., Carlson, R.H., Cole, D.L., Selectivity of anion exchange chromatography and capillary gel electrophoresis for the analysis of phosphorothioate oligonucleotides, *J. Pharm. Biomed. Analysis*, 1997, 16, 619–630.
34. Bourque, A.J., Cohen, A.S., Quantitative analysis of phosphorothioate oligonucleotides in biological fluids using direct injection fast anion-exchange chromatography and capillary gel electrophoresis, *J. Chromatogr. B*, 1994, 662, 343–349.
35. Shanagar, J., Purification of a synthetic oligonucleotide by anion exchange chromatography: method optimisation and scale-up, *J. Biochem. Biophys. Methods*, 2005, 64, 216–225.
36. Deshmukh, R.R., Leitch, W.E., Cole, D.L., Application of sample displacement techniques to the purification of synthetic oligonucleotides and nucleic acids: a mini-review with experimental results, *J. Chromatogr. A*, 1998, 806, 77–92.
37. Temsamani, J., Kubert, M., Agrawal, S., Sequence identity of the n-1 product of a synthetic oligonucleotide, *Nucleic Acids Res.*, 1995, 23, 1841–1844.
38. Chen, D., Yan, Z., Cole, D.L., Srivatsa, G.S., Analysis of internal (n-1)mer deletion sequences in synthetic oligodeoxyribonucleotides by hybridization to an immobilized probe array, *Nucleic Acids Res.*, 1999, 27, 389–395.
39. Gilar, M., Analysis and purification of synthetic oligonucleotides by reversed-phase high-performance liquid chromatography with photodiode array and mass spectrometry detection, *Analyt. Biochem.*, 2001, 298, 196–206.
40. Krotz, A. H., Klopchin, P., Walker, K. L., Srivatsa, S., Cole, D. L., Ravikumar, V. T., On the formation of longmers in phosphorothioate oligodeoxyribonucleotide synthesis, *Tetrahedron*, 1997, 38, 3875–3878.
41. Krotz A.H., Gorman D., Mataruse P., Foster C., Godbout J.D., Coffin C.C., Scozzari A.N., Phosphorothioate oligonucleotides with low phosphate diester content: greater than 99,9% sulfurization efficiency with 'aged' solutions of phenylacetyl disulfide (PADS), *Org. Process Res. Dev.*, 2004, 8, 852–858.
42. Morgan, M.A., Kazakov, S.A., Hecht, S.M., Phosphoryl migration during the chemical synthesis of RNA, *Nucleic Acids Res.*, 1995, 23, 19, 3949–3953.
43. Mikkola, S., Kaukinen, U., Lonnberg, H., The effect of secondary structure on cleavage of the phosphodiester bonds of RNA, *Cell Biochem. Biophys.*, 2001, 34, 95–119.
44. Wincott, F., DiRenzo, A., Shaffer, C., Grimm, S., Tracz, D., Workman, C., Sweedler, D., Gonzalez, C., Scaringe, S., Usman, N., Synthesis, deprotection, analysis and purification of RNA and ribozymes, *Nucleic Acids Res.*, 1995, 23, 2677–2684.
45. Sproat, B.S., Rupp, T., Menhardt, N., Keane, D., Beijer, B., Fast and simple purification of chemically modified hammerhead ribozymes using a lipophilic capture tag, *Nucleic Acids Res.*, 1999, 27, 1950–1955.
46. Noll, B., Seiffert, S., Hertel, F., Debelak, H., Hadwiger, P., Vornlocher, H.-P., Roehl, I., Purification of small interfering RNA using non-denaturing anion-exchange chromatography, *Nucleic Acid Technol.*, 2011, 21, 383–393.
47. Bergot, B.J., Egan, W., Separation of synthetic phosphorothioate oligodeoxynucleotides from their oxygenated (phosphodiester) defect species by strong-anion-exchange high-performance liquid chromatography, *J. Chromatogr.*, 599, 35.
48. Zhang, Y., Cremer, P.S., Interactions between macromolecules and ions: The Hofmeister series, *Curr. Opin.Chem. Biol.*, 2006, 10, 6, 658–663.

49. Yang, X., Hodge, R.P., Luxon, B.A., Shope, R., Gorenstein, D.G., Separation of synthetic oligonucleotide dithioates from monothiophosphate impurities by anion-exchange chromatography on a mono-Q column, *Analyt. Biochem.*, 2002, 306, 92–99.
50. Huber, C.G., Oberacher, H., Analysis of nucleic acids by on-line liquid chromatography-mass spectrometry, *Mass Spectrometry Rev.*, 2001, 20, 5, 310–343.
51. Apffel, A., Chakel, J.A., Fischer, S., Lichtenwalter, K., Hancock, W.S., Analysis of oligonucleotides by HPLC-electrospray ionization mass spectrometry, *Analyt. Chem.*, 1997, 69, 7, 1320–1325.
52. Apffel, A., Chakel, J.A., Fischer, S., Lichtenwalter, K., Hancock, W.S., New procedure for the use of high-performance liquid chromatography-electrospray ionization mass spectrometry for the analysis of nucleotides and oligonucleotides, *J. Chromatogr. A*, 1997, 777, 3–21.
53. Gilar, M., Fountain, K.J., Budman, Y., Holyoke, J.L., Davoudi, H., Gebler J.C., Characterization of therapeutic oligonucleotides using liquid chromatography with on-line mass spectrometry detection, *Oligonucleotides*, 2003, 13, 229–243.
54. Fountain, K.J., Gilar, M., Gebler J.C., Analysis of native and chemically modified oligonucleotides by tandem ion-pair reversed-phase high-performance liquid chromatography/electrospray ionization mass spectrometry, *Rapid Communs Mass Spectrometry*, 2003, 17, 646–653.
55. Beverly, M., Hartsough, K., Machemer, L., Liquid chromatography/electrospray mass spectrometric analysis of metabolites from an inhibitory RNA duplex, *Rapid Communs Mass Spectrometry*, 2005, 19, 1675–1682.
56. Premstaller, A., Oberacher, H., Huber, C.G., High-performance liquid chromatography-electrospray ionization mass spectrometry of single- and double-stranded nucleic acids using monolithic capillary columns, *Analyt. Chem.*, 2000, *72*, 4386–4393.
57. Oberacher, H., Parson, W., Muhlmann, R., Huber, C.G., Analysis of polymerase chain reaction products by on-line liquid chromatography-mass spectrometry for genotyping of polymorphic short tandem repeat loci, *Analyt. Chem.*, 2001, 73, 5109–5115.
58. Gelhaus, S.L., LaCourse, W.R., Hagan, N.A., Amarasinghe, G.K., Fabris, D., Rapid purification of RNA secondary structures, *Nucleic Acids Res.*, 2003, 31, e135.
59. Waghmare, S.P., Pousinis, P., Hornby, D.P., Dickman, M.J., Studying the mechanism of RNA separations using RNA chromatography and its application in the analysis of ribosomal RNA and RNA:RNA interactions, *J. Chromatogr. A*, 2009, 1216, 1377–1382.
60. Holzl, G., Oberacher, H., Pitsch, S., Stutz, A., Huber, C.G., Analysis of biological and synthetic ribonucleic acids by liquid chromatography-mass spectrometry using monolithic capillary columns, *Analyt. Chem.*, 2005, 77, 2, 673–680.
61. Oberacher, H., Niederstatter, H., Casetta, B., Parson, W., Detection of DNA sequence variations in homo- and heterozygous samples via molecular mass measurements by electrospray ionization time-of-flight mass spectrometry, *Analyt. Chem.*, 2005, 77, 15, 4999–5008.
62. Oberacher, H., Pitterl, F., On the use of ESI-QqTOF-MS/MS for the comparative sequencing of nucleic acids, *Biopolymers*, 2009, 91, 6, 401–409.
63. McCarthy, S.M., Gilar, M., Gebler, J., Reversed-phase ion-pair liquid chromatography analysis and purification of small interfering RNA, *Analyt. Biochem.*, 2009, 390, 181–188.
64. Gilar, M., Fountain, K.J., Budman, Y., Neue, U.D., Yardley K.R., Rainville, P.D., Russell, R.J., Gebler, J.C., Ion-pair reversed-phase high-performance liquid chromatography analysis of oligonucleotides: retention prediction, *J. Chromatogr. A*, 2002, 958, 1–2, 167–182.
65. Beverly, M., Hartsough, K., Machemer, L., Pavco, P., Lockridge, J., Liquid chromatography electrospray ionization mass spectrometry analysis of the ocular metabolites from a short interfering RNA duplex, *J. Chromatogr. B*, 2006, 835, 62–70.
66. Bleicher, K., Bayer, E., Various factors influencing the signal intensity of oligonucleotides in electrospray mass spectrometry, *Biol. Mass Spectrometry*, 1994, 23, 6, 320–322.
67. Andrus A., Kuimelis R.G., Overview of purification and analysis of synthetic nucleic acids, *Current Protocols in Nucleic Acid Chemistry*, 2001, Ch. 10, Unit 10.3, John Wiley and Sons, Chichester, UK.
68. Noll, B., Seiffert, S., Vornlocher, H.P., Roehl, I., Characterization of small interfering RNA by

non-denaturing ion-pair reversed-phase liquid chromatography, *J. Chromatogr. A*, 2011, 1218, 33, 5609–5617.
69. McKeown, A.P., Shaw, P.N., Barrett, D.A., Retention behaviour of an homologous series of oligodeoxythmidilic acids using reversed-phase ion-pair chromatography, *Chromatographia*, 2002, 55, 5–6, 271–277.
70. Janson, J.-C., Ryden, L., *Gel Filtration in Protein Purification Principles High Resolution Methods snd Applications*, 1989, VCH Publishers Inc., New York/Weinheim/Cambridge.
71. Chan, D., Roymoulik, I., Purity analyisis and molecular weight determination by size exclusion HPLC. In *Handbook of Analysis of Oligonucleotides and Related Products*, Bonilla, J.V., Srivatsa, G.S. (Eds), 2011, CRC Press, Boca Raton, FL, USA.
72. Li, W., Szoka, F.C. Jr, Lipid-based nanoparticles for nucleic acid delivery, *Pharm. Res.*, 2007, 24, 3, 438–449.
73. Fattal, E., Barratt, G., Nanotechnologies and controlled release systems for the delivery of antisense oligonucleotides and small interfering RNA, *Br. J. Pharmacol.*, 2009, 157, 2, 179–194.
74. De Rosa, G., La Rotonda, M.I., Nano and microtechnologies for the delivery of oligonucleotides with gene silencing properties, *Molecules*, 2009, 14, 2801–2823.
75. Tolley, L., Jorgenson, J.W., Moseley, M.A., Very high pressure gradient LC/MS/MS, *Analyt. Chem.*, 2001, 73, 2985–2991.
76. Neue, U.D., Kele, M., Bunner, B., Kromidas, A., Dourdeville, T., Mazzeo, J.R., Grumbach, E.S., Serpa, S., Wheat, T.E., Hong, P., Gilar, M., Ultra-performance liquid chromatography technology and applications, *Adv. Chromatogr.*, 2010, 48, 99–143.
77. Gilar, M., *Waters Application Notes: UPLC Separation of Oligonucleotides: Method Development*, 2007, 720002383EN.
78. Cramer, H., Finn, K.J., Herzberg, E., Purity analysis and impurities determination by reversed-phase high-performance liquid chromatography. In *Handbook of Analysis of Oligonucleotides and Related Products*, Bonilla, J.V., Srivatsa, G.S. (Eds), 2011, CRC Press, Boca Raton, FL, USA.
79. Agilent Technologies, *Agilent Application Notes*, 2010, 5990–5195EN.
80. Alexandrov, M.L., Gall, L.N., Krasnov, N.V., Nikolaev, V.I., Pavlenko, V.A., Shkurov, V.A., Экстракция ионов из растворов при атмосферном давлении – Метод масс-спектрометрического анализа биоорганических веществ (Extraction of ions from solutions at atmospheric pressure – A method for mass spectrometric analysis of bioorganic substances), Doklady Akademii Nauk SSSR, 1984, 277, 2, 379–383 (in Russian).
81. Yamashita, M., Fenn, J.B., Electrospray ion source. Another variation on the free-jet theme, *J. Phys. Chem.*, 1984, 88, 20, 4451–4459.

4 Analytical Characterisation of Oligonucleotides by Mass Spectrometry

Patrick A. Limbach

4.1 Introduction

During the past decade there has been increased application of oligonucleotides in the biotechnology market. Oligonucleotides are used as diagnostic tools and as a drug development platform allowing for the modulation of gene expression. Unmodified oligodeoxyribonucleotides and oligoribonucleotides typically have low specificity and therapeutic efficacy *in vivo* due to nuclease-mediated degradation. Those problems can be overcome by chemical alteration of their native structure via modifications of the phosphodiester backbone, ribose sugar and the heterocyclic bases [1].

Analytical methods are required to verify both the composition and sequence of synthetic oligonucleotides, especially those destined for therapeutic use. Despite significant progress in solid-phase oligomerisation chemistry, synthetic oligonucleotides still contain multiple classes of low-level impurities (refer to Chapter 2 for more details). These impurities, when present in oligonucleotide therapeutics, can influence not only their efficacy and toxicity but also lead to potential regulatory issues due to potential changes in the impurity profiles. Impurity profiles can be used to better understand solid-phase oligomerisation chemistry, yielding better control over the manufacturing process.

This chapter describes the use of mass spectrometry (MS) for the analysis of synthetic oligonucleotides, including modified oligonucleotides, and the impurities created during their synthesis. The chapter provides a brief overview of the common MS platforms used for oligonucleotide analysis, a discussion of the important analytical considerations during MS method development, a presentation of

Analysis of Oligonucleotides and their Related Substances, edited by George Okafo, David Elder and Mike Webb.

oligonucleotide gas phase properties and selected illustrations of applications of MS for oligonucleotide characterisation.

There are a number of relatively recent review articles and book chapters focused on particular areas of MS as applied to the characterisation of oligonucleotides [2–11], and the interested reader is referred to those for additional details and information. A few of those references deserve highlighting because of the detailed coverage they provide on selected fundamental characteristics or applications of MS. Beverly [10] has recently reviewed the field of MS used for the characterisation of small interfering ribonucleic acids (siRNAs) and provides a nice discussion of the use of MS specifically focused on double-stranded RNA. The ability of MS to obtain quantitative information from oligonucleotide metabolite profiles has been reviewed through 2007 [7]. Among the most comprehensive reviews of the past several years, Banoub *et al.* [4] covered nearly the entire field of nucleoside, oligonucleotide and nucleic acid MS. That review then formed the basis for a recent book dedicated to the use of MS for these biomolecules [9].

This chapter is organised to provide the reader analytically useful information relating to the use of MS for characterising synthetic oligonucleotides. To start, a brief primer on modern MS instrumentation is given. Readers familiar with MS are welcome to omit this section. Next, the gas phase properties of oligonucleotides are discussed. This section primarily focuses on single-stranded oligonucleotides and includes an overview of the known fragmentation behaviour of oligodeoxynucleotides, oligoribonucleotides and selected modified oligonucleotides. That information is followed by a brief overview of analytical method development as applied to oligonucleotide analysis. Methods covered include the more common high-performance liquid chromatography–MS (HPLC-MS) and matrix-assisted laser desorption/ionisation–MS (MALDI-MS) methods used for oligonucleotide characterisation. The remainder of this chapter highlights selected applications from the recent literature. Whilst this section is not comprehensive, owing to space limitations, each application is chosen to provide the interested reader with guidance and insight into the common analytical challenges faced when characterising oligonucleotides, along with successful approaches for identifying the scientific information of interest from such samples.

4.2 MS Instrumentation

MS is a powerful tool for obtaining valuable analytical information such as molecular composition and structural information from biomolecules such as oligonucleotides. MS is most notable for its ability to provide absolute molecular weight measurements with relatively high sensitivity. Historically, however, the application of MS to the analysis of oligonucleotides was limited to nucleosides and low molecular weight oligonucleotides ($<$ 2000 Da) because the polyanionic character of these samples did not allow for the efficient generation of gas-phase ions of such samples.

The development of two so-called 'soft ionisation' techniques, electrospray ionisation (ESI) [12, 13] and MALDI [14], provided practitioners of MS with the means to generate gas-phase ions from oligonucleotides and nucleic acids of moderate to high molecular weights. No clearer measure of the importance of these two ionisation methods for MS can be found than the awarding of the Nobel Prize in chemistry in 2002 to Koichi Tanaka for laser desorption/ionisation MS and the late John Fenn for ESI-MS. Without question, these two ionisation techniques have revolutionised the mass spectrometric analysis of oligonucleotides. In fact, further developments and refinements to ESI-MS and ultra-violet (UV) MALDI-MS have led to the current situation wherein MS is a routine analytical technique for the characterisation of oligonucleotides and their constituents.

Every commercial mass spectrometer will have one (or more) ionisation sources and one (or more) mass analysers. The ionisation source is responsible for generating the gas-phase ions which are then separated based on the ion's mass-to-charge ratio (*m/z*) by the mass analyser. Because these two components are both ubiquitous and important for MS-based characterisation of oligonucleotides, each will be treated in more detail in the sections that follow. In addition to an ionisation source and a mass analyser, commercial mass spectrometers will typically have some means of sample introduction (e.g. HPLC), a detector to register the *m/z* values present in the sample, and software and associated electronics for instrument control, sample acquisition and data analysis.

4.2.1 Ionisation Sources

4.2.1.1 MALDI

MALDI-MS [14–16] is one of the most useful techniques for determining the mass of oligonucleotides. A number of recent reviews can be consulted for more detailed discussions regarding the processes in MALDI-MS [8, 17, 18]. MALDI-MS offers high mass accuracy, good sensitivity, simplicity and speed, all of which make MALDI a suitable technique for the analysis of oligonucleotides. Conceptually, sample analysis using MALDI-MS is a relatively straightforward procedure. As will be discussed in further detail below, the analyte of interest is mixed with an appropriate matrix, spotted on a sample plate and analysed. Normally, singly charged ions of oligonucleotides are observed in MALDI-MS, which makes MALDI mass spectra very easy to interpret and simplifies the analysis of mixtures.

Currently, MALDI-MS can be used to analyse oligonucleotides at the 120mer size and below, although it is most practical and effective when used for the rapid identification and characterisation of synthetic oligonucleotides at the 50mer level and below. In fact, this approach has replaced Maxam–Gilbert chemical degradation followed by polyacrylamide gel electrophoresis for the sequence verification of smaller oligonucleotides [19]. MALDI-MS requires fmol to low pmol per microlitre amounts of oligonucleotides for analysis. Because MALDI is capable of analysing mixtures, synthetic oligonucleotides (usually less than 30mers) can be characterised by their failure sequences [20, 21]. Alternatively, exonuclease digestion has proven to be a versatile approach to the sequence determination of moderate length oligonucleotides [22, 23]. Both of these techniques will be covered in more depth at the end of this chapter.

4.2.1.2 The MALDI Laser

Among the various commercial vendors of MALDI-based mass spectrometers, each system can be equipped with lasers of different laser wavelength, repetition rate and laser spot size. The pulsed nitrogen laser ($\lambda = 337$ nm) and the frequency-tripled neodymium-doped yttrium aluminium garnet (Nd:YAG) laser ($\lambda = 355$ nm) are the most common lasers used in UV-MALDI-MS [24]. Operationally, laser fluences (fluxes) of the order of $10^6 - 10^7$ W/cm^2 are common in MALDI-MS. The best results are achieved by working at or just above the threshold irradiance necessary to generate analyte signal. To achieve the appropriate laser irradiance, some means of attenuating the laser output are typically employed and are common accessories on commercial instruments. Lasers are available in a wide range of repetition rates, ranging from the Hz range for nitrogen lasers up to modern Nd:YAG lasers with rates > 1000 Hz. Typically, sample throughput increases with increasing laser repetition rate as the number of co-added scans per sample can be increased (thereby increasing signal-to-noise (S/N) ratio) without an increase in analysis time. Laser spot sizes from 5 μm up to 150 μm are available, depending on the beam focusing technology used. Except for specialised applications not typically encountered during oligonucleotide analysis, there are no advantages to very tightly focused lasers.

4.2.1.3 Matrix

The MALDI technique involves the process of desorption, dissociation and ionisation of the analyte and the matrix under high laser energy. The essence of MALDI-MS is the matrix. Generally analyte compounds are embedded in a surplus of matrix (around 1000-fold molar excess) and are co-desorbed upon laser excitation. The general features of effective MALDI matrices are well known: they must dissolve (liquid matrix) or co-crystallise (solid matrix) with the sample, strongly absorb the laser light, remain in the condensed phase under high-vacuum conditions, stifle both chemical and thermal degradation of the sample, and promote the ionisation of the sample via any of a number of mechanisms [25].

As with other biomolecule compound classes, a large number of candidate matrices have been investigated in MALDI-MS analyses of oligonucleotides. Over the years, only a handful have found widespread use within the field: picolinic acid, 3-hydroxypicolinic acid (3-HPA), 2-amino thiothymine (ATT), 3,4-diaminobenzophenone (DABP) and 2,4,6-trihydroxyacetophenone (THAP). The *de facto* standard for small oligonucleotides for many years was THAP together with di- and tri-ammonium salts of organic or inorganic acids [26]. A mixture of THAP in acetonitrile and aqueous triammonium citrate in a 1:1 molar proportion was found to be a good matrix for the detection of synthetic oligodeoxynucleotide samples [27]. A high proportion of volatile solvent as well as the high salt content ensured fast co-crystallisation of the matrix, co-matrix and analyte molecules. ATT, co-crystallised with ammonium citrate, has been shown to be a suitable matrix for the analysis of oligonucleotides and short DNA fragments [28]. The major advantages of ATT over other conventional matrices, for example THAP and 3-HPA, were improved resolution and mass accuracy, easy sample preparation, and applicability to crude or partially purified samples.

Although DABP was originally introduced as a novel matrix for the analysis of peptides and proteins by MALDI-MS, it was also demonstrated to be useful in the analysis of oligonucleotides [29]. Intact oligonucleotide ions with this matrix can be readily produced with lower laser powers, resulting in better detection limits, less fragmentation and fewer alkali metal ion adducts compared to conventional matrices, for example 3-HPA and THAP (Figure 4.1). This matrix can be used for the analysis of small oligonucleotides (< 30mers) without any co-matrices, providing the sample is sufficiently desalted. Minimal fragmentation and fewer alkali metal ion adducts were detected even at low concentrations of oligonucleotides. It was also found that samples prepared with DABP are highly homogeneous, which resulted in excellent shot-to-shot reproducibility, better resolution and a higher S/N ratio.

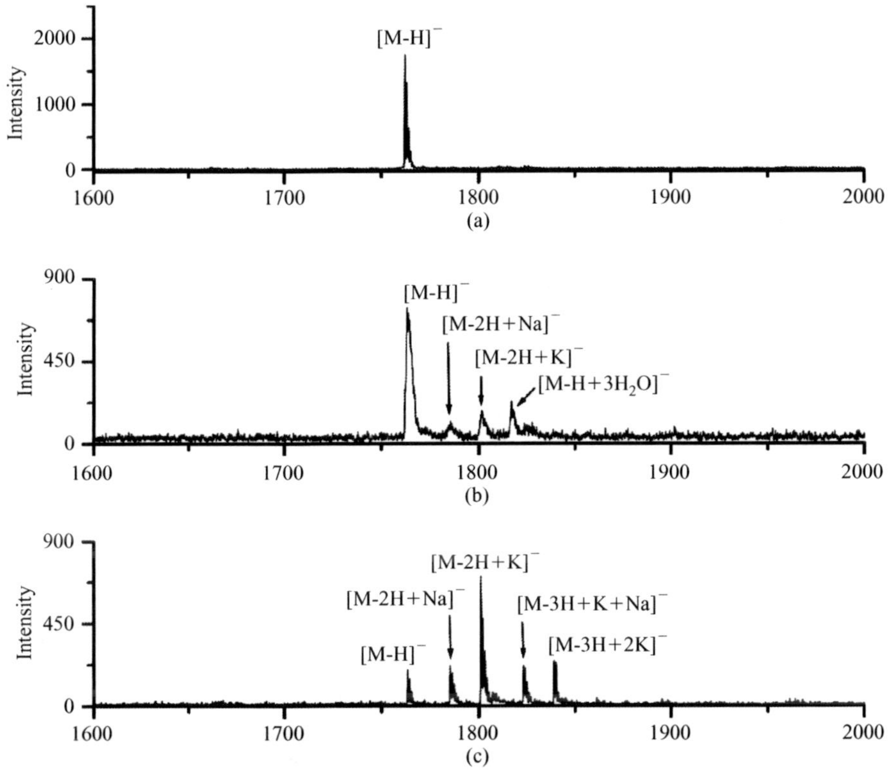

Figure 4.1 MALDI mass spectra of $d(T)_6$ obtained using (a) DABP, (b) 3-HPA and (c) 2,4,6-THAP as the matrix, respectively. The quantity of $d(T)_6$ loaded was 1.4 pmol without adding extra NaCl to the sample solution. (Source: Fu, Y., Xu, S., Pan, C., Ye, M., Zou, H., Guo, B., A matrix of 3,4-diaminobenzophenone for the analysis of oligonucleotides by matrix-assisted laser desorption/ionization time-of-flight mass spectrometry, *Nucleic Acids Res.*, 2006, 34, e94, by permission of Oxford University Press.)

A unique aspect of MALDI-MS analysis of nucleic acid components is the advantages one gains by adding a co-matrix to minimise cation adduction and improve oligonucleotide signals. The effect of ammonium salt in the detection of oligonucleotides was systematically investigated using several matrices with ammonium salt additives [30]. The results showed that the presence of ammonium salt in the matrix had a beneficial effect on protonation and deprotonation of oligonucleotides in addition to suppressing alkali-ion adducts. The addition of a co-matrix of sufficiently high proton affinity can serve as a proton sink to reduce oligonucleotide fragmentation in MALDI-MS [31, 32]. A direct correlation between the co-matrix proton affinity and the oligonucleotide molecular ion stability was found. The polyamine co-matrices (spermine tetrahydrochloride, spermine, spermidine trihydrochloride and spermidine) were evaluated for their effectiveness at enhancing the mass spectral quality of oligonucleotides [33, 34]. In general, polyamine co-matrices were found to be more effective than monofunctional amines for improved mass spectral data.

The oldest and original sample preparation method introduced in 1988 by Karas and Hillenkamp [15], which has remained, with minor modifications, in place for nearly two decades is the 'dried droplet method'. The approach is simple – a freshly prepared saturated solution of matrix compound is mixed with analyte solution and dried at ambient temperature – but aggregation of higher amounts of analyte/matrix crystals in a ring around the edge of the drop yields inhomogeneous and irregularly distributed crystals on the MALDI target. This inhomogeneity requires the search for 'hot' spots to generate good-quality spectra [35, 36].

Various methods were subsequently developed in attempts to reduce the preparation inhomogeneity. The 'crushed crystal' [37], 'fast evaporation' [38], 'overlayer' [36] and 'sandwich' [39, 40] methods are the more common examples. Among these, the two-layer ('overlayer', 'sandwich') approaches wherein matrix is first spotted on the sample target followed by the analyte and additional matrix or analyte–matrix mixture are most commonly used for solid matrix preparations. These approaches provide high detection sensitivity and excellent spot-to-spot reproducibility, which are mainly due to the increase of the matrix-to-analyte ratio and improved isolation between analyte molecules as a result of analyte/matrix deposition features.

Owing to difficulties in preparing homogeneous samples using the solid matrix techniques, alternatives based on the use of liquid matrices have also been investigated. Room-temperature ionic liquids (RTILs) are salts with melting points close to or below room temperature. They look like a classical liquid but they do not contain any molecules: they are made of ions. Typically, a RTIL consists of nitrogen- or phosphorus-containing organic cations and large organic or inorganic anions [41]. They have good thermal stabilities, remain liquid over a range of 200–300°C and have practically no vapour pressure [42]. They produce homogenous solutions with great vacuum stabilities and are good solvents for a variety of organic, inorganic and polymeric substances [43].

It has been demonstrated that ionic liquids and solids make useful MALDI matrices [44]. With ionic matrices, it is possible to combine the beneficial qualities of liquid and solid matrices. Ionic liquids produce a much more homogeneous sample solution (as do all liquid matrices), yet they have greater vacuum stability than most solid matrices. In most cases, an ionic matrix can be found that produces greater spectral peak intensities and lower limits of detection than comparable solid matrices. While most ionic liquids readily dissolve biological oligomers, such as oligonucleotides, they can vary tremendously in their ability to promote analyte ionisation. Therefore, while they have previously been investigated as UV-MALDI matrices for oligonucleotides, their performance has not yet matched the performance already available using conventional solid matrices (Figure 4.2) [45–47]. However, their potential advantages suggest further investigations may be warranted, especially for selected applications of MALDI-MS for modified oligonucleotides.

4.2.2 Electrospray Ionisation

The analysis of biomolecules such as nucleic acids and proteins by MS is a rather difficult process due to the high molecular weight and thermal lability of these molecules. Two of the major problems in such analyses are the need for a 'soft' ionisation method capable of generating intact molecular ions and the limited upper mass range of most mass analysers. In the mid-1980s, Fenn *et al.* [48] demonstrated that ESI could be used to analyse molecules with molecular weights (M_r) larger than the *m/z* limit of the mass analyser. Later work on oligonucleotides opened the door for accurate and high-resolution analysis of these compounds by ESI-MS [49].

ESI allows for the analysis of high molecular weight compounds through the generation of multiply charged ions in the gas phase. Because the basis of the mass spectrometric measurement is the *m/z* value of the molecule, the presence of multiple charges on the molecule will result in a decrease in the *m/z* values and allow characterisation using mass analysers with limited *m/z* ranges. An example of a typical electrospray mass spectrum one obtains for oligonucleotides is shown in Figure 4.3. As is illustrated in the top mass spectrum, each of the oligonucleotide species is detected as multiply charged negative ions. After spectral deconvolution, the molecular weights of each of the three oligonucleotides can be determined, as well as the presence of sodium adducts.

A necessary requirement for ESI is that the analyte molecules be charged in solution. The negatively charged phosphate backbone of oligonucleotides allows for negative-ion mode analysis of ESI-generated ions. Transfer of ions from solution phase to the gas phase is accomplished by generating an electric field between a spraying needle, which is held at a high negative potential, and a counter electrode held at ground or a positive potential some distance from the needle. The solution being sprayed exits the needle as a conical distribution of droplets ('Taylor cone'), each containing excess negative charge. A heated drying gas, such as nitrogen, is typically used to assist evaporation of the solvent sheath from the ion. The desolvated, multiply charged ion is then introduced into the mass spectrometer for analysis [50].

ESI performance is directly related to the inner diameter (i.d.) of the spraying needle (capillary) [51]. Spraying capillaries with a 1–2 μm i.d. spraying orifice, so-called 'nanoelectrospray' sources, have several advantages over the traditional ESI needle, particularly in biomolecular analyses. This small spraying orifice requires low flow rates (nL/min), which aids in sample conservation. In addition, there is an

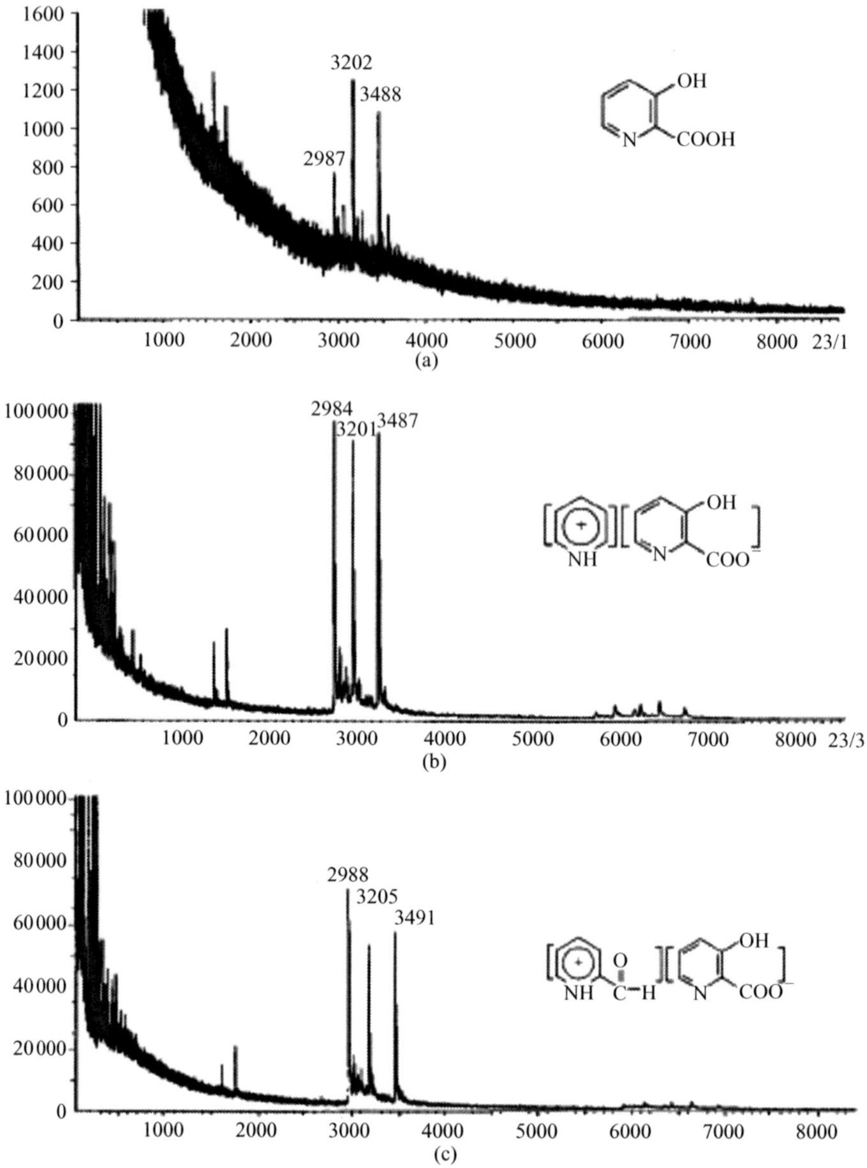

Figure 4.2 MALDI mass spectra of the three oligonucleotides ($d(pT)_{10}$, $d(pC)_{11}$ and $d(pC)_{12}$) in different matrices: (a) 3-HPA – 10% ammonium citrate, (b) ionic solid 21 and (c) ionic solid 26. Spectra obtained cumulating 100 UV 237 nm laser shots. For the three experiments, the oligonucleotide-to-matrix molar ratio was 1:500 000 and the laser fluence was the same (attenuation 10). The signal strength is expressed in arbitrary units corresponding to the accumulation of 100 shots on a good spot. The 3-HPA scale (top spectrum) differs eight times from that for the two salts (bottom spectra). (Source: Carda-Broch, S., Berthod, A., Armstrong, D.W., Ionic matrices for matrix-assisted laser desorption/ionization time-of-flight detection of DNA oligomers, *Rapid Communs Mass Spectrom.*, 2003, 17, 553–560, by permission of John Wiley and Sons.)

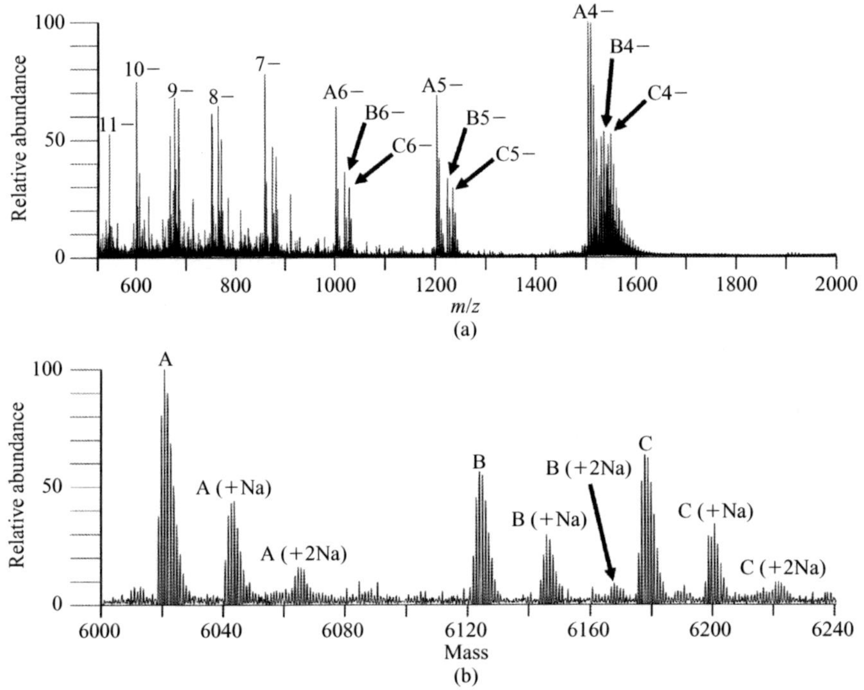

Figure 4.3 (a) ESI mass spectrum of three oligonucleotides (**A**: d(pT)$_{20}$, **B**: T7 promoter (5′-TAATACGACTCACTATAGGG-3′), and **C**: T7 terminator (5′-TAGTTATTGCT-CAGCGGTGG-3′) illustrating typical charge state distribution for oligonucleotides that are generated during electrospray. (b) Deconvoluted mass spectral data from (a). The parent ions plus sodium adducts are detected for each oligonucleotide.

increased overall efficiency in desolvation, ionisation and transfer efficiency. The small size of the droplets generated results in one analyte molecule per droplet. Because the size of the droplets is significantly reduced, desolvation is easily achieved without the need of a drying gas. The overall charge-to-volume ratio of the droplets is also higher than in conventional ESI, thereby improving ionisation. Transfer of ions into the mass analyser is improved because the spraying tip can be placed close to the orifice of the analyser (1–2 mm). Other important advantages of the nanoelectrospray source are that it has a higher tolerance to salt adduction than conventional ESI and can be operated under high pH conditions. As described in the sample preparation section below, both of these characteristics can aid in the analysis of oligonucleotides.

ESI is a gentler ionisation technique compared to MALDI and generally produces multiply charged ions of intact molecules. Non-volatile cation adduction (e.g. Na^+ and K^+) is a major problem with ESI-MS analysis of oligonucleotides. The negative charge on the phosphate backbone of oligonucleotides results in a large degree of Coulombic strain. In solution, solvent molecules help to reduce these Coulombic interactions. In the gas phase, where solvent molecules are absent, relief of this strain is achieved by neutralisation or cation adduction. These adduct peaks reduce the sensitivity of mass measurement as the ion current is dispersed among multiple cationised ions and can also result in peak broadening in spectra if the mass analyser has insufficient resolving power. Without question, sample preparation becomes the key to obtaining reliable and accurate analytical information when performing ESI-MS of oligonucleotides [52].

4.2.3 ESI-MS or MALDI-MS for Oligonucleotides?

A brief comparison of the two approaches for oligonucleotide analysis is presented, and a summary is provided in Table 4.1. The advantages of ESI-MS include ease of coupling to on-line separation methods, and the ability to characterise non-covalent interactions. It is also the preferred method for gas-phase MS sequencing (tandem MS). MALDI-MS is more tolerant of sample contaminants, can handle complex mixture analysis and is the preferred method for sequencing by an exonuclease digestion protocol. In general, both methods are capable of providing molecular weight and sequence information from oligonucleotides, and the choice of ionisation method depends on the available instrumentation, type of analyses desired and user preference.

4.2.4 Mass Analysers

The most common instrument configuration is the MALDI ion source coupled to a time-of-flight (TOF) mass analyser. Separation of ions of different *m/z* values in a TOF mass analyser is accomplished by accelerating the ions through a short 20–

Table 4.1 Summary of characteristics of MALDI-MS versus ESI-MS.

Parameter or characteristic	**MALDI-MS**	**ESI-MS**
Mass errors	0.05–1.0%	0.005–0.1%
Upper mass limit	ca. 50 000 Da	ca. 120 000 Da
Ionisation (polarity/distribution)	Positive/primarily singly charged ions	Negative/primarily multiply charged ions
Feasibility of:		
Mixture analysis	Yes	Difficult (without prior LC separation)
Easy coupling to separation methods	Difficult	Yes
Exonuclease sequencing	Preferred	Possible
Sequencing via gas-phase dissociation (MS/MS)	Possible (with MALDI-TOF/TOF)	Preferred (with tandem MS platform)
Characterisation of noncovalent interactions	No	Yes

Note: TOF denotes time of flight.

30 kV electric field, allowing these ions to drift through a field free region, and measuring the total flight time from ion formation to impact on an electron multiplier detector. Lower-mass ions have shorter flight times than do higher-mass ions and thus, by calibration of the ions' flight times through the instrument with standards, a mass spectrum can be obtained.

The sensitivity, resolution and mass accuracy depend on the particular TOF utilised. MALDI was first coupled with a basic linear TOF (L-TOF) instrument. In L-TOF, ions are accelerated for less than a few hundred nanoseconds. Once ions leave the acceleration region and enter the field-free region, any fragmentation that occurs will not be detected. In this case, less fragmentation is observed in L-TOF as this instrumental configuration cannot differentiate the parent and fragment ions. A significant disadvantage of the L-TOF is that it cannot reduce peak broadening occurring from the initial velocity distribution of ions. Time-lag focusing, commonly referred to as delayed-extraction (DE), was coupled to TOF as a means of reducing the initial velocity spread of ions while they are still in the accelerating region of the mass spectrometer [53–55]. In the DE mode of operation, a second electrostatic gate is included in the ion source region. During the laser pulse, this electrode is held at a potential sufficient to prevent the acceleration of the laser-desorbed ions into the field-free region of the mass spectrometer. After a suitable delay period, which is mass dependent [55], the ions are then accelerated into the field-free region and analysed as usual. DE L-TOF instruments achieve extremely reasonable resolution (ca. 1000 FWHM (full width at half maximum)) and lower mass errors (ca. 0.01–0.1%) for oligonucleotides up to the 50mer level [56].

A common instrumental approach for increasing the flight path, and an approach that also addresses a higher order ion focusing problem, is the addition of an electrostatic mirror at the end of the flight tube that reverses the ion's direction and refocuses the ion toward the detector. This 'reflectron' TOF (re-TOF) mass analyser effectively doubles the path length of the ion, yielding longer flight times and, hence, higher mass spectral resolution. The electrostatic ion mirror compensates for the initial energy spread of the ions thereby improving focusing of ions of a single *m/z* more effectively at the detector. In addition, DE can be combined with a re-TOF mass analyser to yield a 'high-performance' MALDI mass spectrometer with resolution approaching 15 000 FWHM and ppm mass errors. The use of a reflectron can resolve metastable dissociations, but unless certain parameters are adjusted [57], these ions appear as broad uninformative peaks.

ESI can be coupled with a large variety of mass analysers. Table 4.2 lists the analysers most often coupled to an ESI source. Quadrupole ion trap (QIT) and linear ion trap (LIT) mass analysers are the most popular configuration, owing to their ease of operation and low cost. Triple quadrupole (QqQ) instruments are used primarily for quantitative studies. Fourier transform-based mass spectrometers (Fourier transform ion cyclotron resonance (FTICR) or Orbitrap) are configurations which offer the highest resolution and mass accuracy. ESI-TOF and ESI Q-TOF mass spectrometers offer high sensitivity, extended upper *m/z* range, adequate resolution and mass accuracy, along with a high duty cycle compatible with liquid chromatography.

Table 4.2 Mass analysers typically coupled to ESI source.

Analyser	Resolution	Sensitivity	Mass accuracy	Upper *m/z* limit	Tandem MS capability
TOF	Good	High	High	ca. 50 000	No
Quadrupole	Fair	Good	Fair	3000	No
QqQ	Fair	High	Fair	12 000	Yes
QIT	Fair	Good	Fair	4000	Yes
LIT	Good	High	Good	4000	Yes
Q-TOF	High	High	High	ca. 50 000	Yes
Orbitrap	Very high	High	Very high	4000	Yes
FTICR	Very high	Good	Very high	15 000	Yes

4.2.5 Tandem MS

Tandem MS, or MS-MS, is a now generic description of an instrumental method for obtaining structural information from gas-phase ions. In this section, the discussion will be limited to the various options in instrument configuration that allow for MS-MS analysis. Later sections will describe the fragmentation processes that can occur with oligonucleotides during MS-MS analysis, as well as selected applications in oligonucleotide characterisation by tandem MS approaches.

In MS-MS experiments, the ion of interest (i.e. the precursor ion) is first separated or isolated by a mass analyser and then internally excited (by various means) to enable bond dissociation reactions to occur. The charged products of those bond dissociation events (i.e. product ions) are subsequently mass analysed and the information from the precursor ion mass along with the product ions can be used to characterise the structure of the precursor ion. Thus, MS-MS experiments require two separate steps of mass analysis with an intermediate step of ion dissociation. Instrumentally, a number of options exist for mass analysis and ion dissociation, and while the discussion here will be limited to those approaches most commonly found in oligonucleotide analysis, the interested reader is referred to a variety of resources for more detailed information on MS-MS instrumentation, ion activation and MS-MS methods [2, 9, 58–69].

MS-MS can be implemented using a variety of different mass analysers. Tandem 'in space' instruments require multiple mass analysers. Conceptually, the simplest example is the QqQ instrument, where Q(1) and Q(3) are quadrupole mass analysers and q(2) is a quadrupole-based collision cell where the fragmentation process occurs. A more common platform used in oligonucleotide analysis is the hybrid mass spectrometer, the Qq-TOF, where Q(1) is a quadrupole mass analyser, q(2) is a quadrupole-based collision cell, and the TOF serves as the second mass analyser.

With tandem 'in time' instruments, only one mass analyser is required and the steps of precursor ion selection, ion dissociation and product ion analysis occur within that single mass analyser sequentially in time. The most common example here is the ion trap, either a QIT or an LIT. Other examples include the FT-based mass analysers,

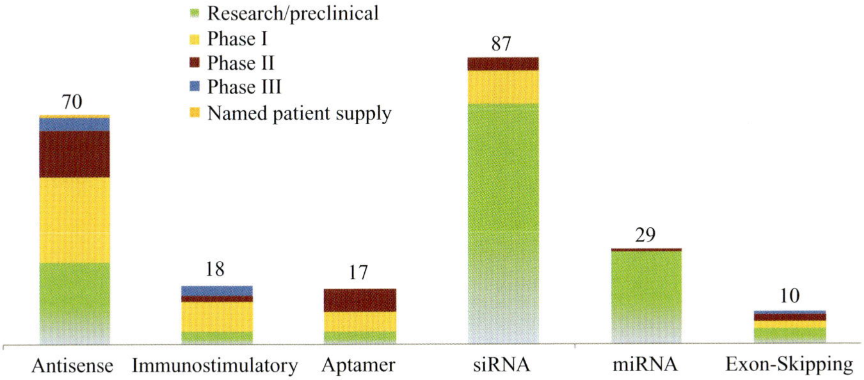

Figure 1.9 Oligonucleotides in research and development by type and clinical phase in 2012. (Source: © Agilent Technologies, Inc 2013. Reproduced with permission, courtesy of Agilent Technologies, Inc.).

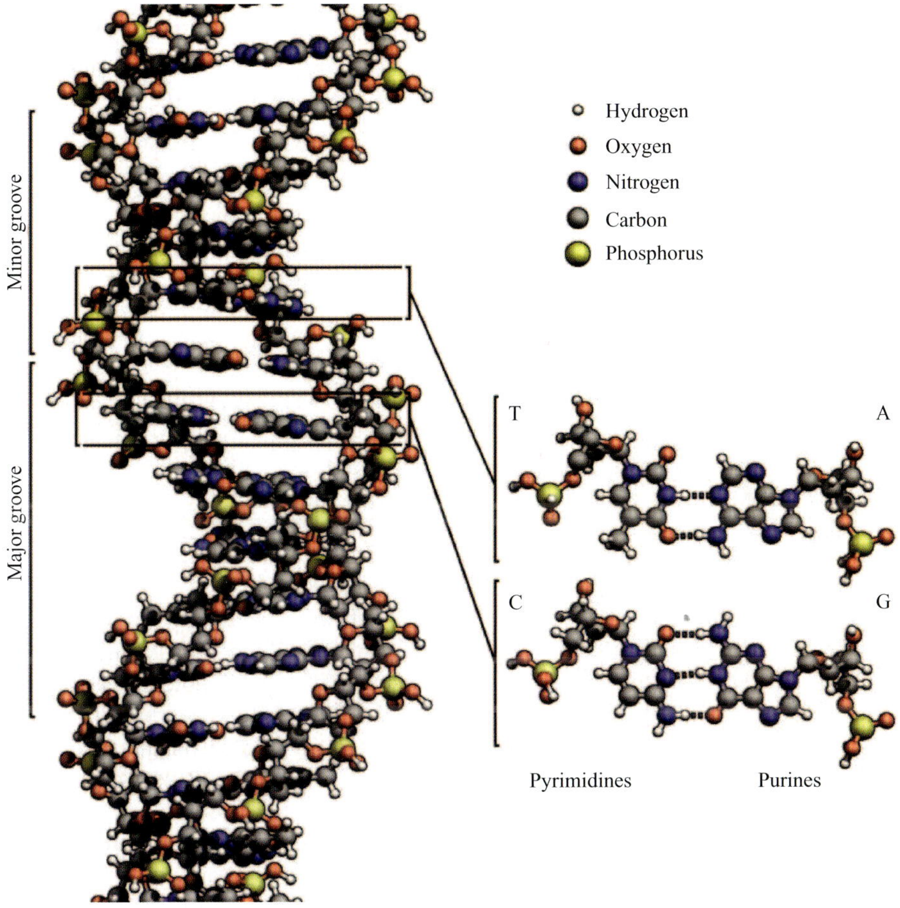

Figure 3.1 Structure of double-stranded (or duplex) nucleic acids.

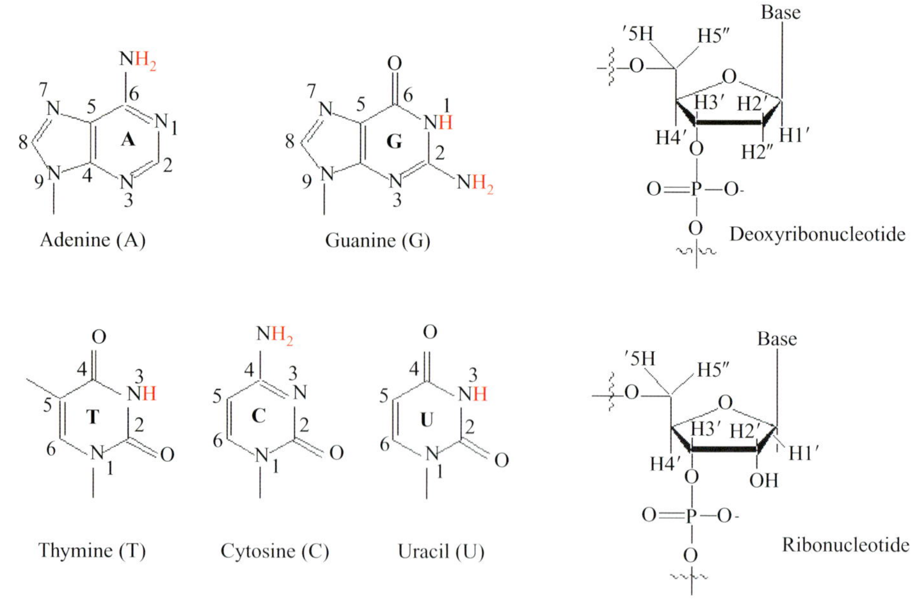

Figure 5.4 Chemical structures of DNA (A, G, C and T) and RNA (A, G, C and U) nucleotide bases along with the deoxyribose and ribose sugar ring fragments. Exchangeable protons are indicated in red.

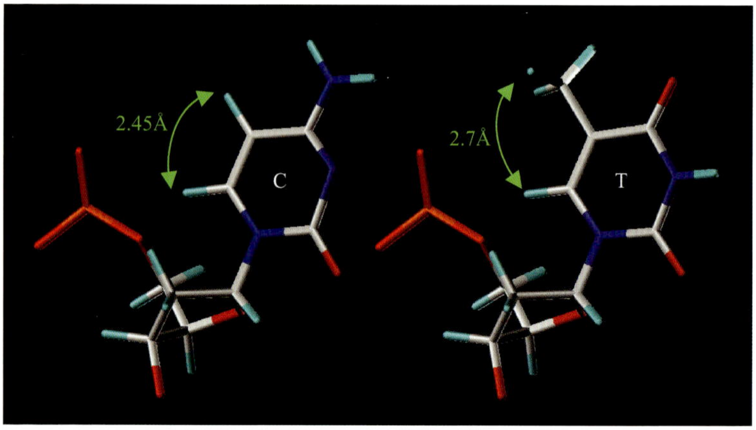

Figure 5.12 3D structures of cytidine and thymidine nucleotide residues. Arrows indicate those protons of the aromatic rings (C(H5)–C(H6) and T(CH_3)–T(H6)), which are involved in J-coupling interactions via chemical bonds as well as in through-space dipole–dipole interactions due to their close location.

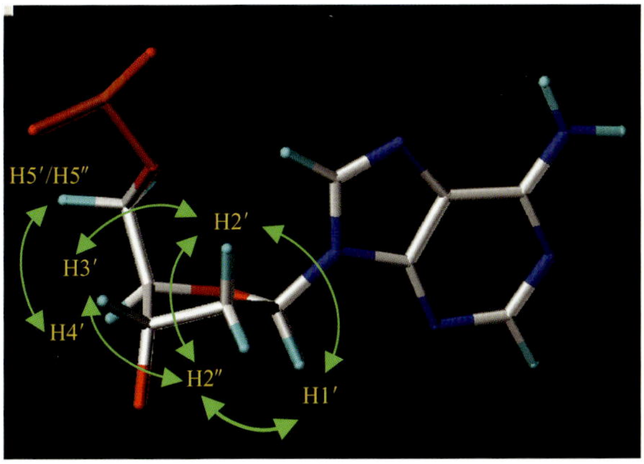

Figure 5.13 3D structure of the guanosine nucleotide residues demonstrating the sugar ring spin system. Deoxyribose protons, which are normally involved in J-coupling interactions, are indicated by arrows.

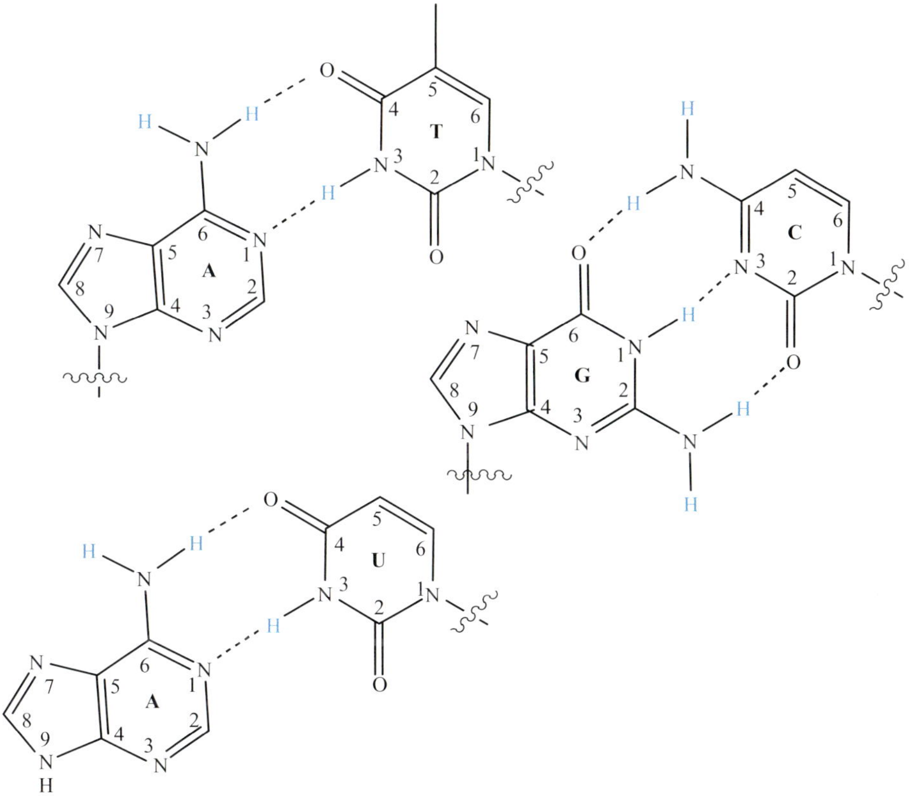

Figure 5.15 Schematic representation of Watson–Crick base pairs A-T, G-C and A-U present in nucleic acids. Imino and amino protons involved in the hydrogen bond formation are highlighted in blue.

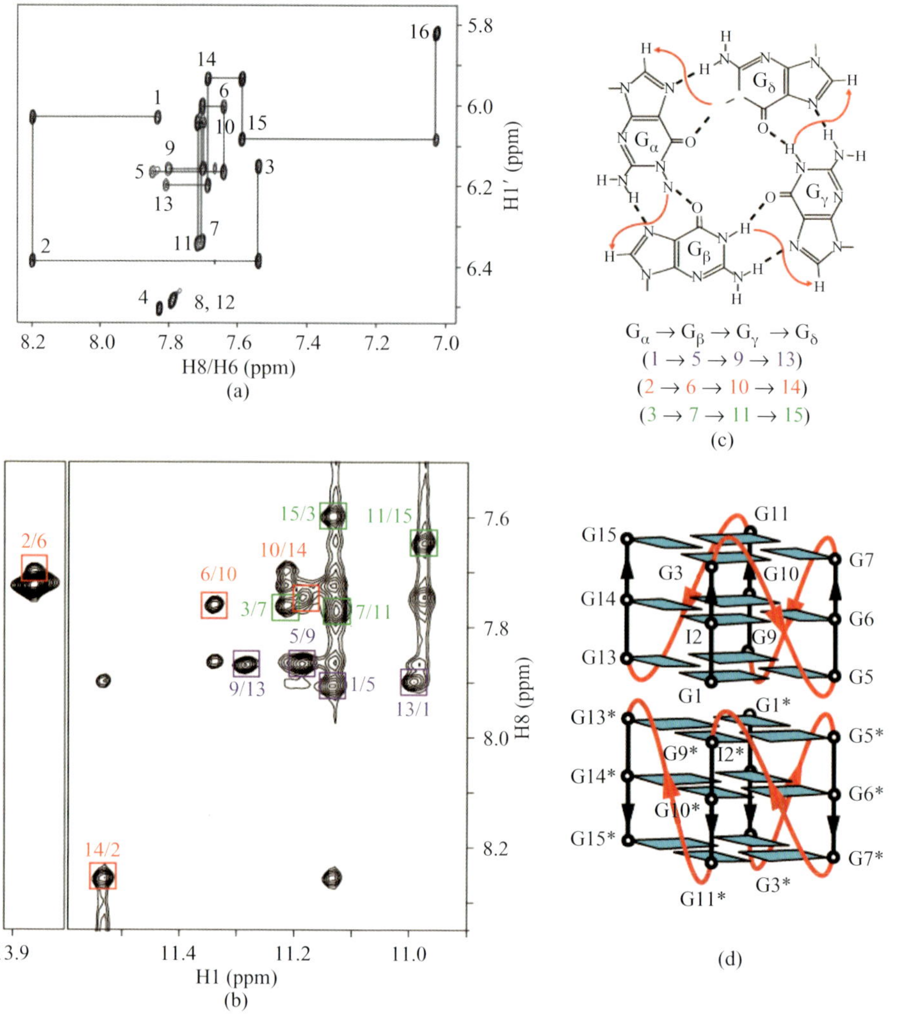

Figure 5.27 Determination of G-quadruplex folding topology of J19 in K+ solution. (a) H8/H6–H10 region of NOESY spectrum (mixing time, 300 ms) at 25°C. The assignments and NOE sequential connectivities are shown. (b) Imino-H8 region of NOESY spectrum (mixing time, 400 ms) at 25°C. The characteristic guanine imino-H8 cross-peaks for G-tetrads are framed and labelled with the imino proton assignment in the first position and that of the H8 proton in the second position. (c) Specific imino-H8 connectivity pattern around a G-tetrad (Ga_Gb_Gg_Gd) indicated with arrows (connectivity between Gd and Ga implied). The connectivities observed for G1_G5_G9_G13 (purple), I2_G6_G10_G14 (red) and G3_G7_G11_G15 (green) tetrads are shown below. (d) The schematic structure of J19 dimeric parallel-stranded G-quadruplex. (Source: Do, N.Q., Lim, K.W., Teo, M.H., Heddi, B., Phan, A.T., Stacking of G-quadruplexes: NMR structure of a G-rich oligonucleotide with potential anti-HIV and anticancer activity, *Nucleic Acids Res.*, 2011, 39, 9448–9457, by permission of Oxford University Press.)

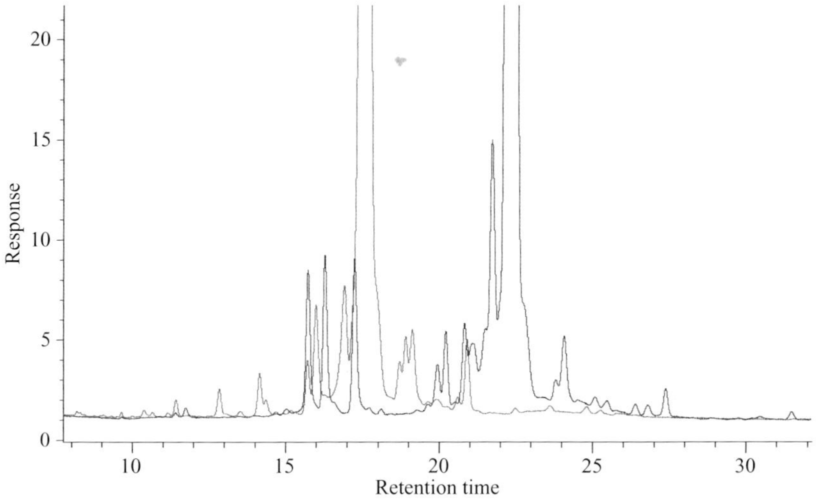

Figure 7.2 RP-IP chromatography overlay of individual sense (green) and antisense strands (blue) of a 21mer siRNA, 0.25 mg/ml each. Overlay demonstrates co-elution of impurities between the two strands and why impurities analysis on the individual strands prior to annealing into the duplex may provide improved analysis. Method conditions: Waters Acquity UPLC system, OST C18 2.1 × 100 mm 1.7 μm column, 20 mM triethylammonium (TEA) mobile phase with 0.25% methanol/min gradient, 0.25 ml/min flow rate, 60°C column temperature.

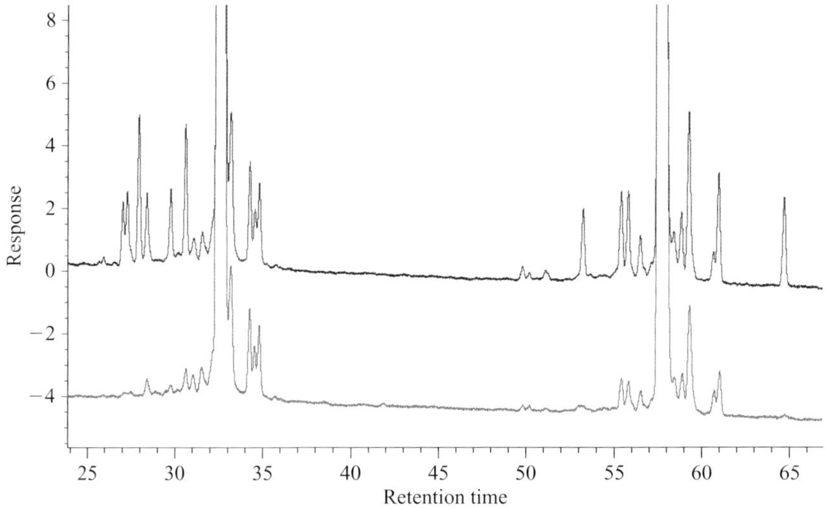

Figure 7.3 Denaturing RP-IP chromatography of a 21mer siRNA at initial (green) and after 2 months of storage in neat water at 25°C (blue). Chromatography demonstrates complete separation between the sense and antisense groups of impurities. Method conditions: Waters Acquity UPLC system, ethylene bridged hybrid (BEH) shield RP18 2.1 × 150 mm 1.7 μm column, 10 mM TEA/10 mM propylamine mobile phase with 0.17% methanol/min gradient, 0.175 ml/min flow rate, 55°C column temperature [50].

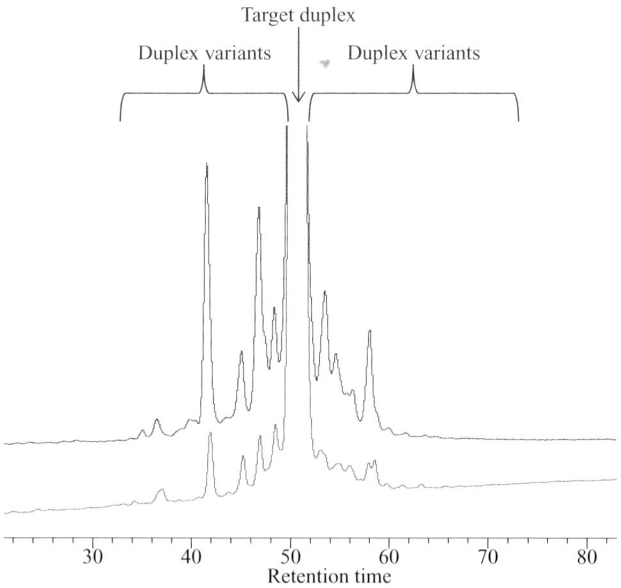

Figure 7.4 Chromatography for non-denaturing SAX method showing lyophilised siRNA API at initial (green) and after 6 months' storage at 25°C with 20% RH headspace (blue). Method conditions: Agilent 1100 HPLC system, Dionex DNAPac PA200 2 × 250 mm column, 12.5 mM mobile phase A: tris base, 1 M NaCl (pH 8.0 with 1M HCl), B: water, gradient: 58% A at initial linear to 40.7% A at 80 min, 0.5 ml/min flow rate, 15°C column temperature, 5 µL injection, 0.25 mg/ml siRNA concentration.

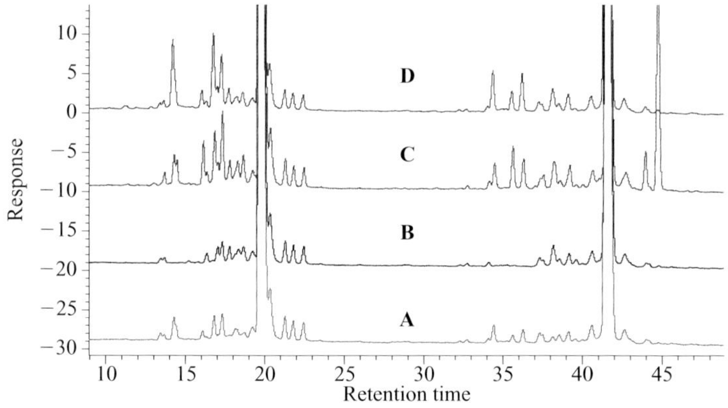

Figure 7.9 Denaturing RP-IP UPLC impurity profile chromatograms of siRNA **B** (middle T_m) at initial (**A**, green) and after 2 months' storage at 25°C in 1X PBS (**B**, black), water (**C**, red), or in the solid state with sealed 20% RH headspace (**D**, blue). Method conditions: Waters Acquity UPLC system, BEH shield RP18 2.1 × 150 mm 1.7 µm column, 10 mM TEA/10 mM propylamine mobile phase with 0.17% methanol/min gradient, 0.175 ml/min flow rate, 55°C column temperature [50].

either FTICR or orbitraps, although both of these are more commonly used in a hybrid configuration where the isolation and dissociation steps can occur in separate mass analysers.

In addition to the variety of instrumental configurations that are available for tandem MS, a number of options exist to generate the bond dissociation processes required to fragment the precursor ion so that structural information can be obtained. The most commonly encountered dissociation technique is collision-induced dissociation (CID). In this technique, the precursor ion of interest is accelerated and undergoes one or more collisions with neutral gas atoms or molecules. These collisions result in the conversion of translational energy into internal energy. When sufficient internal energy is deposited within the ion, bond dissociation can occur leading to the fragmentation of the precursor ion. All of the various tandem mass spectrometer configurations described above allow for CID, typically using nitrogen or argon as the collision gas.

4.3 Oligonucleotides in the Gas Phase

4.3.1 Types and Structure

The various types and nomenclatures used to represent oligodeoxynucleotide (DNA), oligoribonucleotide (RNA) and modified oligonucleotides (e.g. phosphorothioate oligonucleotides) are discussed in Chapter 1. Of particular interest here is the nomenclature used to characterise oligonucleotides of any type during mass spectral analysis, specifically during sequencing by CID or other fragmentation techniques. McLuckey *et al.* [70] proposed a systematic nomenclature for the identification of oligonucleotide dissociation products that is similar to the nomenclature proposed by Roepstorff and Fohlman [71] for peptide fragmentation, and which is now universally used to identify specific fragment ions from oligonucleotides (Figure 4.4).

Within the McLuckey nomenclature, fragment ions are denoted as a, b, c, or d if the charge remains on the 5′-terminus of the oligonucleotide, and w, x, y or z if the charge remains on the 3′-terminus of the oligonucleotide. The specific letters used to identify fragment ions represent different bond dissociation, with complementary pairs being created (e.g. a and w, b and y, etc.) depending on the location of the charge. Each specific fragment ion is also annotated with a subscript following the identifying

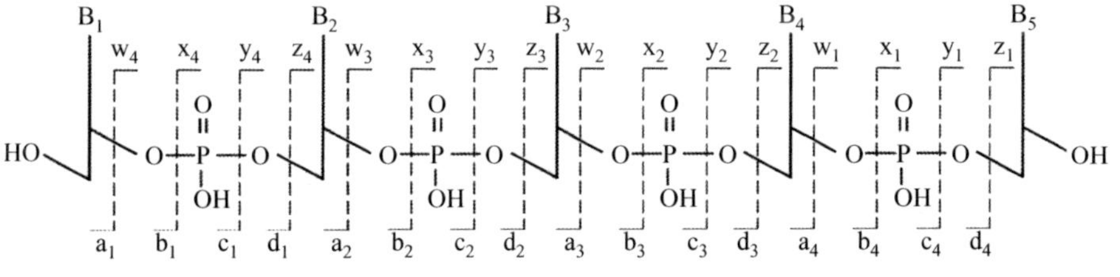

Figure 4.4 Oligonucleotide fragment ion nomenclature of McLuckey *et al.* [70].

letter. This subscript identifies the location in the primary sequence that the bond dissociation occurs.

Presentation of oligonucleotide dissociation assignments within the literature is often simplified using the shorthand notation shown in Figure 4.5. In this example, the assigned fragment ions (c_1-c_3 and y_1-y_4) are identified on the oligoribonucleotide sequence without including the phosphodiester linkages. The McLuckey nomenclature can be adapted to any modified oligonucleotides where the linkages between nucleotide residues remain consistent with the phosphodiester linkage of standard oligonucleotides.

4.3.2 Fragmentation Mechanisms

A significant number of studies have been published on the gas-phase fragmentation mechanisms of oligonucleotides and the interested reader is referred to recent publications that summarise or review these mechanisms [73–76]. While a consistent nomenclature is used in the field of MS to identify oligonucleotide fragment ions, the specific types of fragment ions one will actually detect during MS depends in great part on the type of oligonucleotide being characterised. It is important to note that fragmentation mechanisms can be influenced not only by the composition of the oligonucleotide (that is, whether it is an oligodeoxynucleotide or an oligoribonucleotide), but is also influenced by the instrumental and experimental conditions used during analysis. As such, several general rules are described below, but the fragmentation of model oligonucleotides of interest should always be examined on the

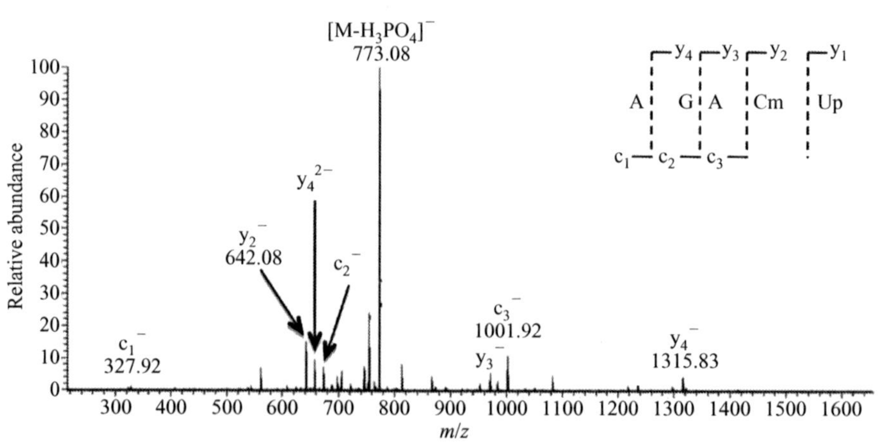

Figure 4.5 Representative example of shorthand notation to identify oligonucleotide fragment ions and their corresponding sequence assignment. (Source: Adapted from Krivos, K.L., Addepalli, B., Limbach, P.A., Removal of 3'-phosphate group by bacterial alkaline phosphatase improves oligonucleotide sequence coverage of RNase digestion products analyzed by collision-induced dissociation mass spectrometry, *Rapid Communs Mass Spectrom.*, 2011, 25, 3609–3616, by permission of John Wiley and Sons.)

instrument of choice to best identify what types of fragment ions are to be expected during analysis.

The fragmentation behaviour of ESI-generated oligodeoxynucleotides was first studied in detail, resulting in the conclusion that DNA-based oligonucleotides fragment via nucleobase loss. The nucleobase can be lost as a neutral or a charged ion, depending on the overall charge state of the oligodeoxynucleotide being fragmented. Generally, neutral loss occurs at low parent ion charge state, with charged ions lost as the parent ion charge state increases [77]. Loss of the nucleobase is then followed by cleavage of the 3′ C–O bond of the deoxyribose at the site of nucleobase loss. This cleavage leads to the generation of (a-Base)- and w-type ions. These complementary fragment ion products are very useful in assigning sequence products for oligodeoxynucleotide dissociations.

Oligoribonucleotides were found to exhibit different fragmentation behaviour, most often attributed to the presence of the 2′-OH group [74, 76, 78–80]. Those reports concur that oligoribonucleotide fragmentation, unlike oligodeoxyribonucleotide fragmentation, is not influenced by the nucleobase. Rather, the predominant pathway arises from cleavage of the 5′-P-O bond to generate c- and y-type complementary ions. In addition, cleavage of the 3′-P-O bond after nucleobase loss can also occur to generate (a-B)- and w-type complementary ions, although these are far less prevalent in oligoribonucleotides than in oligodeoxyribonucleotides. McLuckey and co-workers [79] also have shown that the relative abundances of c/y and (a-B)/w fragment ions depend on the overall charge state of the parent ion, with c/y-ions being more prevalent at low to intermediate charge states.

Because chemistry plays on important role in oligonucleotide fragmentation, it is not surprising that the various modified oligonucleotides can exhibit vastly different fragmentation pathways – dependent on the chemical effects of the modification. Phosphorothioate oligonucleotides have been shown to fragment via the same pathways as the unmodified analogue [81], which is not unexpected due to the minimal change in replacing a backbone oxygen with a sulfur that has similar chemical properties. In contrast, methylphosphonates, which eliminate the charge along the oligonucleotide backbone, can result in rather significant changes to the fragmentation pathways [75].

Locked nucleic acids (LNAs) were found to generate all possible 5′-P-O and base-loss initiated 3′-P-O fragments yielding the complementary pairs (a-B)/w, b/x, c/y and d/z [82]. Oligonucleotides containing a mixture of LNAs and DNA yielded significantly different behaviour, which is discussed in more detail in Section 4.5.5.3. Similarly, 2′-fluoro- and 2′-*O*-methyl modified oligonucleotides were also found to exhibit no dissociation pathway preference and so, too, generated the same four complementary pairs as LNAs [74].

Regardless of the type of oligonucleotide being analysed, fragmentation of ESI-generated oligonucleotides is most effective when the size of the oligonucleotide is at the 15mer or below [83]. Sequence information from oligonucleotides larger than this size can be obtained by MS-MS, although as the size increases the likelihood that sufficient fragmentation will occur to yield complete 5′- and/or 3′-sequencing reads decreases.

Because MALDI generates primarily singly charged ions (of either polarity), there are some differences observed in the fragmentation behaviour of oligonucleotides when using this ionisation source. In contrast to ESI, the desorption and ionisation steps in MALDI can lead to energy being deposited into the oligonucleotide, which may lead to bond cleavage and fragmentation, even in the absence of any additional collisional activation steps. This source-derived fragmentation has been classified into four differing timescales: prompt, fast, fast metastable and metastable. Prompt dissociations occur on a timescale equal to or less than the desorption event. Fast dissociations occur after the desorption event, but before or at the beginning of the acceleration event. Fast metastable decays occur on the timescale of the acceleration event. Metastable decays occur after the acceleration event, during the field-free flight time of the ion. While a richer fragmentation spectrum is typically observed in MALDI-MS, the predominant fragment products for oligodeoxynucleotides remain the (a-Base)- and w-type ions that are also seen in ESI-CID fragmentation studies. Although no detailed investigations have been undertaken, it appears that source-generated dissociation of oligonucleotides is more effective at presenting complete sequence information for longer oligonucleotides than is possible via MS-MS methods of ESI-generated oligonucleotides.

4.4 Method Development

4.4.1 Sample Preparation

One reason most other ionisation methods are not useful for oligonucleotides and nucleic acids is the substantial difficulty associated with generating a gas-phase ion from such a polar analyte. Indeed, even with ESI or MALDI methods, a key concern in generating gas-phase ions involves the charge neutralisation that occurs by adduction of cations during ionisation. Cations are adducted because oligonucleotides possess a negatively charged phosphate backbone at neutral pH. In solution, Coulombic repulsion due to adjacent charges is reduced by the presence of solvent. However, once oligonucleotides are taken to the gas phase, solvent molecules are no longer present. To reduce charge repulsion, neutralisation of the backbone charges occurs through cation adduction. Cation adducts, except H^+ or NH_4^+, are undesirable as they reduce the sensitivity by spreading the ion species in multiple mass values and also cause peak broadening. Thus, a significant and general key to ionising oligonucleotides efficiently is the need to reduce or eliminate the presence of cation adducts from the sample or solution prior to analysis.

Because the cationisation of oligonucleotides deteriorates the quality of their resultant mass spectra, sample preparation becomes critical to avoid cationisation. A major goal of all the oligonucleotide sample preparation protocols is to remove or minimise the presence of Na^+ or K^+ or other metal ions within the sample. The common protocols for metal cation removal include cation-exchange resin beads, solid-phase microextraction pipette tips, HPLC, microdialysis or precipitation from organic solvents [84–86]. Water or other solvents used in sample preparation must also have minimal amounts of cations, and in the case of RNA samples, must be RNase free. Detailed protocols for oligonucleotide sample preparation prior to mass spectrometric analysis are available [87, 88].

4.4.2 MALDI-MS Sample Preparation

Sample preparation/handling is often considered the most important step in MALDI-MS. The quality of the preparation has a direct impact on the quality of the analytical results. Different types of sample preparations are available for various types of analytes and purposes [40, 89, 90]. Moreover, most instrument vendors now provide pre-made MALDI sample targets to reduce both the labour involved in sample preparation and to increase the reproducibility of analysis. A few of the more common protocols used in sample preparation for MALDI-MS are briefly discussed. The interested reader is referred to the following expanded discussion of MALDI matrix preparations (both historical and modern) for more details [17].

Cations can arise from any source the oligonucleotide sample comes into contact with, such as solvents, buffers, the sample probe, solution storage containers or pipette tips. Doubly distilled and deionised ('nanopure') water should always be used to reduce cation contamination. Organic solvents used in the preparation of the matrix solutions may need to be desalted by distillation or by activated ion-exchange resin beads (e.g. BioRad AG501-X8(d) resin beads) prior to their use [32]. Many buffers simply interfere with ion formation without forming intense adduct peaks. The tolerance limits for several common buffers have been investigated [91]. It was determined that oligonucleotides could be analysed in the presence of up to 10 mM concentration of salt or 500 mM concentration of buffer in the sample solution. It was found also that positive ions are not affected by impurities as much as negative ions. A simple method to remove minor cations (at concentrations less than 10 mM) is to add an ammonium salt, such as ammonium citrate, tartrate or fluoride to the matrix solution [26, 30, 92, 93]. Ammonium salt concentrations should be 10 mM or less except with THAP; otherwise, the matrix does not crystallise well, limiting the ability to find 'hot spots' that yield abundant ion signal.

Often the sample holder can be a source of contaminants. If the sample probe is not cleaned, cross-contamination with the previous sample will interfere with the newly deposited oligonucleotides. Therefore, sample plates should be cleaned carefully prior to use. Sonicating and washing in organic solvents, such as methanol and acetone, effectively removes most matrices and analytes. Further, stainless steel sample plates can be cleaned by polishing followed by acid oxidation in a dilute inorganic solution (e.g. 0.1 M nitric acid). The sample plates should be rinsed with nanopure water to eliminate the residual cations.

A popular and useful method for sample purification is to use a solid-phase microextraction approach prior to sample and matrix spotting, based on commercially available ZipTips (Millipore). A number of vendors (Millipore, Pharmacia, Boehringer Mannheim, Amicon) offer prepackaged purification systems. These purification devices are typically 0.5–2.5 mL microcentrifuge tubes which contain coated particles or filters that typically retain the oligonucleotide and allow contaminants to be washed off the sample. C18 reversed-phase (RP) cartridges utilise the strong binding affinity of oligonucleotides to the stationary-phase packing material in the presence of 2M triethylamine as the means of trapping the sample for purification. The oligonucleotides are eluted with 20% aqueous acetonitrile. Sephadex G-25 spin column purification is a very useful and quick method (ca. 5 min) to eliminate most salts and buffers if the oligonucleotide is larger than a 10mer. Equilibration of spin columns in nanopure water before desalting is necessary. Generally, one pass through a spin column yields satisfactory results. Molecular weight cut-off filters or membranes retain oligonucleotides above a certain molecular weight and allow the contaminants to be washed through the membrane or filter. The membrane or filter is then inverted and the oligonucleotide is eluted and lyophilised. The drawback to these protocols is that the volume must be reduced prior to combination with matrix and spotting on the MALDI sample target.

4.4.3 ESI-MS Sample Preparation

Nearly all of the approaches described above can also be used in sample preparation for ESI-MS. In addition, Muddiman and co-workers [94–99] have utilised microdialysis for efficient sample clean-up. They initially demonstrated the applicability of this approach by characterising an 89 base pair polymerase chain reaction product with ESI-FTICR-MS. HPLC with a C18 RP column has also been used for desalting larger oligonucleotides including intact transfer RNAs [100]. Gradient elution using triethylammonium acetate (buffer A) and acetonitrile (buffer B) allowed for the acquisition of a mass spectrum with a high S/N ratio and only minor salt adducts.

Because electrospray is a solution-based ionisation method, the solvents used in the analysis of oligonucleotides and nucleic acids are important for the success of the method. Typical solvents used for oligonucleotides analysis consist of a mixture of nanopure water and an organic solvent. The most common organic solvents used are methanol, isopropanol and acetonitrile. While the ESI process prefers solvents of low surface tension (i.e. not water) [50], the upper limit to the organic solvent content is often determined by the oligonucleotide solubility. Typically a stock solution of the oligonucleotide in water is prepared and then this stock solution is diluted into the appropriate volume of organic solvent.

In addition to the off-line approaches to sample purification mentioned above, organic additives (typically organic bases) have also been used for the suppression of cation adducts, the enhancement of the S/N ratio, and for the reduction of the charge state distribution [85, 94, 101–103]. Piperidine and triethylamine strongly suppress cation adduction, but the increase in the overall pH of the sample solution can lead to a decrease in the total ion current from the oligonucleotide sample. Imidazole has been found to yield increases in the total ion current from oligonucleotide samples and, thus, is often used in combination with piperidine or triethylamine allowing for efficient cation reduction without loss in total ion current.

An illustration of the effectiveness of organic base solutions for cation reduction is shown in Figure 4.6. Figure 4.6(a) is the electrospray mass spectrum of a phosphorothioate 20mer obtained after ethanol precipitation. As noted in this spec-

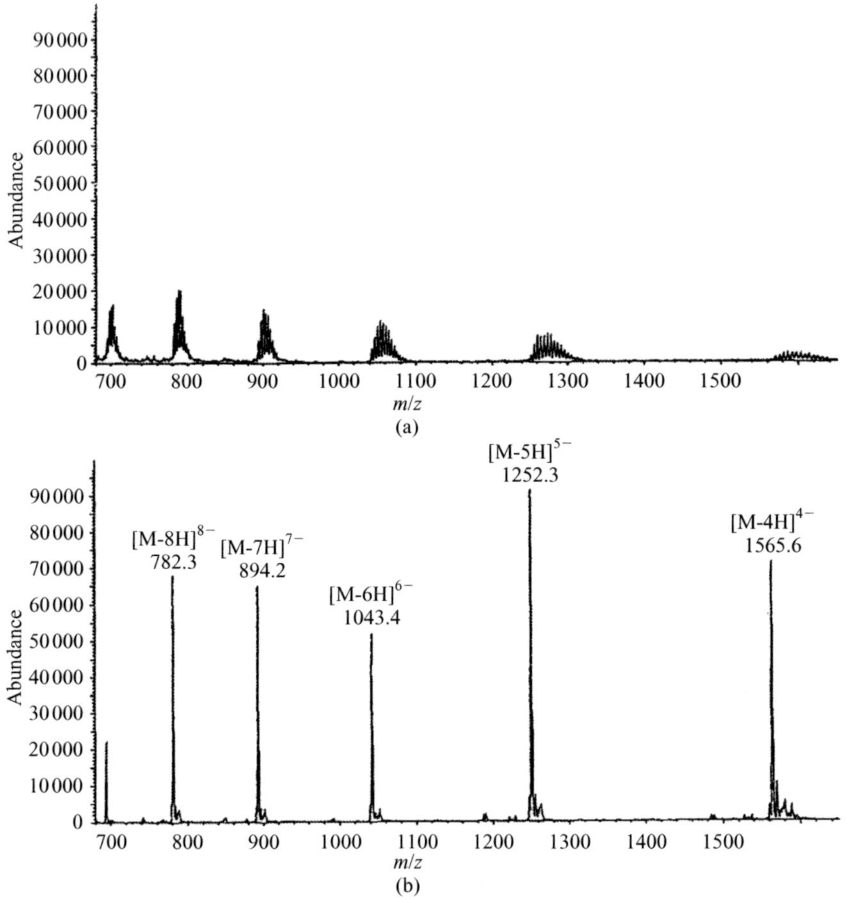

Figure 4.6 Electrospray mass spectra of DNA oligomer 1 precipitated from 10 M ammonium acetate: (a) control; (b) with dissolution in a buffer containing 25 mM imidazole and 25 mM piperidine. (Source: Greig, M., Griffey, R.H., Utility of organic bases for improved electrospray mass spectrometry of oligonucleotides, *Rapid Communs Mass Spectrom.*, 1995, 9, 97–102, by permission of John Wiley and Sons.)

trum, the ion abundance is low and there are a large number of cation adducts. Figure 4.6(b) is the electrospray mass spectrum of the same sample now obtained with the co-addition of 25 mM imidazole and 25 mM piperidine. As seen, the co-addition of imidazole and piperidine yields a 15-fold increase in ion abundance and a dramatic reduction in cation adduction [101].

Care must be exercised if organic bases are utilised with oligoribonucleotide samples, such as siRNAs. RNA is less stable in basic solutions and is subject to strand cleavage upon the addition of imidazole. Furthermore, many RNAs have a strong affinity for divalent metal ions such as Mg^{2+}. To remove divalent metal ions and to reduce alkali metal cations, a combination of a chelating agent, such as trans-1,2-

diaminocyclohexane-*N,N,N′,N′*-tetraacetic acid (CDTA), can be used with low amounts of triethylamine [85]. CDTA (or ethylenediamine *N,N,N′,N′*-tetraacetic acid (EDTA)) has a high binding affinity for Mg^{2+} and other divalent metal ions, and addition of CDTA results in a substantial improvement in the analysis of intact transfer RNAs and larger nucleic acids.

4.4.4 HPLC-MS

Probably the most common hyphenated instrumental platform used in oligonucleotide analysis is HPLC-MS, which is possible using an ESI source. While the ESI source can be readily coupled to other separation techniques such as capillary electrophoresis [104] and capillary electrochromatography [105], HPLC-MS has proven to be not only the most popular combination but also among the easiest to implement.

The most popular HPLC-MS method for the analysis of oligonucleotides involves the use of ion-pairing agents in the HPLC mobile phase and non-polar stationary phases, such as C18 (see Chapter 3 for more details) [2, 106]. When coupling separation techniques to electrospray ionisation sources for the analysis of oligonucleotide or nucleic acid mixtures, particular attention must be paid to the buffer and/or solvent systems utilised in the separation step [107–109]. Many stand-alone chromatographic methods utilise salt gradients for oligonucleotide separation. These salt gradients are typically incompatible with the electrospray technique due to substantial cation adduction, as previously discussed. The most commonly used separation buffers will utilise volatile ammonium salts (e.g. triethylammonium acetate (TEAA)) [110], or hexafluoroisopropanol with triethylamine (HFIP/TEA) [111, 112] as ion-pairing agents in an IP-RP HPLC method.

There are a large number of published chromatographic methods and approaches for the analysis of oligonucleotides, each having their own strengths and weaknesses [2, 111, 113–119]. RP chromatography can be used for the rapid cleaning and desalting of oligonucleotides in addition to being used in the IP-RP-HPLC mode [120]. Different ion pairing agents with a RP column successfully separate oligonucleotides and the choice of IP agent and its concentration should be assessed from a mass spectral perspective. For example, a direct comparison of the separation of polythymidylic acids reveals that the IP agent triethylammonium acetate (TEAA) was more effective at separating small oligonucleotides than is the HFIP/TEA IP agent [87]. Conversely, mixtures of larger oligonucleotides will be more effectively separated by the use of HFIP/TEA than the TEAA ion pairing agent [107]. Given the large number of publications describing HPLC-MS for oligonucleotide separation, it is often easiest to start at HPLC conditions that have proven appropriate previously and then optimise the buffer, gradient and flow rates appropriately for the experiment or analytes of interest.

Optimal HPLC-MS methods provide baseline separation of components via chromatography with minimal increase in mass spectral background, ion suppression, cation adducts to analytes, and damage to the instrument. Online HPLC-MS analysis with ESI provides an inherent advantage by concentrating analytes as they elute from the column. However, the continuous introduction of the column eluent lends an inherent disadvantage of the coupling of these systems, as buffer salts and IP agents sprayed into the instrument, especially at high amounts, increase the need for frequent cleaning, degrade instrument components and decrease instrument performance.

4.5 Applications

4.5.1 Synthesis Confirmation

One of the simplest applications of MS to determine indirectly the sequence of an oligonucleotide is to use the accurate measurement of mass to define the base composition, which may be compatible with an expected sequence [121]. Although an accurate mass measurement does not permit the determination of the oligonucleotide sequence, it is a very powerful way to confirm rapidly whether a predicted or an expected sequence may be present [87, 88]. The quality of the molecular weight information obtained during MS depends both on the sample purification step and the performance of the mass analyser.

Either MALDI-MS or ESI-MS are appropriate platforms for confirming oligonucleotide synthesis [107, 122–126]. As MALDI-MS is predominately implemented using TOF mass analysers, differences in mass accuracy will depend on whether linear, linear with delayed extraction, or reflectron with delayed extraction instruments are used. Mass accuracy increases in the order listed directly above with reflectron instruments equipped with delayed extraction providing molecular weight information with mass measurement errors at the ppm level for smaller oligonucleotides increasing to the 0.5–1.0% level for intact nucleic acids. For ESI-MS, the mass analyser will also determine the mass accuracy of the measurement, and typically hybrid Q-TOF or FTMS (ICR or Orbitrap) mass analysers are preferred for synthesis confirmation [126, 127].

4.5.2 Synthesis Purity and Optimisation

As discussed in detail in Chapter 2, the solid-phase synthesis of oligonucleotides is known to result in formation of the failure sequences and synthesis impurities in addition to the desired full-length product (FLP) (Figure 4.7) [128, 129]. MS is an extremely useful analytical method for the identification and characterisation of such components. The most common method involves the use of HPLC-MS or HPLC-MS-MS for the selective detection and/or characterisation of synthesis impurities. Representative examples are discussed below.

4.5.2.1 Failure Sequences

Internal (n – 1) deletion sequences (Figure 4.7(a)) belong to a class of impurities typically linked to non-quantitative removal of the dimethyl trityl (DMTr) group or inefficient coupling followed by inefficient capping [130]. These impurities can be mitigated by adjustments to the synthesis process parameters (contact time and the reagent composition). Such impurities are among the easiest to identify by MS, as

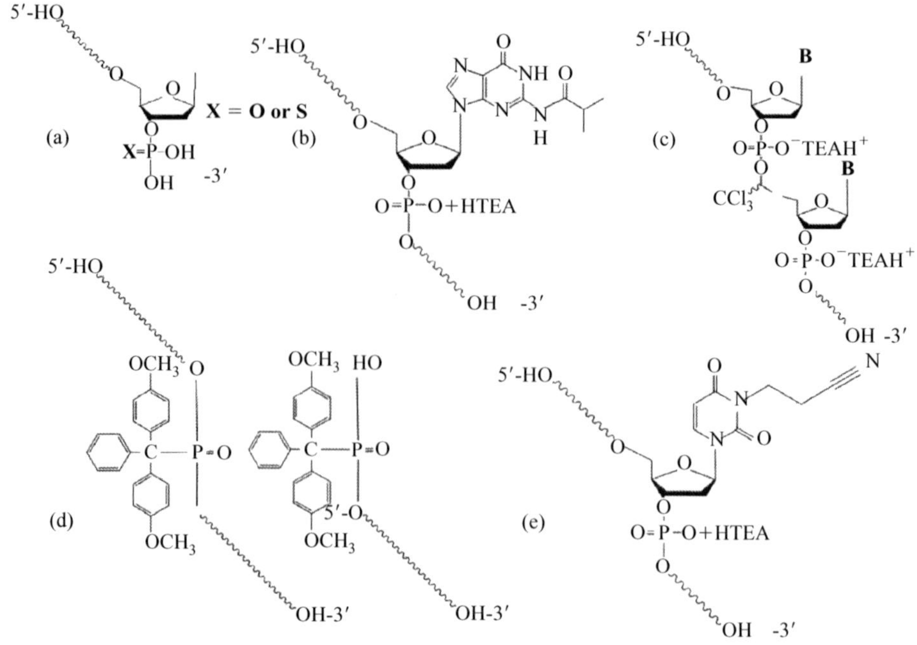

Figure 4.7 Structures of representative synthesis impurities that can be detected by MS. (a) Failure sequences carrying 3'-terminal mono-and monothiophosphates (*n*−1) PO or (*n*−1) PS. (b) FLP isobutyryl adduct. (c) FLP chloral adduct. (d) 5'-O-4,4-dimethoxytrityl-protected oligonucleotide or 5'-O-4,4-dimethoxytrityl-protected *n*−1 deletion sequences. (e) FLP N^3 (2-cyanoethyl) adduct.

they result in a sequence ladder of products with each successive peak differing by the mass of the nucleoside lost in the step-wise synthesis [131].

4.5.2.2 Synthesis Adducts

Detection of isobutyryl protected FLPs is a signature that synthesis deprotection conditions were less than optimal. This adduct (Figure 4.7(b)) results in a 57 Da mass increase over the expected molecular weight, and therefore is also readily identified by standard MS approaches. The trichloroacetaldehyde modified FLP indicates the presence of trichloroacetaldehyde in the detritylating solution, which can lead to chloral (Cl_3CCHO) adducts (Figure 4.7(c)) along the oligonucleotide backbone. MS in combination with nuclear magnetic resonance was used by Capaldi and co-workers [132] to characterise the presence and location of such adducts (Figure 4.8).

N^3-cyanoethyl adducts (Figure 4.7(e)) are related to an inefficient removal of the cyanoethyl protecting groups from the phosphorothioate triester linkages followed by generation of acrylonitrile and its subsequent addition to thymidine residues. Capaldi *et al.* [133] used MS to identify the presence of acrylonitrile on thymidine from the synthesis of a 20mer phosphorothioate. As this adduct results in a 53 Da mass increase

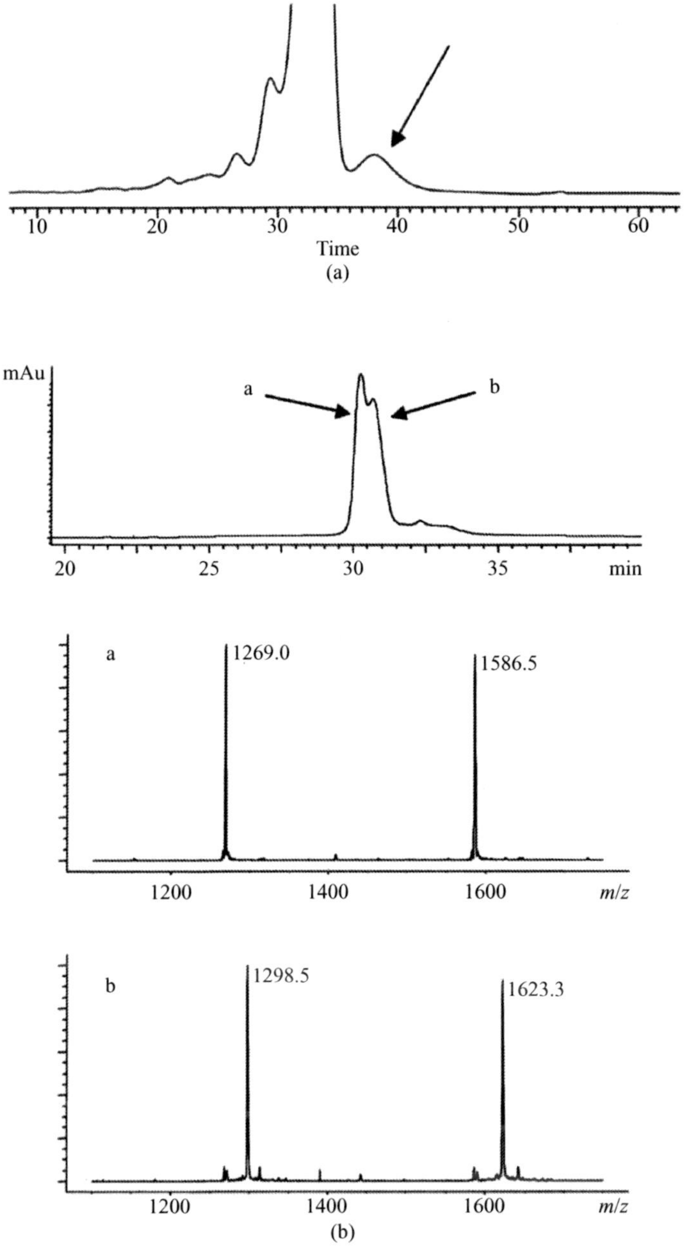

Figure 4.8 (a) Strong anion exchange chromatogram of a batch of 1. (b) Upper panel: HPLC–UV trace of a sample enriched in the 147 Da impurity. Middle panel: the average mass spectrum under peak 'a'. Lower panel: the average mass spectrum under peak 'b'. (Source: Gaus, H., Olsen, P., Sooy, K.V., Rentel, C., Turney, B., Walker, K.L., McArdle, J.V., Capaldi, D.C., Trichloroacetaldehyde modified oligonucleotides, *Bioorg. Med. Chem. Lett.*, 2005, 15, 4118–4124 with permission from Elsevier.)

over the expected molecular weight of the final product, identification by standard MALDI- or ESI-MS is straightforward.

4.5.2.3 Polymerisation

Synthesis impurities that are significantly higher in mass than expected from the FLP can arise due to polymerisation and/or cross-linking reactions [119]. In general, such products are easily amenable to MS characterisation, and it should be possible to use any of a number of sequencing strategies (discussed in subsequent sections) to identify these products.

4.5.2.4 Depurination/Deamination

Depurination and/or deamination can result from improper synthesis conditions, sample purification conditions, or sample storage. Base depurination of the FLP will generate secondary peaks 135 Da (Ade) or 151 Da (Gua) less than the major peak along with dehydrated versions of the same. These effects are easily identified by MS, either ESI- or MALDI-MS (Figure 4.9) [125, 134]. Care must be taken when examining depurination peaks in ESI-MS. Nucleobase loss from negatively charged oligodeoxynucleotides or phosphorothioates can occur within the electrospray source region, and such depurinated products could be interpreted as arising from synthesis or sample handling. Often source-derived depurination can be identified by varying the instrument tuning parameters to reduce any in-source CID of the sample [135].

4.5.2.5 Phosphorothioate Impurities

Phosphorothioates have been extensively characterised, and, in addition to the various impurities described above, these oligonucleotides yield additional impurities (see Chapter 2 for more details). Such impurities include n − 1 terminal thio-monophosphates, which arise from the direct coupling of nucleoside phosphoramidites to the silanol groups of the solid support or reaction of the residual phosphitylating agent present in the phosphoramidite synthons, and desulfurised FLP (containing one or multiple insertions of the phosphodiester linkages). These impurities again can be

Figure 4.9 Experimental mass spectra for unpurified 5-hydroxyuracil (HOU)-containing oligonucleotide, purified oligonucleotide, and oligonucleotides damaged by depurination and oxidation. (a) Oligo sequence can be read from the crude DMT-off 12mer HOU with the failure sequence ladder. Note that the numbering of the sequence is from the 30 end, opposite from the standard 50 numbering but reflecting the direction of synthesis. (b) Pure 12mer HOU. (c) Depurination of the oligo was detected by MALDI-TOFMS. Peak 1: 12mer with HOU; peak 2: 12mer with HOU minus A; peak 3: 12mer with HOU minus G; peak 4: 12mer with HOU minus G minus H_2O; peak 5: 12mer with HOU minus G minus A; peak 6: 12mer with HOU minus G minus A minus H_2O; peak 7: 12mer with HOU minus G minus A minus 2H2O. (d) Oxidation product of 12mer HOU was detected by MALDI-MS. The peak that is 12 Da smaller than 12mer HOU corresponds to 5-hydroxyhydantoin. The −68-Da peak corresponds to the ring-fragmented urea derivative. (Source: Cui, Z., Theruvathu, J.A., Farrel, A., Burdzy, A., Sowers, L.C., Characterization of synthetic oligonucleotides containing biologically important modified bases by matrix-assisted laser desorption/ionization time-of-flight mass spectrometry, *Anal. Biochem.*, 2008, 379, 196–207 with permission from Elsevier.)

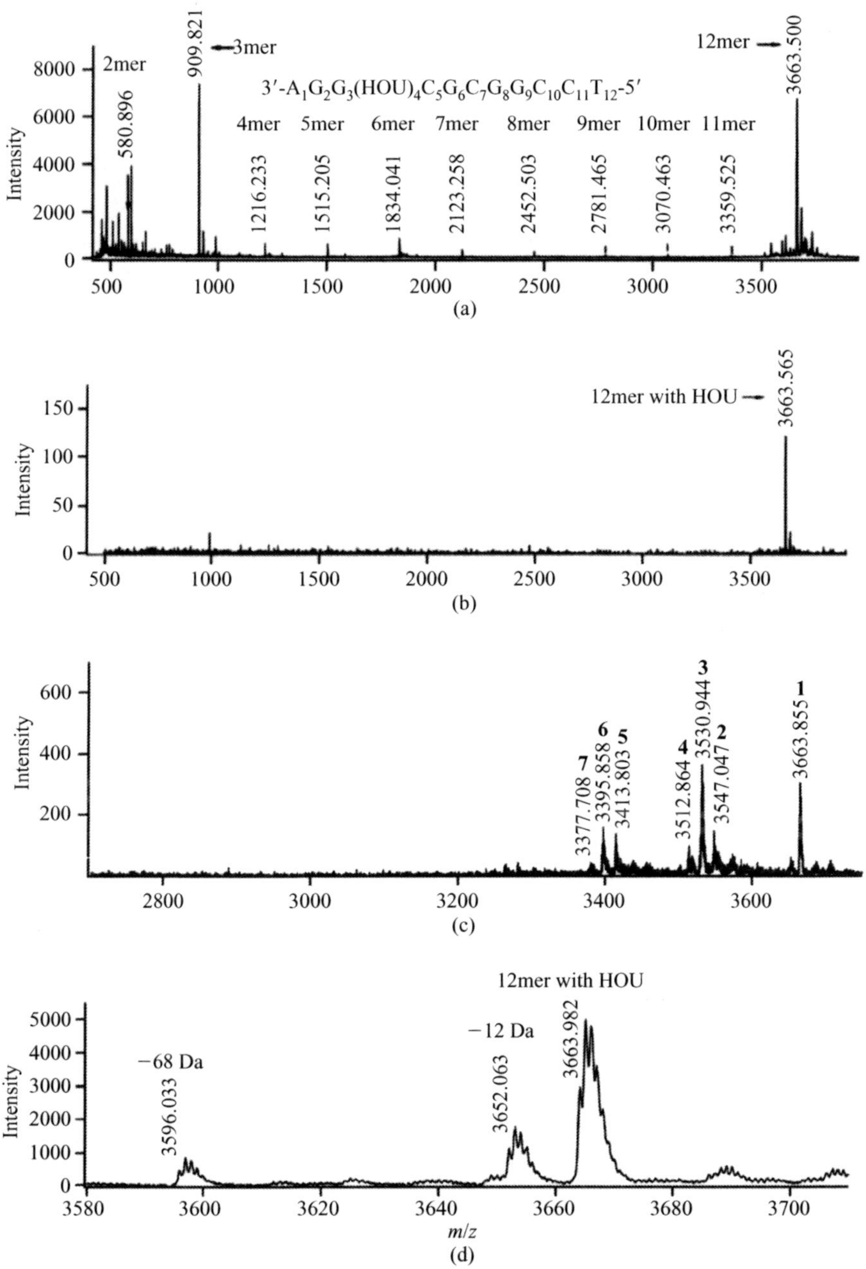
2mer
580.896
909.821
3mer
3'-A1G2G3(HOU)4C5G6C7G8G9C10C11T12-5'
4mer
1216.233
5mer
1515.205
6mer
1834.041
7mer
2123.258
8mer
2452.503
9mer
2781.465
10mer
3070.463
11mer
3359.525
12mer
3663.500
Intensity
(a)
12mer with HOU
3663.565
Intensity
(b)
7
3377.708
6
3395.858
5
3413.803
4
3512.864
3
3530.944
2
3547.047
1
3663.855
Intensity
(c)
12mer with HOU
−68 Da
3596.033
−12 Da
3652.063
3663.982
Intensity
m/z
(d)

characterised by either MALDI- or ESI-MS (with or without on-line HPLC separation). For example, Capaldi *et al.* [118] used HPLC-MS to identify a series of 5′-phosphonate modified phosphorothioates. MS analysis was straightforward due to the presence of sequencing ladders that could be used to localise sites of phosphorothioate and phosphonate linkages (Figure 4.10 and Table 4.3).

4.5.3 Impurity Profiling

In addition to using MS to identify specific classes of synthesis impurities, recently there have been several reports describing the combination of HPLC with high-resolution MS for impurity profiling [135, 136]. In these reports, the use of a higher performance mass spectrometer makes it possible to identify synthesis impurities in a single HPLC-MS or HPLC-MS-MS experiment. Profiling is possible because the accurate mass measurement of impurity peaks allows such peaks to be identified on the basis of their elemental formulas (Table 4.4), which can then be confirmed when necessary by MS-MS data.

4.5.4 Quantitative Analysis of Impurities

In general, MS methods are not recommended for the quantitative analysis of synthesis impurities. In MS, ion abundance is related to both the concentration of the analyte in the sample and the ionisation efficiency of the analyte. As ionisation efficiencies are influenced by the chemical nature of the analyte, impurities that are, for example, more (or less) hydrophobic than the FLP will generate different ion responses than the FLP, even when present at the same concentration [126, 137]. If

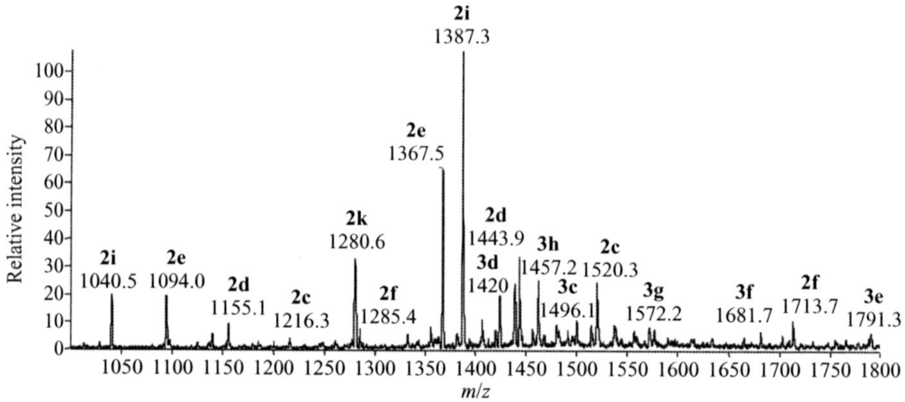

Figure 4.10 Electrospray ionisation mass spectrum of late-eluters 2 and 3. Shown are -5, -4 and -3 charge states. Specific sequences of oligonucleotides are listed in Table 4.3. (Source: Capaldi, D.C., Gaus, H.J., Carty, R.L., Moore, M.N., Turney, B.J., Decottignies, S.D., McArdle, J.V., Scozzari, A.N., Ravikumar, V.T., Krotz, A.H., Formation of 4,4'-dimethoxy-trityl-C-phosphonate oligonucleotides, *Bioorg. Med. Chem. Lett.*, 2004, 14, 4683–4690 with permission from Elsevier.)

Table 4.3 Comparison of masses found in the late-eluters **2** and **3** and deletion sequences of PS-d(GCCCAAGCTGGCATCCGTCA) (**1**). (Source: Capaldi, D.C., Gaus, H.J., Carty, R.L., Moore, M.N., Turney, B.J., Decottignies, S.D., McArdle, J.V., Scozzari, A.N., Ravikumar, V.T., Krotz, A.H., Formation of 4,4'-dimethoxytrityl-C-phosphonate oligonucleotides, *Bioorg. Med. Chem. Lett.*, 2004, 14, 4683–4690 with permission from Elsevier.)

Compound, mass	**1 and its 5′-deletion sequences, mass (calc'd)**	**Δ*m***
2a, (not found)	GCCCAAGCTGGCATCCGTCA, 6368	
3a, (not found)*		271
2b, (not found)	CCCAAGCTGGCATCCGTCA, 6023	
3b, (not found)		
2c, 6085	CCAAGCTGGCATCCGTCA, 5718	367
3c, 5990		272
2d, 5780	CAAGCTGGCATCCGTCA, 5412	368
3d, 5684		272
2e, 5474	AAGCTGGCATCCGTCA, 5107	367
3e, 5375		268
2f, 5144	AGCTGGCATCCGTCA, 4778	366
3f, 5048		270
2g, (not found)	GCTGGCATCCGTCA, 4449	
3g, 4720		271
2h, (not found)	CTGGCATCCGTCA, 4103	
3h, 4373		270
2i, 4166	TGGCATCCGTCA, 3798	367
2k, 3845	GGCATCCGTCA, 3478	367

*Note: 5′-*O*-DMTr-3a (6941 Da) is the most abundant DMTr-*C*-phosphonate diester **3** in crude product.

Table 4.4. Impurities found for FLP PS-d(TCGTCGTTTTGTCGTTTTGTCGTT) during single-run HPLC-MS impurity profiling by high-resolution MS. While all short-mers were detected, only (n − 1) included as representative in the table. DMTr: 4,4-dimethoxytrityl; CE: 2-cyanoethyl; IBU: isobutyryl. (Source: Nikcevic, I., Wyrzykiewicz, T.K., Limbach, P.A., Detecting low-level synthesis impurities in modified phosphorothioate oligonucleotides using liquid chromatography-high resolution mass spectrometry, *Int. J. Mass Spectrom.*, 2011, 304, 98–104 with permission from Elsevier.)

Compound	***m/z* (observed)**	***m/z* (theoretical)**
n − 1 short-mer	1842.15354	1842.15287
(n − 1) PS	1866.13788	1866.13872
FLP depurination product	1888.39372	1888.39506
FLP with PO linkage	1918.17083	1918.16439
FLP CE adduct	1935.66233	1935.66722
FLP IBU adduct	1939.93134	1939.92979
FLP chloral adduct	1958.90085	1958.88796
5′-DMTr OGN	1997.94883	1997.94321

quantitative analysis of synthesis impurities is desired, then appropriate internal standards [138] or calibration curves [139] must be developed. A more detailed discussion of the use of MS for the quantitative analysis of oligonucleotides is presented later in this chapter.

4.5.5 Oligonucleotide Sequencing

Sequence information from synthetic oligonucleotides can be obtained using several different approaches. Two of the easier approaches to implement are failure sequence analysis and exonuclease digestion. These approaches are well suited for MALDI-MS as they depend on the formation of a 'mass ladder' for determining oligonucleotide sequence [21, 140]. However, when the original synthesis product has been purified or when it contains backbone or other modifications that inhibit enzymatic digestion, alternative instrumental methods for oligonucleotide sequencing, such as CID-MS, are required.

4.5.5.1 Failure Sequence Analysis

One of the simplest methods for characterising the chain length and sequence of synthetic oligonucleotides is the detection of failure sequences from the original synthesis step [20, 124, 141, 142]. This procedure takes advantage of the fact that automated solid-phase synthesis of oligonucleotides, especially those that contain modified internucleotide linkages such as methylphosphonates or phosphorothioates, is not 100% efficient. Sequence determination of oligonucleotides by an analysis of the failure sequences is an extremely simple and straightforward method. The mass spectrum will contain a series of peaks that corresponds to the final product and to each one of the failure sequences, each of which differs in mass by the appropriate nucleotide residue value (Figure 4.11). The sequence of the oligonucleotide is determined in the 5′ to 3′ direction from the mass ladder of the synthesis failure products.

Sequence information arising from the formation of a mass ladder is obtained by determining the mass difference between successive peaks in the mass spectrum. As is the case for all sequence determination methods that rely on the mass measurements of successive *n*-mers, oligodeoxynucleotides are easier to characterise due to the relatively large differences in mass among the four oligodeoxynucleotide residues. The ribonucleotide residues are more difficult to sequence by creating mass ladders due to the small mass difference between Ura(U) and Cyt(C) containing nucleotides (a mass difference of only 1 Da). Here, methods take advantage of higher mass accuracy measurement, uridine-specific cleavage [143], or differences in ion abundances [144] to correctly distinguish Us from Cs. Furthermore, all mass ladder methods have a distinct advantage for sequence determination because it is the difference in two mass measurements that results in the desired information – the identity of the nucleotide residue.

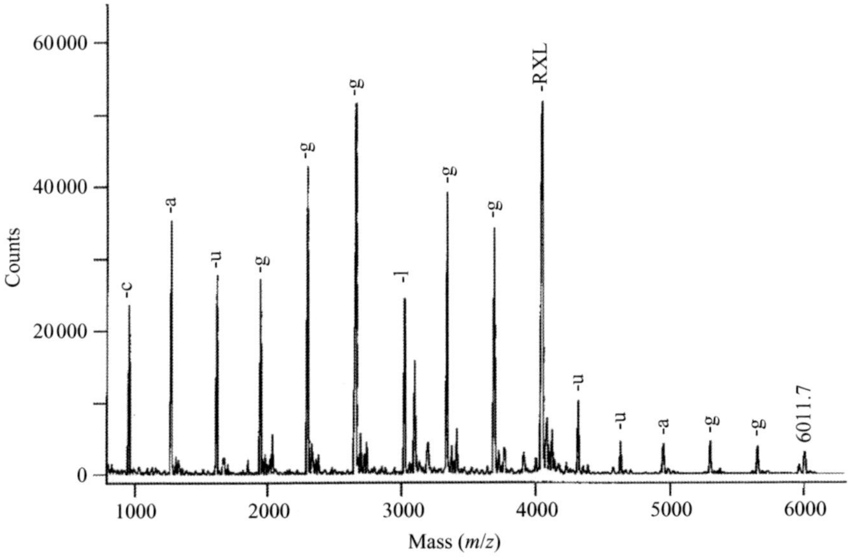

Figure 4.11 Positive ion MALDI-TOF mass spectrum of extracted failure sequences from the crude synthesis of a 17mer (5′ ggauu(RXL)gglggguacagT 3′) composed primarily of 2′-*O*-methylribonucleotides and a single non-nucleosidic linker. Note: upper case letters denote deoxyribonucleotides, lowercase letters denote 2′-*O*-methylribonucleotides, and RXL denotes non-nucleosidic linker. (Source: Alazard, D., Filipowsky, M., Raeside, J., Clarke, M., Majilessi, M., Russell, J., Weisburg, W., Sequencing of production-scale synthetic oligonucleotides by enriching for coupling failures using matrix-assisted laser desorption/ionization time-of-flight mass spectrometry, *Anal. Biochem.*, 2002, 301, 57–64 with permission from Elsevier.)

4.5.5.2 Exonuclease Digestion

The use of exonucleases to generate mass ladders of oligonucleotides that are suitable for analysis by MS is now a standard method for sequencing small to moderate length unmodified oligonucleotides (up to 30-mer for L-TOF without DE and 50-mer for DE-TOF) [21, 23]. Unlike the failure sequence analysis method described above, this approach is also suitable for the sequence analysis of naturally occurring oligonucleotides. Snake venom phosphodiesterase (SVP) is a 3′-5′ exonuclease and will generate 3′-5′ sequence information. Calf spleen phosphodiesterase (CSP) is a 5′-3′ exonuclease and will generate 5′-3′ sequence information.

SVP digestion of an oligonucleotide is generally more rapid than CSP digestion and is not inhibited by the presence of base- or sugar-modified nucleosides. CSP is inhibited by 2′-*O*-methyl-modified nucleosides [26]. Such inhibition may be advantageous when base- and sugar-methyl-modifications must be distinguished in the overall sequence. Digestion with SVP would locate methyl-modified nucleosides, and digestion with CSP would distinguish between a base- or sugar-modification. Typically, complete sequence coverage of moderate length (15–50 nucleotide) oligonucleotides can be achieved within a 20–40 min reaction period, depending on the enzyme and digestion conditions used [143, 145].

4.5.5.3 CID Tandem MS (CID-MS-MS)

In MS-MS, molecules are allowed to dissociate in the gas phase, and analysis of the resulting fragmentation pattern yields insight into the structure and, in the case of oligonucleotides, sequence of the molecule. Thus, CID-MS-MS is a routine approach to characterise synthetic oligonucleotides, both unmodified as well as modified, providing instrumentation with MS-MS capabilities is available [81–83, 146–148]. The fundamentals of oligonucleotide fragmentation have already been covered in an earlier section of this chapter, with an emphasis on fragmentation mechanisms identified based on the type of modification present. Hence, it is now described how such information was used in a specific case to obtain more complete sequencing information from a modified synthetic oligonucleotide.

McLuckey and co-workers [82] have demonstrated the utility of MS-MS for sequencing LNAs, and in the process have explored the influence of parent ion charge state on fragmentation behaviour. Model LNAs could be readily sequenced by CID-MS-MS with no influence of the parent ion charge state on fragmentation. However, when chimeric LNA-DNA oligonucleotide mixmers were analysed, these researchers found that MS-MS information obtained from low parent ion charge states (e.g. -1 or -2) was limited due to fragmentation occurring at the 3′-C-O bond(s) of the deoxynucleotides in the oligonucleotide. Using a specialised MS approach, they then demonstrated that by dissociation of parent ions at higher charge state (> -5) of these same mixmers, more complete fragmentation from both 5′- and 3′-bond cleavages could be obtained. The specialised MS approach involved reducing the charge states of the fragment ions to -1 or -2, which enabled their identification and interpretation on the Q-TOF mass spectrometer used in this study. While such capabilities are not available routinely on most commercial MS instruments, the use of a high-resolution MS platform, such as an FTMS, should also make it possible to identify fragment ion charge states and *m/z* values sufficiently that this general approach could be used by other investigators.

4.5.5.4 Sequencing Modified Oligonucleotides

Several methods have been developed for the sequence characterisation of modified oligonucleotides [75, 107, 125, 149–154]. The approaches used include chemical digestion, enzymatic digestion, CID or various combinations of two or more of these approaches. Because synthetic oligonucleotides for therapeutic applications are often modified to increase biological stability, enzymatic digestion is often of limited utility or used to confirm sequence location of known, unmodified nucleosides. CID, as discussed previously in this chapter, can be used to generate sequence-specific fragment ions from modified oligonucleotides. However, CID spectra from heterogeneously modified oligonucleotides are significantly more difficult to interpret than spectra obtained from homogeneously modified oligonucleotides, such as phosphorothioates or methylphosphonates.

Presently, the most useful approach for sequencing heterogeneously modified oligonucleotides requires a combination of approaches, with chemical digestion being a necessary step for generating sequence-specific information from the sample [152–154]. Farand and co-workers [153, 154] have illustrated the utility of such a combined approach by characterising a set of oligonucleotides that contain a mixture of 2′-deoxy, 2′-fluoro, 2′-*O*-methyl, abasic and ribonucleotides. In their initial publication, a double-stranded-siRNA with these modifications was characterised using a variety of chemicals either alone or in combination as noted in Table 4.5 [153]. By characterising the known and expected reactivity of these chemicals to the specific modified nucleosides present in the sample, a series of digestion products could be generated and analysed using MS. Because MS yields base composition information and mass differences in digestion products yield nucleoside identities, the resulting digestion products could be re-assembled into final sequences for both the guide (sense) and passenger (anti-sense) strands of the ds-siRNA.

This general approach was then expanded to utilise both chemical digestion and precise analysis of nucleoside composition to achieve the *de novo* sequencing of two heavily modified oligonucleotides [154]. As before, chemical digestion along with MS-MS were used to generate a proposed oliognucleotide sequence for the unknown

Table 4.5 Chemical reactivity of modified oligonucleotide (iB-dG-fluU-dG-fluU-fluU-fluU-dG-dG-fluC-dA-dG-fluC-dG-dG-fluC-dA-dA-fluC-fluU-dT-dT-iB) to chemical digestion. iB: reverse abasic nucleotide; dN: deoxynucleotide; fluN: 2′-fluoro nucleotide. *a* denotes no H_2PO_3 at 5′-end; *b* denotes with H_2PO_3 at 5′-end. (Source: Reprinted with permission from Farand, J., Beverly, M., Sequence confirmation of modified oligonucleotides using chemical degradation, electrospray ionization, time-of-flight, and tandem mass spectrometry, *Anal. Chem.*, 80, 7414–7421. Copyright (2008) American Chemical Society.)

N-*x* (5′→3′)	Digested nt (5′→3′)	$C_5H_{11}N$	NaOH	$C_2H_6O_4S$, $C_5H_{11}N$	DEPC, $C_5H_{11}N$	NH_2OH, $C_5H_{11}N$	Aniline
N-1	iB		*a*				
N-2	dG				*b*		*b*
N-3	fluU	*b*	*b*		*b*		
N-4	dG	*b*			*b*		*b*
N-5	fluU	*b*	*b*		*b*	*b*	
N-6	fluU	*b*	*b*		*b*	*b*	
N-7	fluU	*b*	*b*		*b*	*b*	
N-8	dG	*b*			*b*	*b*	*b*
N-9	dG	*b*			*b*	*b*	
N-10	fluC		*b*			*b*	
N-11	dA				*b*		
N-12	dG			*b*	*b*		*b*
N-13	fluC		*b*			*b*	
N-14	dG	*b*		*b*	*b*	*b*	*b*
N-15	dG	*b*		*b*	*b*	*b*	*b*
N-16	fluC		*b*			*b*	
N-17	dA			*b*	*b*		*b*
N-18	dA			*b*	*b*		*b*
N-19	fluC		*b*			*b*	
N-20	fluU	*b*	*b*		*b*	*b*	*b*

sample (Figure 4.12). Then, after digesting the oligonucleotide to nucleosides for analysis by ultra-performance liquid chromatography, quantitative confirmation of the amounts of each individual nucleoside was obtained. These amounts could be converted into relative numbers for each nucleoside, and that information was used to confirm the proposed sequence obtained by chemical digestion and MS-MS. Although not examined in their work, this latter step should be amenable to the absolute quantitation methods described below, which would allow one both to characterise the sequence of modified oligonucleotides and determine the absolute concentration of the oligonucleotide in the sample.

While the approaches described above were conducted using HPLC-MS and HPLC-MS-MS, a similar method for sequencing siRNAs has been reported by Bahr *et*

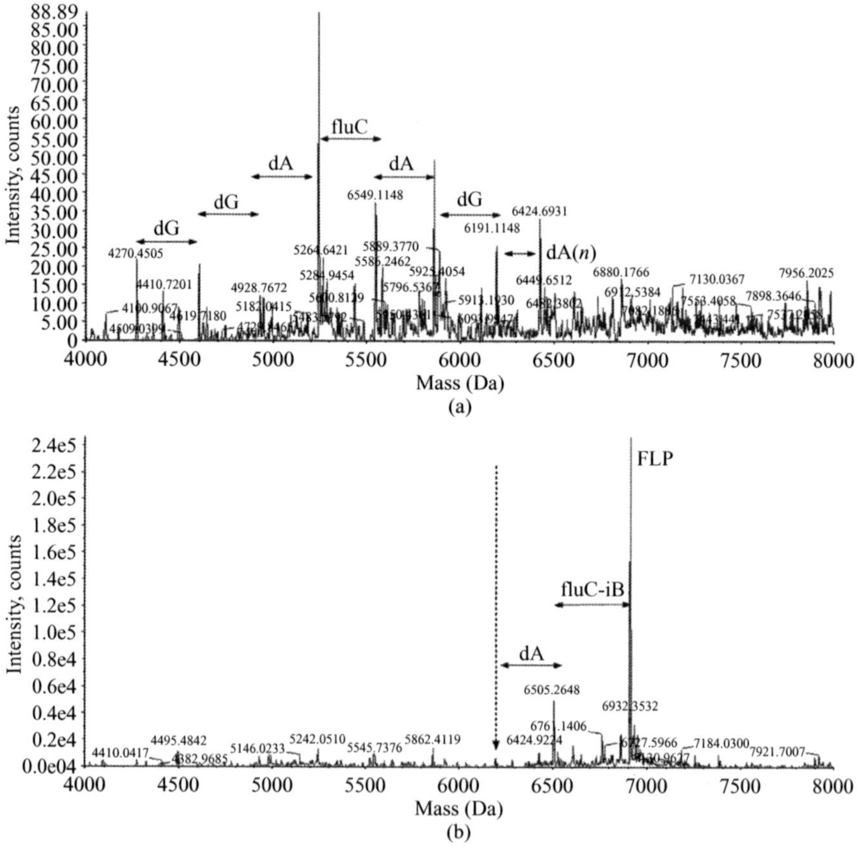

Figure 4.12 Deconvoluted fragments between 4000 and 6912 u after chemical digestion with NH_2OH/piperidine (a,b). Masses corresponding to loss of nucleotides and nucleosides (n, without 5′ phosphate) were observed. (Source: Reprinted with permission from Farand, J., Gosselin, F., De novo sequence determination of modified oligonucleotides, *Anal. Chem.*, 81, 3723–3730. Copyright (2009) American Chemical Society.)

al. [152] using MALDI-MS and MALDI-MS-MS. In that report, only acid hydrolysis was used to chemically digest the siRNA samples. Because MALDI generates singly charged ions, mass ladders could be obtained that enable a straightforward interpretation of the oligonucleotide sequence. However, while this approach is certainly easy to implement and yielded high-quality sequencing ladders (Figure 4.13), the model siRNAs were not heavily modified as compared to those analysed by Farand and co-workers [153, 154], so the overall utility of an acid-only chemical digestion approach is more limited than the various chemical digestion schemes used by those researchers.

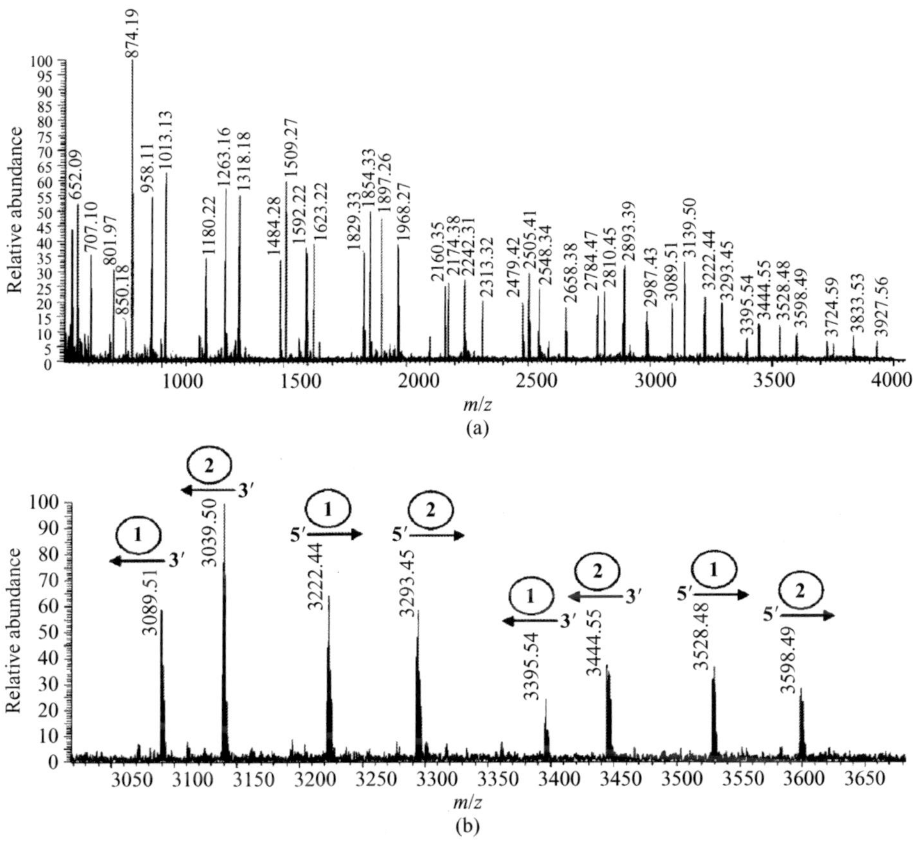

Figure 4.13 (a) MALDI mass spectrum of the 21mer double-strand ON-2AB after acid hydrolysis. (b) Section of the mass spectrum: ① assigns fragments from single-strand ON-2B, ② from ON-2A of the double strand; the arrows indicate 5′-fragments or 3′-fragments, respectively. (Source: Reprinted with permission from Bahr, U., Aygun, H., Karas, M., Sequencing of single- and double-stranded RNA oligonucleotides by acid hydrolysis and MALDI mass spectrometry, *Anal. Chem.*, 81, 3173–3179. Copyright (2009) American Chemical Society.)

4.6 Quantitative Analysis

While the MALDI process itself is not inherently suited for quantitative measurements, quantitative analysis of low molecular weight oligonucleotides has been demonstrated [155–157]. The key to quantitative analysis is the choice of appropriate internal standards or use of isotope labelling. The ideal internal standard would have the same solubility, desorption and ionisation efficiencies as the sample of interest. For those reasons, known, stable oligonucleotides are the best internal standards to use for quantitative analysis of oligonucleotides.

By far the most common MS method for quantitative analysis of oligonucleotides involves the use of targeted scans (e.g. selected reaction monitoring (SRM)) during HPLC-MS-MS with quantitation typically achieved by use of calibration standards [7]. The basis for targeted scan quantification is the identification of an oligonucleotide fragmentation pathway that is unique to the analyte of interest. Using a triple-quadrupole mass spectrometer (typically), SRM assays using the specific fragmentation pathway are readily able to provide limits of quantitation in the range of 1–100 ng/mL. While this approach is typically utilised in the pharmaceutical industry for analyte and metabolite characterisation of oligonucleotides, it can also be used for oligonucleotide quality control when necessary, provided that the appropriate HPLC-MS-MS instrumentation is available.

Other approaches exist for using MS for the relative quantification of oligonucleotides [6, 45, 158–160]. One useful example is the work by Bahr *et al.* [160] who described the relative quantification of siRNA double strands by MALDI-MS. Using ATT as the matrix with at least 100 mM but not more than 150 mM diammonium hydrogen citrate as the co-matrix, native ds-RNAs were readily detected by MALDI-MS. Importantly, because ATT is a neutral, water-soluble matrix, the sample preparation conditions necessary for high-quality MALDI could be adapted to maintain non-covalent interactions that maintain the double-stranded nature of the sample. Other matrices or sample preparation conditions led to denaturing of the ds-RNA. Quantification could then be achieved by using a single-stranded oligonucleotide reference, which was added to the ds-RNA of interest. Analysis of the mixture under native and denaturing conditions enabled the relative quantification of the ds-RNA from the original sample. Because MALDI-MS is often used to confirm the purity and quality of synthetic oligonucleotides, this method could be easily incorporated into analytical laboratories that address ds-RNA products as well as single-stranded oligonucleotides.

Isotope dilution MS (IDMS) has been used for the absolute quantification of oligonucleotides [159, 161, 162]. The general protocol has involved digestion of the target oligonucleotide using phosphodiesterases and/or nucleases with phosphatase treatment to generate a mixture of deoxynucleoside monophosphate or deoxynucleosides that can be separated and analysed using HPLC-MS. IDMS-based quantification requires the addition of a known, isotopically labelled standard that is then used to quantify the desired unknown sample. O'Connor *et al.* [161] described the initial proof-of-concept experiments for this quantitative method. That initial publication focused on the quantitative analysis of deoxynucleoside monophosphates that were generated after enzymatic digestion of a target oligonucleotide (Figure 4.14).

More recently, Kinumi *et al.* [159] compared the results obtained for quantifying a 21mer failure sequence product in a mixture with a 20mer synthetic oligonucleotide target sequence using a variety of analytical methods. They reported that the IDMS technique provided good agreement with quantitative data obtained by either inductively coupled plasma-optical emission spectroscopy (measuring total phosphorus content) or ion chromatography. They also noted that the IDMS technique was more accurate than conventional UV-based quantification approaches.

4.7 Future Developments

Ion mobility-MS (IM-MS) is a relatively recent commercialised instrument that has proven to provide unique benefits for the separation and analysis of complex mixtures of biomolecules [163, 164]. Unlike more conventional chromatographic separations, IM provides for the separation of gas-phase ions based roughly on the size and shape of the ion. Separation is effected by moving the ion through a neutral background gas. Ions with low surface area (i.e. small or compact ions) will undergo fewer collisions

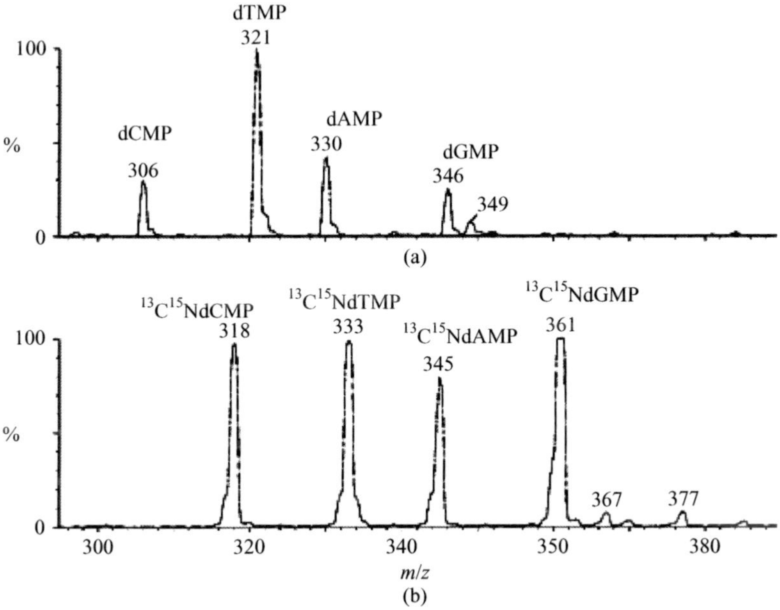

Figure 4.14 Electrospray mass spectrum of (a) natural deoxynucleotides and (b) isotopically enriched deoxynucleotides. (Source: Reprinted with permission from O'Connor, G., Dawson, C., Woolford, A., Webb, K.S., Catterick, T., Quantitation of oligonucleotides by phosphodiesterase digestion followed by isotope dilution mass spectrometry: proof of concept, *Anal. Chem.*, 74, 3670–3676. Copyright (2002) American Chemical Society.)

with the background gas than ions with high surface area (i.e. large or denatured ions). Because the time-frame for IM separation is compatible with both upstream chromatographic separation and downstream MS analysis, HPLC-IM-MS-(MS) offers unique possibilities for multi-dimensional separation and characterisation of complex mixtures.

To date, the use of IM-MS for analysing oligonucleotides has been limited [165–169], and primarily these reports are focused more on the capabilities of IM-MS, *per se*, than on any applications related to the analysis of synthetic oligonucleotides. Among the most recent publications, Fenn *et al.* [170] demonstrated that the collision cross-section of oligonucleotides (deoxynucleotide trimers and tetramers of various isomeric sequences) is sufficiently different from other biomolecules to enable their separation via IM-MS. Moreover, certain isomers (e.g. 5′-d(CGAT)-3′ and 5′-d(TGCA)-3′) adopted different gas-phase conformations as they yielded different collisional cross sections within IM. These results are promising, as they suggest that isomeric oligonucleotides (which cannot be differentiated by MS alone) may be separated under appropriate IM-MS conditions. Even more intriguing is the possibility of using IM-MS to characterise double-stranded oligonucleotides, as well as its potential for characterising polyplexes used as drug delivery vehicles for nucleic acids [171].

Acknowledgments

The author thanks K.W. Gaston for generating and providing Figure 4.3. Financial support of the author's research in oligonucleotide mass spectrometry is provided by the National Institutes of Health (GM58843) and the National Science Foundation (CHE0910751).

References

1. Wilson, C., O'Keefe, A.D., Building oligonucleotide therapeutics using non natural chemistries, *Curr. Opin. Chem. Biol.*, 2006, 10, 607–614.
2. Huber, C.G., Oberacher, H., Analysis of nucleic acids by on-line liquid chromatography-mass spectrometry, *Mass Spectrom. Rev.*, 2001, 20, 310–343.
3. Kenseth, J.R., He, Y., Tallman, D., Pang, H.M., Coldiron, S.J., Techniques for high-throughput characterization of peptides, oligonucleotides and catalysis efficiency, *Curr. Opin. Chem. Biol.*, 2004, 8, 327–333.
4. Banoub, J.H., Newton, R.P., Esmans, E., Ewing, D.F., Mackenzie, G., Recent developments in mass spectrometry for the characterization of nucleosides, nucleotides, oligonucleotides, and nucleic acids, *Chem. Rev.*, 2005, 105, 1869–1915.
5. Hofstadler, S.A., Sannes-Lowery, K.A., Hannis, J.C., Analysis of nucleic acids by FTICR MS, *Mass Spectrom. Rev.*, 2005, 24, 265–285.
6. Willems, A.V., Deforce, D.L., Van Peteghem, C.H., Van Bocxlaer, J.F., Analysis of nucleic acid constituents by on-line capillary electrophoresis-mass spectrometry, *Electrophoresis*, 2005, 26, 1221–1253.

7. Lin, Z., Li, W., Dai, G., Application of LC-MS for quantitative analysis and metabolite identification of therapeutic oligonucleotides, *J. Pharm. Biomed. Anal.*, 2007, 44, 330–341.
8. Sauer, S., The essence of DNA sample preparation for MALDI mass spectrometry, *J. Biochem. Biophys. Methods*, 2007, 70, 311–318.
9. Banoub, J.H., Limbach, P.A., *Mass Spectrometry of Nucleosides and Nucleic Acids*, 2010, p. 492, CRC Press, Taylor & Francis Group, Boca Raton, Florida, USA.
10. Beverly, M.B., Applications of mass spectrometry to the study of siRNA, *Mass Spectrom. Rev.*, 2011, 30, 979–998.
11. Oberacher, H., Frontiers of mass spectrometry in nucleic acids analysis, *Eur. J. Mass Spectrom.*, 2010, 16, 351–365.
12. Yamashita, M., Fenn, J.B., Electrospray ion source. Another variation on the free-jet theme, *J. Phys. Chem.*, 1984, 88, 4451–4459.
13. Yamashita, M., Fenn, J.B., Negative ion production with the electrospray ion source, *J. Phys. Chem.*, 1984, 88, 4671–4675.
14. Karas, M., Bachmann, D., Bahr, U., Hillenkamp, F., Matrix-assisted ultraviolet laser desorption of non-volatile compounds, *Int. J. Mass Spectrom. Ion Processes*, 1987, 78, 53–68.
15. Karas, M., Hillenkamp, F., Laser desorption ionization of proteins with molecular masses exceeding 10,000 daltons, *Anal. Chem.*, 1988, 60, 2299–2301.
16. Tanaka, K., Waki, H., Ido, Y., Akita, S., Yoshida, Y., Yoshida, T., Protein and polymer analyses up to *m/z* 100,000 by laser ionization time-of-flight mass spectrometry, *Rapid Communs Mass Spectrom.*, 1988, 2, 151–153.
17. Hossain, M., Limbach, P.A., A comparison of MALDI matrices. In *Electrospray and MALDI Mass Spectrometry: Fundamentals, Instrumentation, Practicalities, and Biological Applications*, Cole, R.B. (Ed.), 2010, pp. 215–244, John Wiley, Hoboken, NJ, USA.
18. Knochenmuss, R., Ion formation mechanisms in UV-MALDI, *Analyst*, 2006, 131, 966–986.
19. Maxam, A.M., Gilbert, W., Sequencing end-labeled DNA with base-specific chemical cleavages. In *Methods in Enzymology*, Grossman, L., Moldave, K., (Eds), 1980, pp. 499–560, Academic Press, London/New York.
20. Keough, T., Baker, T.R., Dobson, R.L.M., Lacey, M.P., Riley, T.A., Hasselfield, J.A., Hesselberth, P.E., Antisense DNA oligonucleotides II: The use of matrix-assisted laser desorption/ionization mass spectrometry for the sequence verification of methylphosphonate oligodeoxyribonucleotides, *Rapid Communs Mass Spectrom.*, 1993, 7, 195–200.
21. Limbach, P.A., Indirect mass spectrometric methods for characterizing and sequencing oligonucleotides, *Mass Spectrom. Rev.*, 1996, 15, 297–336.
22. Smirnov, I.P., Roskey, M.T., Juhasz, P., Takach, E.J., Martin, S.A., Haff, L.A., Sequencing oligonucleotides by exonuclease digestion and delayed extraction matrix-assisted laser desorption ionization time-of-flight mass spectrometry, *Anal. Biochem.*, 1996, 238, 19–25.
23. Zhang, L., Ren, Y., Rempel, D., Taylor, J., Gross, M., Determination of photomodified oligodeoxynucleotides by exonuclease digestion, matrix-assisted laser desorption/ionization and post-source decay mass spectrometry, *J. Am. Soc. Mass Spectrom.*, 2001, 12, 1127–1135.
24. Holle, A., Haase, A., Kayser, M., Hohndorf, J., Optimizing UV laser focus profiles for improved MALDI performance, *J. Mass Spectrom.*, 2006, 41, 705–716.
25. Knochenmuss, R., Dubois, F., Dale, M.J., Zenobi, R., The matrix suppression effect and ionization mechanisms in matrix-assisted laser desorption/ionization, *Rapid Communs Mass Spectrom.*, 1996, 10, 871–877.
26. Pieles, U., Zurcher, W., Schar, M., Moser, H.E., Matrix-assisted laser desorption ionization time-of-flight mass spectrometry: a powerful tool for the mass and sequence analysis of natural and modified oligonucleotides, *Nucleic Acids Res.*, 1993, 21, 3191–3196.
27. Lavanant, H., Lange, C., Sodium-tolerant matrix for matrix-assisted laser desorption/ionization mass spectrometry and post-source decay of oligonucleotides, *Rapid Commun. Mass Spectrom.*, 2002, 16, 1928–1933.
28. Lecchi, P., Le, H., Pannell, L., 6-Aza-2-thiothymine: a matrix for MALDI spectra of oligonucleotides, *Nucleic Acids Res.*, 1995, 23, 1276–1277.
29. Fu, Y., Xu, S., Pan, C., Ye, M., Zou, H., Guo, B., A matrix of 3,4-diaminobenzophenone for the

analysis of oligonucleotides by matrix-assisted laser desorption/ionization time-of-flight mass spectrometry, *Nucleic Acids Res.*, 2006, 34, e94.

30. Zhu, Y.F., Taranenko, N.I., Allman, S.L., Martin, S.A., Haff, L., Chen, C.H., The effect of ammonium salt and matrix in the detection of DNA by MALDI-TOF mass spectroscopy, *Rapid Communs Mass Spectrom.*, 1996, 10, 1591–1596.
31. Simmons, T.A., Limbach, P.A., The use of a co-matrix for improved MALDI-TOFMS analysis of oligonucleotides, *Rapid Communs Mass Spectrom.*, 1997, 11, 567–572.
32. Simmons, T.A., Limbach, P.A., Influence of co-matrix proton affinity on oligonucleotide ion stability in matrix-assisted laser desorption/ionization time-of-flight mass spectrometry, *J. Am. Soc. Mass Spectrom.*, 1998, 9, 668–675.
33. Asara, J.M., Allision, J., Enhanced detection oligonucleotides in UV MALDI MS using the tetraamine spermine as a matrix additive, *Anal. Chem.*, 1999, 71, 2866–2870.
34. Vandell, V.E., Limbach, P.A., Polyamine co-matrices for matrix-assisted laser desorption/ionization mass spectrometry of oligonucleotides, *Rapid Communs Mass Spectrom.*, 1999, 13, 2014–2021.
35. Stahl, B., Steup, M., Karas, M., Hillenkamp, F., Analysis of neutral oligosaccharides by matrix-assisted laser desorption/ionization mass spectrometry, *Anal. Chem.*, 1991, 63, 1463–1466.
36. Dai, Y., Whittal, R., Li, L., Confocal fluorescence microscopic imaging for investigating the analyte distribution in MALDI matrices, *Anal. Chem.*, 1996, 68, 2494–2500.
37. Xiang, F., Beavis, R., A method to increase contaminant tolerance in protein matrix-assisted laser desorption/ionization by the fabrication of thin protein-doped polycrystalline films, *Rapid Communs Mass Spectrom.*, 1994, 8, 199–204.
38. Vorm, O., Roepstorff, P., Mann, M., Improved resolution and very high sensitivity in MALDI TOF of matrix surfaces made by fast evaporation, *Anal. Chem.*, 1994, 66, 3281–3287.
39. Li, L., Golding, R., Whittal, R., Analysis of single mammalian cell lysates by mass spectrometry, *J. Am. Chem. Soc.*, 1996, 118, 11662–11663.
40. Kussmann, M., Nordhoff, E., Rahbek-Nielsen, H., Haebel, S., Rossel-Larsen, M., Jakobsen, L., Gobom, J., Mirgorodskaya, E., Kroll-Kristensen, A., Palm, L., Roepstorff, P., Matrix-assisted laser desorption/ionization mass spectrometry sample preparation techniques designed for various peptide and protein analytes, *J. Mass Spectrom.*, 1997, 32, 593–601.
41. Wasserscheid, P., Keim, W., Ionic liquids – New 'solutions' for transition metal catalysis, *Angew. Chem. Int. Ed.*, 2000, 39, 3773–3789.
42. Welton, T., Room-temperature ionic liquids. solvents for synthesis and catalysis, *Chem. Rev.*, 1999, 99, 2071–2083.
43. Armstrong, D., Zhang, L., He, L., Gross, M., Ionic liquids as matrixes for matrix-assisted laser desorption/ionization mass spectrometry, *Anal. Chem.*, 2001, 73, 3679–3686.
44. Tholey, A., Heinzle, E., Ionic (liquid) matrices for matrix-assisted laser desorption/ionization mass spectrometry-applications and perspectives, *Anal. Bioanal. Chem.*, 2006, 386, 24–37.
45. Li, Y.L., Gross, M.L., Ionic-liquid matrices for quantitative analysis by MALDI-TOF mass spectrometry, *J. Am. Soc. Mass Spectrom.*, 2004, 15, 1833–1837.
46. Jones, J.J., Batoy, S.M., Wilkins, C.L., Liyanage, R., Lay, J.O., Jr, Ionic liquid matrix-induced metastable decay of peptides and oligonucleotides and stabilization of phospholipids in MALDI FTMS analyses, *J. Am. Soc. Mass Spectrom.*, 2005, 16, 2000–2008.
47. Carda-Broch, S., Berthod, A., Armstrong, D.W., Ionic matrices for matrix-assisted laser desorption/ionization time-of-flight detection of DNA oligomers, *Rapid Communs Mass Spectrom.*, 2003, 17, 553–560.
48. Fenn, J.B., Mann, M., Meng, C.K., Wong, S.F., Whitehouse, C.M., Electrospray ionization for mass spectrometry of large biomolecules, *Science*, 1989, 246, 64–71.
49. Covey, T.R., Bonner, R.F., Shushan, B.I., Henion, J., The determination of protein, oligonucleotide, and peptide molecular weights by ion-spray mass spectrometry, *Rapid Communs Mass Spectrom.*, 1988, 2, 249–256.
50. Gaskell, S.J., Electrospray: Principles and practice, *J. Mass Spectrom.*, 1997, 32, 677–688.
51. Wilm, M., Mann, M., Analytical properties of the nanoelectrospray ion source, *Anal. Chem.*, 1996, 68, 1–8.

52. Manduzio, H., Martelet, A., Ezan, E., Fenaille, F., Comparison of approaches for purifying and desalting polymerase chain reaction products prior to electrospray ionization mass spectrometry, *Anal. Biochem.*, 2010, 398, 272–274.
53. Brown, R.S., Lennon, J.J., Mass resolution improvement by incorporation of pulsed ion extraction in a matrix-assisted laser desorption ionization time-of-flight mass spectrometer, *Anal. Chem.*, 1995, 67, 1998–2003.
54. Colby, S.M., King, T.B., Reilly, J.P., Improving the resolution of matrix-assisted laser desorption/ionization time-of-flight mass spectrometry by exploiting the correlation between ion position and velocity, *Rapid Communs Mass Spectrom.*, 1994, 8, 865–868.
55. Vestal, M.L., Juhasz, P., Martin, S.A., Delayed extraction matrix-assisted laser desorption time-of-flight mass spectrometry, *Rapid Communs Mass Spectrom.*, 1995, 9, 1044–1050.
56. Langley, G., Herniman, J., Davies, N., Brown, T., Simplified sample preparation for the analysis of oligonucleotides by matrix-assisted laser desorption/ionisation time-of-flight mass spectrometry, *Rapid Communs Mass Spectrom.*, 1999, 13, 1717–1723.
57. Spengler, B., Post-source decay analysis in matrix-assisted laser desorption/ionization mass spectrometry of biomolecules, *J. Mass Spectrom.*, 1997, 32, 1019–1036.
58. de Hoffman, E., Stroobant, V., *Mass Spectrometry: Principles and Applications*, 2007, p. 502, John Wiley, Chichester, West Sussex.
59. Dass, C., *Fundamentals of Contemporary Mass Spectrometry*, 2007, p. 608, John Wiley, Hoboken, NJ, USA.
60. Perry, R.H., Cooks, R.G., Noll, R.J., Orbitrap mass spectrometry: instrumentation, ion motion and applications, *Mass Spectrom. Rev.*, 2008, 27, 661–699.
61. Vestal, M.L., Campbell, J.M., Tandem time-of-flight mass spectrometry, *Methods Enzymol.*, 2005, 402, 79–108.
62. Cooper, H.J., Hakansson, K., Marshall, A.G., The role of electron capture dissociation in biomolecular analysis, *Mass Spectrom. Rev.*, 2005, 24, 201–222.
63. Ens, W., Standing, K.G., Hybrid quadrupole/time-of-flight mass spectrometers for analysis of biomolecules, *Methods Enzymol.*, 402, 49–78.
64. Glish, G.L., Burinsky, D.J., Hybrid mass spectrometers for tandem mass spectrometry, *J. Am. Soc. Mass Spectrom.*, 2008, 19, 161–172.
65. Laskin, J., Futrell, J.H., Activation of large ions in FT-ICR mass spectrometry, *Mass Spectrom. Rev.*, 2005, 24, 135–167.
66. Pittenauer, E., Allmaier, G., High-energy collision induced dissociation of biomolecules: MALDI-TOF/RTOF mass spectrometry in comparison to tandem sector mass spectrometry, *Comb. Chem. High Throughput Screen*, 2009, 12, 137–155.
67. Reilly, J.P., Ultraviolet photofragmentation of biomolecular ions, *Mass Spectrom. Rev.*, 2009, 28, 425–447.
68. Rompp, A., Taban, I.M., Mihalca, R., Duursma, M.C., Mize, T.H., McDonnel, L.A., Heeren, R.M., Examples of Fourier transform ion cyclotron resonance mass spectrometry developments: from ion physics to remote access biochemical mass spectrometry, *Eur. J. Mass Spectrom.*, 2005, 11, 443–456.
69. Zubarev, R.A., Electron-capture dissociation tandem mass spectrometry, *Curr. Opin. Biotechnol.*, 2004, 15, 12–16.
70. McLuckey, S.A., Berkel, G.J.V., Glish, G.L., Tandem mass spectrometry of small, multiply charged oligonucleotides, *J. Am. Soc. Mass Spectrom.*, 1992, 3, 60–70.
71. Roepstorff, P., Fohlman, J., Proposal for a common nomenclature for sequence ions in mass spectra of peptides, *Biomed. Mass Spectrom.*, 1984, 11, 601.
72. Krivos, K.L., Addepalli, B., Limbach, P.A., Removal of 3'-phosphate group by bacterial alkaline phosphatase improves oligonucleotide sequence coverage of RNase digestion products analyzed by collision-induced dissociation mass spectrometry, *Rapid Communs Mass Spectrom.*, 2011, 25, 3609–3616.
73. Wu, J., McLuckey, S.A., Gas-phase fragmentation of oligonucleotide ions, *Int. J. Mass Spectrom.*, 2004, 237, 197–241.
74. Tromp, J.M., Schurch, S., Gas-phase dissociation of oligoribonucleotides and their analogs

studied by electrospray ionization tandem mass spectrometry, *J. Am. Soc. Mass Spectrom.*, 2005, 16, 1262–1268.
75. Monn, S.T., Schurch, S., New aspects of the fragmentation mechanisms of unmodified and methylphosphonate-modified oligonucleotides, *J. Am. Soc. Mass Spectrom.*, 2007, 18, 984–990.
76. Andersen, T.E., Kirpekar, F., Haselmann, K.F., RNA fragmentation in MALDI mass spectrometry studied by H/D-exchange: mechanisms of general applicability to nucleic acids, *J. Am. Soc. Mass Spectrom.*, 2006, 17, 1353–1368.
77. McLuckey, S.A., Viadyanathan, G., Habibi-Goudarzi, S., Charged vs. neutral Nucleobase loss from multiply charged oligonucleotide anions, *J. Mass Spectrom.*, 1995, 30, 1222–1229.
78. Schurch, S., Bernal-Mendez, E., Leumann, C.J., Electrospray tandem mass spectrometry of mixed-sequence RNA/DNA oligonucleotides, *J. Am. Soc. Mass Spectrom.*, 2002, 13, 936–945.
79. Huang, T.Y., Kharlamova, A., Liu, J., McLuckey, S.A., Ion trap collision-induced dissociation of multiply deprotonated RNA: c/y-ions versus (a-B)/w-ions, *J. Am. Soc. Mass Spectrom.*, 2008, 19, 1832–1840.
80. Nyakas, A., Stucki, S.R., Schurch, S., Tandem mass spectrometry of modified and platinated oligoribonucleotides, *J. Am. Soc. Mass Spectrom.*, 2011, 22, 875–887.
81. Ivleva, V.B., Yu, Y.Q., Gilar, M., Ultra-performance liquid chromatography/tandem mass spectrometry (UPLC/MS/MS) and UPLC/MS(E) analysis of RNA oligonucleotides, *Rapid Communs Mass Spectrom.*, 2010, 24, 2631–2640.
82. Huang, T.Y., Kharlamova, A., McLuckey, S.A., Ion trap collision-induced dissociation of locked nucleic acids, *J. Am. Soc. Mass Spectrom.*, 2010, 21, 144–153.
83. Pomerantz, S.C., McCloskey, J.A., Tarasow, T.M., Eaton, B.E., Deconvolution of combinatorial oligonucleotide libraries by electrospray ionization tandem mass spectrometry, *J. Am. Chem. Soc.*, 1997, 119, 3861–3867.
84. Limbach, P.A., Matrix-assisted laser desorption-ionization mass spectrometry: an overview, *Spectroscopy*, 1998, 13, 17–27.
85. Limbach, P.A., Crain, P.F., McCloskey, J.A., Molecular mass measurement of intact ribonucleic acids via electrospray ionization quadrupole mass spectrometry, *J. Am. Soc. Mass Spectrom.*, 1995, 6, 27–39.
86. Ragas, J.A., Simmons, T.A., Limbach, P.A., A comparative study on methods of optimal sample preparation for the analysis of oligonucleotides by matrix-assisted laser desorption/ionization mass spectrometry, *Analyst*, 2000, 125, 575–581.
87. Castleberry, C.M., Rodicio, L.P., Limbach, P.A., Electrospray ionization mass spectrometry of oligonucleotides, *Curr. Protoc. Nucleic Acid Chem.*, 2008, Chapter 10, Unit 10 12.
88. Castleberry, C.M., Chou, C.W., Limbach, P.A., Matrix-assisted laser desorption/ionization time-of-flight mass spectrometry of oligonucleotides, *Curr. Protoc. Nucleic Acid Chem.*, 2008, Chapter 10, Unit 10 11.
89. Nordhoff, E., Schürenberg, M., Thiele, G., Lübbert, C., Kloeppel, K.-D., Theiss, D., Lehrach, H., Gobom, J., Sample preparation protocols for MALDI-MS of peptides and oligonucleotides using prestructured sample supports, *Int. J. Mass Spectrom.*, 2003, 226, 163–180.
90. Xu, Y., Bruening, M.L., Watson, J.T., Non-specific, on-probe cleanup methods for MALDI-MS samples, *Mass Spectrom. Rev.*, 2003, 22, 429–440.
91. Shaler, T.A., Wickham, J.N., Sannes, K.A., Wu, K.J., Becker, C.H., Effect of impurities on the matrix-assisted laser desorption mass spectra of single-stranded oligodeoxynucleotides, *Anal. Chem.*, 1996, 68, 576–579.
92. Currie, G.J., Yates, J.R., Analysis of oligonucleotides by negative-ion matrix-assisted laser desorption mass spectrometry, *J. Am. Soc. Mass Spectrom.*, 1993, 4, 955–963.
93. Cheng, S.-W., Chan, T.W.D., Use of ammonium halides as co-matrices for matrix-assisted laser desorption/ionization studies of oligonucleotides, *Rapid Communs Mass Spectrom.*, 1996, 10, 907–910.
94. Muddiman, D.C., Cheng, X., Udseth, H.R., Smith, R.D., Charge-state reduction with improved signal intensity of oligonucleotides in electrospray ionization mass spectrometry, *J. Am. Soc. Mass Spectrom.*, 1996, 7, 697–706.
95. Liu, C., Muddiman, D.C., Tang, K., Smith, R.D., Improving the microdialysis procedure for

electrospray ionization mass spectrometry of biological samples, *J. Mass Spectrom.*, 1997, 32, 425–431.
96. Liu, C., Wu, Q., Harms, A., Smith, R., On-line microdialysis sample cleanup for electrospray ionization mass spectrometry of nucleic acids, *Anal. Chem.*, 1996, 68, 3295–3299.
97. Muddiman, D.C., Anderson, G.A., Hofstadler, S.A., Smith, R.D., Length and base composition of PCR-amplified nucleic acids using mass measurements from electrospray ionization mass spectrometry, *Anal. Chem.*, 1997, 69, 1543–1549.
98. Null, A.P., Hannis, J.C., Muddiman, D.C., Preparation of single-stranded PCR products for electrospray ionization mass spectrometry using the DNA repair enzyme lambda exonuclease, *Analyst*, 2000, 125, 619–625.
99. Hannis, J.C., Muddiman, D.C., Characterization of a microdialysis approach to prepare polymerase chain reaction products for electrospray ionization mass spectrometry using on-line ultraviolet absorbance measurements and inductively coupled plasma-atomic emission spectroscopy, *Rapid Communs Mass Spectrom.*, 1999, 13, 323–330.
100. Little, D.P., Thannhauser, T.W., McLafferty, F.W., Verification of 50- to 100-mer DNA and RNA sequences with high-resolution mass spectrometry, *Proc. Natl Acad. Sci. USA*, 1995, 92, 2318–2322.
101. Greig, M., Griffey, R.H., Utility of organic bases for improved electrospray mass spectrometry of oligonucleotides, *Rapid Communs Mass Spectrom.*, 1995, 9, 97–102.
102. Greig, M.J., Gaus, H.J., Griffey, R.H., Negative-ionization micro electrospray mass spectrometry of oligonucleotides and their complexes, *Rapid Communs Mass Spectrom.*, 1996, 10, 47–50.
103. Griffey, R.H., Sasmor, H., Greig, M.J., Oligonucleotide charge states in negative ionization electrospray-mass spectrometry are a function of solution ammonium ion concentration, *J. Am. Soc. Mass Spectrom.*, 1997, 8, 155–160.
104. Barry, J.P., Muth, J., Law, S.-J., Karger, B.L., Vouros, P., Analysis of modified oligonucleotides by capillary electrophoresis in a polyvinylpyrrolidone matrix coupled with electrospray mass spectrometry, *J. Chromatogr. A*, 1996, 732, 159–166.
105. Ding, J., Vouros, P., Capillary electrochromatography and capillary electrochromatography–mass spectrometry for the analysis of DNA adduct mixtures, *Anal. Chem.*, 1997, 69, 379–384.
106. Oberacher, H., Niederstätter, H., Casetta, B., Parson, W., Some guidelines for the analysis of genomic DNA by PCR-LC-ESI-MS, *J. Am. Soc. Mass Spectrom.*, 2006, 17, 124–129.
107. Fountain, K.J., Gilar, M., Gebler, J.C., Analysis of native and chemically modified oligonucleotides by tandem ion-pair reversed-phase high-performance liquid chromatography/electrospray ionization mass spectrometry, *Rapid Communs Mass Spectrom.*, 2003, 17, 646–653.
108. Gilar, M., Fountain, K.J., Budman, Y., Holyoke, J.L., Davoudi, H., Gebler, J.C., Characterization of therapeutic oligonucleotides using liquid chromatography with on-line mass spectrometry detection, *Oligonucleotides*, 2003, 13, 229–243.
109. Gilar, M., Fountain, K.J., Budman, Y., Neue, U.D., Yardley, K.R., Rainville, P.D., Russell Ii, R. J., Gebler, J. C., Ion-pair reversed-phase high-performance liquid chromatography analysis of oligonucleotides: – Retention prediction, *J. Chromatogr. A*, 2002, 958, 167–182.
110. Polo, L.M., Limbach, P.A., Analysis of oligonucleotides by electrospray ionization mass spectrometry. In *Current Protocols in Nucleic Acid Chemistry*, Beaucage, S., Bergstrom, D.E., Glick, G.D., Jones, R.A. (Eds), 2000, pp. 10.12.11–10.12.20, John Wiley, New York, USA.
111. Apffel, A., Chakel, J.A., Fischer, S., Lichtenwalter, K., Hancock, W.S., Analysis of oligonucleotides by HPLC–electrospray ionization mass spectrometry, *Anal. Chem.*, 1997, 69, 1320–1325.
112. Apffel, A., Chakel, J.A., Fischer, S., Lichtenwalter, K., Hancock, W.S., New procedure for the use of high-performance liquid chromatography-electrospray ionization mass spectrometry for the analysis of nucleotides and oligonucleotides, *J. Chromatogr. A*, 1997, 777, 3–21.
113. Berger, B., Holzl, G., Oberacher, H., Niederstatter, H., Huber, C.G., Parson, W., Single nucleotide polymorphism genotyping by on-line liquid chromatography-mass spectrometry in forensic science of the Y-chromosomal locus M9, *J. Chromatogr. B Analyt. Technol. Biomed. Life Sci.*, 2002, 782, 89–97.
114. Holzl, G., Oberacher, H., Pitsch, S., Stutz, A., Huber, C.G., Analysis of biological and synthetic

ribonucleic acids by liquid chromatography-mass spectrometry using monolithic capillary columns, *Anal. Chem.*, 2005, 77, 673–680.
115. Oberacher, H., Huber, C.G., Oefner, P.J., Mutation scanning by ion-pair reversed-phase high-performance liquid chromatography-electrospray ionization mass spectrometry (ICEMS), *Hum. Mutat.*, 2003, 21, 86–95.
116. Oberacher, H., Niederstatter, H., Parson, W., Liquid chromatography-electrospray ionization mass spectrometry for simultaneous detection of mtDNA length and nucleotide polymorphisms, *Int. J. Legal Med.*, 2007, 121, 57–67.
117. Premstaller, A., Oberacher, H., Huber, C.G., High-performance liquid chromatography-electrospray ionization mass spectrometry of single- and double-stranded nucleic acids using monolithic capillary columns, *Anal. Chem.*, 2000, 72, 4386–4393.
118. Capaldi, D.C., Gaus, H.J., Carty, R.L., Moore, M.N., Turney, B.J., Decottignies, S.D., McArdle, J.V., Scozzari, A.N., Ravikumar, V.T., Krotz, A.H., Formation of 4,4'-dimethoxytrityl-C-phosphonate oligonucleotides, *Bioorg. Med. Chem. Lett.*, 2004, 14, 4683–4690.
119. Kurata, C., Bradley, K., Gaus, H., Luu, N., Cedillo, I., Ravikumar, V.T., Van Sooy, K., McArdle, J.V., Capaldi, D.C., Characterization of high molecular weight impurities in synthetic phosphorothioate oligonucleotides, *Bioorg. Med. Chem. Lett.*, 2006, 16, 607-614.
120. Fountain, K.J., Gilar, M., Gebler, J.C., Electrospray ionization mass spectrometric analysis of nucleic acids using high-throughput on-line desalting, *Rapid Communs Mass Spectrom.*, 2004, 18, 1295–1302.
121. Pomerantz, S.C., Kowalak, J.A., McCloskey, J.A., Determination of oligonucleotide composition from mass spectrometrically measured molecular weight, *J. Am. Soc. Mass Spectrom.*, 1993, 4, 204–209.
122. Ball, R.W., Packman, L.C., Matrix-assisted laser desorption/ionization time-of-flight mass spectrometry as a rapid quality control method in oligonucleotide synthesis, *Anal. Biochem.*, 1997, 248, 185–194.
123. Shah, S., Friedman, S.H., An ESI-MS method for characterization of native and modified oligonucleotides used for RNA interference and other biological applications, *Nat. Protoc.*, 2008, 3, 351–356.
124. Butler, J.M., Jiang-Baucom, P., Huang, M., Belgrader, P., Girard, J., Peptide nucleic acid characterization by MALDI-TOF mass spectrometry, *Anal. Chem.*, 1996, 68, 3283–3287.
125. Cui, Z., Theruvathu, J.A., Farrel, A., Burdzy, A., Sowers, L.C., Characterization of synthetic oligonucleotides containing biologically important modified bases by matrix-assisted laser desorption/ionization time-of-flight mass spectrometry, *Anal. Biochem.*, 2008, 379, 196–207.
126. Oberacher, H., Niederstätter, H., Parson, W., Characterization of synthetic nucleic acids by electrospray ionization quadrupole time-of-flight mass spectrometry, *J. Mass Spectrom.*, 2005, 40, 932–945.
127. Meng, Z., Limbach, P.A., RNase mapping of intact nucleic acids by electrospray ionization Fourier transform ion cyclotron resonance mass spectrometry (ESI-FTICRMS) and 18O labeling, *Int. J. Mass Spectrom.*, 2004, 234, 37–44.
128. Reese, C.B., Oligo- and poly-nucleotides: 50 years of chemical synthesis, *Org. Biomol. Chem.*, 2005, 3, 3851–3868.
129. Sinha, N.D., Michaud, D.P., Recent developments in the chemistry, analysis and control for the manufacture of therapeutic-grade synthetic oligonucleotides, *Curr. Opin. Drug Discov. Devel.*, 2007, 10, 807–818.
130. Chen, D., Yan, Z., Cole, D.L., Srivatsa, G.S., Analysis of internal (n-1)mer deletion sequences in synthetic oligodeoxyribonucleotides by hybridization to an immobilized probe array, *Nucleic Acids Res.*, 1999, 27, 389–395.
131. Alazard, D., Russell, J., Sequencing oligonucleotides by enrichment of coupling failures using matrix-assisted laser desorption/ionization time-of-flight mass spectrometry, *Curr. Protoc. Nucleic Acid Chem.*, 2006, Chapter 10, Unit 10 10.
132. Gaus, H., Olsen, P., Sooy, K.V., Rentel, C., Turney, B., Walker, K.L., McArdle, J.V., Capaldi, D.C., Trichloroacetaldehyde modified oligonucleotides, *Bioorg. Med. Chem. Lett.*, 2005, 15, 4118–4124.

133. Capaldi, D.C., Gaus, H., Krotz, A.H., Arnold, J., Carty, R.L., Moore, M.N., Scozzari, A.N., Lowery, K., Cole, D.L., Ravikumar, V.T., Synthesis of high-quality antisense drugs. addition of acrylonitrile to phosphorothioate oligonucleotides: adduct characterization and avoidance, *Org. Process. Res. Dev.,* 2003, 7, 832–838.
134. Rentel, C., Wang, X., Batt, M., Kurata, C., Oliver, J., Gaus, H., Krotz, A.H., McArdle, J.V., Capaldi, D.C., Formation of modified cytosine residues in the presence of depurinated DNA, *J. Org. Chem.*, 2005, 70, 7841-7845.
135. Nikcevic, I., Wyrzykiewicz, T.K., Limbach, P.A., Detecting low-level synthesis impurities in modified phosphorothioate oligonucleotides using liquid chromatography-high resolution mass spectrometry, *Int. J. Mass Spectrom.*, 2011, 304, 98–104.
136. Smith, M., Characterisation of a modified oligonucleotide together with its synthetic impurities using accurate mass measurements, *Rapid Communs Mass Spectrom.*, 2011, 25, 511–525.
137. Schneider, K., Chait, B.T., Matrix-assisted laser desorption mass spectrometry of homopolymer oligodeoxyribonucleotides. influence of base composition on the mass spectrometric response, *Org. Mass Spectrom.*, 1993, 28, 1353–1361.
138. Bruenner, B.A., Yip, T.T., Hutchens, T.W., Quantitative analysis of oligonucleotides by matrix-assisted laser desorption/ionization mass spectrometry, *Rapid Communs Mass Spectrom.*, 1996, 10, 1797–1801.
139. Addepalli, B., Limbach, P.A., Mass spectrometry-based quantification of pseudouridine in RNA, *J. Am. Soc. Mass Spectrom.*, 2011, 22, 1363–1372.
140. Nordhoff, E., Kirpekar, F., Roepstorff, P., Mass spectrometry of nucleic acids, *Mass Spectrom. Rev.*, 1996, 15, 67–138.
141. Juhasz, P., Roskey, M.T., Smirnov, I.P., Haff, L.A., Vestal, M.L., Martin, S.A., Applications of delayed extraction matrix-assisted laser desorption ionization time-of-flight mass spectrometry to oligonucleotide analysis, *Anal. Chem.*, 1996, 68, 941–946.
142. Alazard, D., Filipowsky, M., Raeside, J., Clarke, M., Majilessi, M., Russell, J., Weisburg, W., Sequencing of production-scale synthetic oligonucleotides by enriching for coupling failures using matrix-assisted laser desorption/ionization time-of-flight mass spectrometry, *Anal. Biochem.*, 2002, 301, 57–64.
143. Tolson, D.A., Nicholson, N.H., Sequencing RNA by a combination of exonuclease digestion and uridine specific chemical cleavage using MALDI-TOF, *Nucleic Acids Res.*, 1998, 26, 446–451.
144. Faulstich, K., Worner, K., Brill, H., Engels, J. W., A sequencing method for RNA oligonucleotides based on mass spectrometry, *Anal. Chem.*, 1996, 68, 4349–4353.
145. Bentzley, C.M., Johnston, M.V., Larsen, B.S., Gutteridge, S., Oligonucleotide sequence and composition determined by matrix-assisted laser desorption/ionization, *Anal. Chem.*, 1996, 68, 2141–2146.
146. Dai, G., Wei, X., Liu, Z., Liu, S., Marcucci, G., Chan, K.K., Characterization and quantification of Bcl-2 antisense G3139 and metabolites in plasma and urine by ion-pair reversed phase HPLC coupled with electrospray ion-trap mass spectrometry, *J. Chromatogr. B Analyt. Technol. Biomed. Life Sci.*, 2005, 825, 201–213.
147. Kretschmer, M., Lavine, G., McArdle, J., Kuchimanchi, S., Murugaiah, V., Manoharan, M., An automated algorithm for sequence confirmation of chemically modified oligonucleotides by tandem mass spectrometry, *Anal. Biochem.*, 2010, 405, 213–223.
148. Bartlett, M.G., McCloskey, J.A., Manalili, S., Griffey, R.H., The effect of backbone charge on the collision-induced dissociation of oligonucleotides, *J. Mass Spectrom.*, 1996, 31, 1277–1283.
149. Polo, L.M., McCarley, T.D., Limbach, P.A., Chemical sequencing of phosphorothioate oligonucleotides using matrix-assisted laser desorption/ionization time-of-flight mass spectrometry, *Anal. Chem.*, 1997, 69, 1107–1112.
150. Gao, H., Liu, Y., Rumley, M., Yuan, H., Mao, B., Sequence confirmation of chemically modified RNAs using exonuclease digestion and matrix-assisted laser desorption/ionization time-of-flight mass spectrometry, *Rapid Communs Mass Spectrom.*, 2009, 23, 3423–3430.
151. Gao, L., Zhang, L., Cho, B.P., Chiarelli, M.P., Sequence verification of oligonucleotides containing multiple arylamine modifications by enzymatic digestion and liquid chromatography mass spectrometry (LC/MS), *J. Am. Soc. Mass Spectrom.*, 2008, 19, 1147–1155.

152. Bahr, U., Aygun, H., Karas, M., Sequencing of single and double stranded RNA oligonucleotides by acid hydrolysis and MALDI mass spectrometry, *Anal. Chem.*, 2009, 81, 3173–3179.
153. Farand, J., Beverly, M., Sequence confirmation of modified oligonucleotides using chemical degradation, electrospray ionization, time-of-flight, and tandem mass spectrometry, *Anal. Chem.*, 2008, 80, 7414–7421.
154. Farand, J., Gosselin, F., De novo sequence determination of modified oligonucleotides, *Anal. Chem.*, 2009, 81, 3723–3730.
155. Sarracino, D., Richert, C., Quantitative MALDI-TOF MS of oligonucleotides and a nuclease assay, *Bioorg. Med. Chem. Lett.*, 1996, 6, 2543–2548.
156. Bleczinski, C.F., Richert, C., Monitoring the hybridization of the components of oligonucleotide mixtures to immobilized DNA via matrix-assisted laser desorption/ionization time-of-flight mass spectrometry, *Rapid Communs Mass Spectrom.*, 1998, 12, 1737–1743.
157. Meng, Z., Limbach, P.A., Quantitation of ribonucleic acids using O18 labeling and mass spectrometry, *Anal. Chem.*, 2005, 77, 1891–1895.
158. Ding, C., Qualitative and quantitative DNA and RNA analysis by matrix-assisted laser desorption/ionization time-of-flight mass spectrometry, *Methods Mol. Biol.*, 2006, 336, 59–71.
159. Kinumi, T., Narukawa, T., Fujii, S., Eyama, S., Saeki, M., Takatsu, A., Quantification of oligonucleotide containing sequence failure product: comparison of isotope dilution mass spectrometry with other quantification methods, *Eur. J. Mass Spectrom.*, 2009, 15, 399–407.
160. Bahr, U., Aygun, H., Karas, M., Detection and relative quantification of siRNA double strands by MALDI mass spectrometry, *Anal. Chem.*, 2008, 80, 6280–6285.
161. O'Connor, G., Dawson, C., Woolford, A., Webb, K.S., Catterick, T., Quantitation of oligonucleotides by phosphodiesterase digestion followed by isotope dilution mass spectrometry: proof of concept, *Anal. Chem.*, 2002, 74, 3670–3676.
162. Donald, C.E., Stokes, P., O'Connor, G., Woolford, A.J., A comparison of enzymatic digestion for the quantitation of an oligonucleotide by liquid chromatography-isotope dilution mass spectrometry, *J. Chromatogr. B Analyt. Technol. Biomed. Life Sci.*, 2005, 817, 173–182.
163. Bohrer, B.C., Merenbloom, S.I., Koeniger, S.L., Hilderbrand, A.E., Clemmer, D.E., Biomolecule analysis by ion mobility spectrometry, *Ann. Rev. Anal. Chem.*, 2008, 1, 293–327.
164. Kanu, A.B., Dwivedi, P., Tam, M., Matz, L., Hill, H.H., Jr, Ion mobility-mass spectrometry, *J. Mass Spectrom.*, 2008, 43, 1–22.
165. Balbeur, D., Widart, J., Leyh, B., Cravello, L., De Pauw, E., Detection of oligonucleotide gas-phase conformers: H/D exchange and ion mobility as complementary techniques, *J. Am. Soc. Mass Spectrom.*, 2008, 19, 938–946.
166. Koomen, J.M., Ruotolo, B.T., Gillig, K.J., McLean, J.A., Russell, D.H., Kang, M., Dunbar, K.R., Fuhrer, K., Gonin, M., Schultz, J.A., Oligonucleotide analysis with MALDI-ion-mobility-TOFMS, *Anal. Bioanal. Chem.*, 2002, 373, 612–617.
167. Rosu, F., Gabelica, V., Poncelet, H., De Pauw, E. Tetramolecular G-quadruplex formation pathways studied by electrospray mass spectrometry, *Nucleic Acids Res.*, ????, 38, 5217–5225.
168. Williams, J.P., Lough, J.A., Campuzano, I., Richardson, K., Sadler, P.J., Use of ion mobility mass spectrometry and a collision cross-section algorithm to study an organometallic ruthenium anticancer complex and its adducts with a DNA oligonucleotide, *Rapid Communs Mass Spectrom.*, 2009, 23, 3563–3569.
169. Woods, A.S., Ugarov, M., Egan, T., Koomen, J., Gillig, K.J., Fuhrer, K., Gonin, M., Schultz, J.A., Lipid/peptide/nucleotide separation with MALDI-ion mobility-TOF MS, *Anal. Chem.*, 2004, 76, 2187–2195.
170. Fenn, L. S., Kliman, M., Mahsut, A., Zhao, S.R., McLean, J.A., Characterizing ion mobility-mass spectrometry conformation space for the analysis of complex biological samples, *Anal. Bioanal. Chem.*, 2009, 394, 235–244.
171. Smiljanic, D., Wesdemiotis, C., Non-covalent complexes between single-stranded oligodeoxynucleotides and poly(ethylene imine), *Int. J. Mass Spectrom.*, 2011, 304, 148–153.

5 Analytical Characterisation of Oligonucleotides by NMR Spectroscopy

Elena Bichenkova

5.1 Introduction

Nuclear magnetic resonance (NMR) spectroscopy is recognised as one of the most powerful and versatile analytical tools, which is broadly used for identification and characterisation of both small molecules and functionally significant biopolymers.

In oligonucleotide chemistry, biochemistry and structural biology this method offers a broad variety of strategies that can be applied to confirm the identity and purity of oligonucleotides and their pharmaceutically relevant analogues. NMR spectroscopy offers many different levels for probing oligonucleotide samples, ranging from very simple 'fingerprint' analysis of complicated oligonucleotide mixtures to comprehensive study of oligonucleotide structure, dynamics and interactions. In contrast to other analytical approaches that are routinely used for elucidation of structural properties of nucleic acids (e.g. X-ray crystallography or neutron diffraction), NMR offers the advantage of structural analysis in solution, thus more closely resembling biological systems. In addition, NMR is the only method capable of providing a complete interpretation of a chemical structure and conformation in a single experiment, which became possible after the development of sequence-specific signal assignment strategies. The knowledge of three-dimensional (3D) structures of folded oligonucleotide conformations, duplexes and their structure–function correlations offers a basis for manipulation with these functionalities and can potentially be exploited for the design of novel therapeutic and diagnostic tools, especially if NMR structural studies are complemented by high-level computational techniques.

The aim of this chapter is to provide an overview of the most recent NMR-based strategies that can be utilised by analytical chemists and used as guidance for probing primary, secondary and tertiary structures of synthetic oligonucleotides. By providing some practical examples, this chapter demonstrates how different one-dimensional (1D) and two-dimensional (2D) NMR formats and techniques can be applied for each particular case in order to gain a desirable level of information. These examples and strategic approaches can then be extended to other applications in the area of oligonucleotide chemistry, biochemistry and structural biology.

5.1.1 Basic Concepts of NMR Spectroscopy

The NMR phenomenon is based on the magnetic properties of atomic nuclei that can be used to provide some essential information about chemical and structural properties of oligonucleotides and their analogues. Sub-atomic particles such as protons, neutrons and electrons have magnetic properties. The magnetic moment of the nucleus depends on the quantum properties of the atom. Any atom with an odd number of neutrons and/or an odd number of protons has magnetic properties (i.e. nuclear spin and magnetic moment) [1, 2].

5.1.1.1 Different Nuclei Used in NMR of Oligonucleotides

The most important isotopic species that are present in nucleic acids, oligonucleotides and their derivatives include ^{1}H, ^{13}C, ^{15}N, ^{19}F and ^{31}P, all of which have a non-zero nuclear spin of $\frac{1}{2}$ that can be detected by NMR. In contrast, ^{12}C and ^{16}O nuclei, which are highly abundant in oligonucleotides, possess zero nuclear spin and are considered to be 'NMR silent' [2].

Spinning nuclei possess an intrinsic angular momentum, P, and, as with any charged object, the spinning nucleus (e.g. ^{1}H, ^{13}C, ^{15}N. ^{19}F or ^{31}P) generates a magnetic field around it, giving rise to an associated magnetic moment μ

$$\mu = \gamma \cdot P \tag{5.1}$$

where γ is the magnetogyric ratio, which is a specific constant for any particular nucleus and is responsible for its magnetic properties [1, 2] (see Table 5.1).

Table 5.1 NMR properties of selected isotopes with spin $\frac{1}{2}$, which are often used for characterisation of nucleic acids and their synthetic analogues [1, 3].

Isotope	Nuclear spin	Natural abundance (%)	NMR frequency[a] (MHz)	Magnetogyric ratio, $\gamma \times 10^7$ (rad/s per T)	Sensitivity[b]
^{1}H	$\frac{1}{2}$	99.9885	400.00	26.7522	1.0
^{13}C	$\frac{1}{2}$	1.07	100.6	6.7283	1.76×10^{-4}
^{15}N	$\frac{1}{2}$	0.37	40.5	−2.7126	3.85×10^{-6}
^{19}F	$\frac{1}{2}$	100	376.3	25.1623	0.83
^{31}P	$\frac{1}{2}$	100	161.9	10.8394	6.63×10^{-2}

[a]Resonance frequencies are given for a 9.4 Tesla magnet (400 MHz spectrometer).
[b]Relative sensitivities are given at constant field and for an equal number of nuclei relative to ^{1}H observation.

5.1.1.2 Key Principles of the Resonance Experiment

When a nucleus with an angular moment of P and a magnetic moment of μ is placed in an external static magnetic field B_0, the nuclear magnetic moments align themselves relative to the externally applied magnetic field in a discrete number of orientations [1, 2]. For the nucleus possessing a nuclear spin quantum number I there are $2I + 1$ possible spin states. For example, for a single nucleus with nuclear spin quantum number of $\frac{1}{2}$ (e.g. ^{1}H, ^{13}C, ^{31}P, ^{19}F or ^{15}N) there are two possible spin orientations relative to the external magnetic field, namely $+\frac{1}{2}$ and $-\frac{1}{2}$ (Figure 5.1). Nuclei corresponding to the $+\frac{1}{2}$ spin state, which are aligned with the applied external field, have the energetically preferred orientation, whereas nuclei with $-\frac{1}{2}$ spin state are energetically unfavourable due to their orientation against the field [2]. These spin states are separated in energy given by

$$\Delta E = h \cdot \nu = \frac{h \cdot \gamma \cdot B_0}{2\pi} \tag{5.2}$$

where ΔE is the difference in energy between two states, h is Planck's constant, γ is the magnetogyric ratio and B_0 is the strength of the applied magnetic field [1, 4].

The ratio between the number of the aligned nuclei (N_α) corresponding to a low energy state and those aligned against the applied magnetic field (N_β) associated with a high energy state is given by the Boltzmann distribution (equation (5.3))

$$N_\alpha / N_\beta = \exp^{(DE/kT)} \tag{5.3}$$

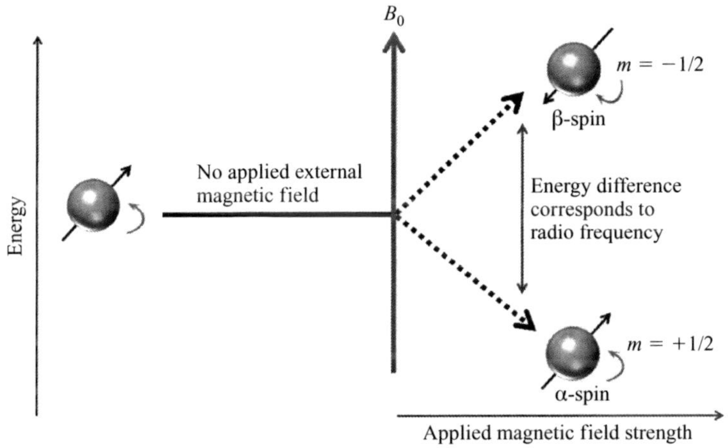

Figure 5.1 Schematic diagram showing the energy levels of a nucleus with spin quantum number $\frac{1}{2}$. When there is no applied static external magnetic field, there is no energy difference between the two states. However, when the external magnetic field (B_0) is applied, nuclear spins align themselves relative to the externally applied magnetic field in a discrete number of orientations corresponding to two different energy levels. Paler, curved arrows represent the spinning directions of the nucleus around itself and the black arrows represent the alignment with or against the applied magnetic field.

This indicates that at equilibrium there will be an excess of nuclei with the spin orientated with the applied magnetic field as compared with those orientated against the field.

According to a model of classical mechanics [1, 2], the magnetic moment μ associated with a spinning spherical charge will precess in an external magnetic field around the direction of the applied magnetic field B_0 at specific angular velocity (Figure 5.2). The rate of the precession corresponds to the angular velocity ω_0 (rad/s) and is proportional to the strength of the magnetic field B_0 in accordance with equation (5.4) [1, 4]

$$\omega_0 = -\gamma B_0 \qquad \text{(rad/s)} \tag{5.4}$$

The frequency ω_0 is called the Larmor frequency of the nucleus and can also be expressed in Hz via equation (5.5) [1, 4]

$$\nu_0 = \frac{\varpi_0}{2\pi} = \frac{\gamma \cdot B_0}{2\pi} \qquad \text{(Hz)} \tag{5.5}$$

The direction of the precession is determined by the sign of magnetogyric ratio γ.

Nuclear magnetic resonance occurs when the nucleus changes its spin state as a result of the absorption of a quantum of energy following electromagnetic radiation, whose frequency exactly matches the Larmor frequency ν_0 of the nucleus. The sample will generate an NMR signal only at a particular combination of radio frequency and magnetic field strength to satisfy resonance conditions described by equation (5.5).

This absorption of energy during the transition between spin states forms the basis of NMR spectroscopy.

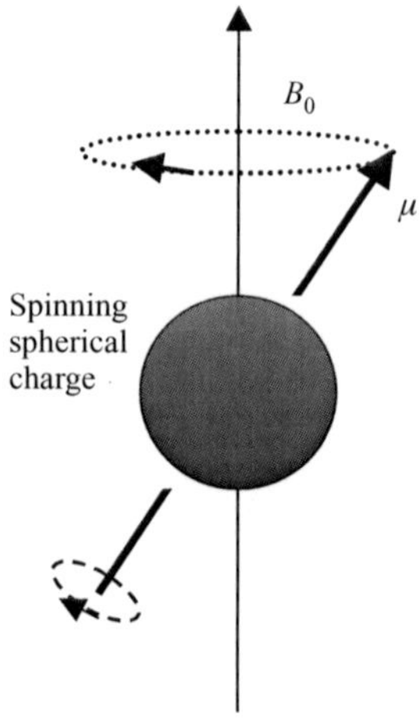

Figure 5.2 Schematic representation of a precession of the magnetic moment μ of a spinning spherical charge in an external magnetic field B_0 around the direction of the applied magnetic field.

5.1.1.3 The Vector Model of NMR Experiments

The behaviour of nuclear spins in pulsed NMR experiments can be more easily understood using the vector model of NMR. Since the energy difference between the two spin states is very small, their corresponding population differences are also very small, as described by equation (5.3). However, this excess in population at the lower energy level gives rise to a resultant 'bulk' magnetisation vector M_0 along the direction of the magnetic field B_0 (Figure 5.3), which behaves according to the rules of classical mechanics [1, 2].

During the standard 1D NMR experiment a sample is subjected to irradiation by a non-selective 90° pulse of radio-frequency field B_1, which is applied along the X axis to rotate the bulk magnetisation vector M_0 into the X, Y plane perpendicular to B_0. The transverse magnetisation in the X, Y plane precesses under the influence of static magnetic field B_0 at Larmor frequency, which corresponds to the resonance frequency ν_0 (see equation (5.5)). This induces an electric current in the detection coil of the NMR probe and thus produces the NMR signal. Since the system tends to return to thermodynamic equilibrium with the bulk magnetisation vector M_0 to be parallel to the direction of the static magnetic field B_0, the transverse magnetisation decays with time as a result of spin relaxation to produce free induction decay (FID). The frequency-domain NMR spectrum is then generated by Fourier transformation of the corresponding time-domain data.

5.2 Different Formats of NMR Used for Nucleic Acids

5.2.1 Key Aspects of 1D NMR Spectroscopy

Conventional 1D NMR spectra could be sufficient to provide some essential information about the identity and purity of synthesised single-stranded oligonucleotides and their derivatives, which can be achieved *via* careful identification of the ^{1}H, ^{31}P and (if necessary) ^{13}C signals. Also, a complete sequence-specific resonance assignment of the oligonucleotide protons is the first step in the subsequent structural analysis of nucleic acids.

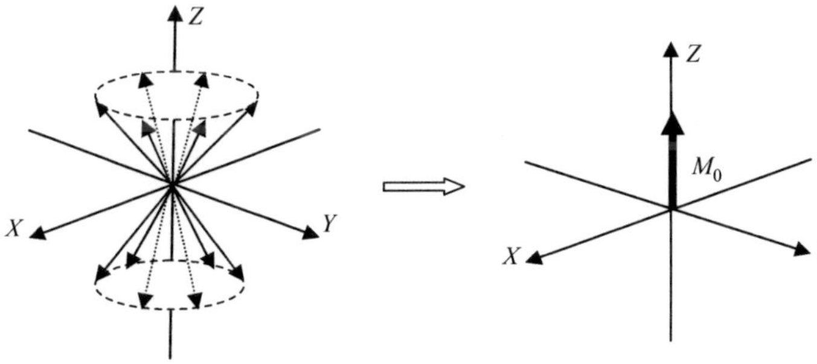

Figure 5.3 The orientations of the resultant bulk magnetisation vector M_0 along the direction of the external magnetic field B_0.

There are a number of important points that should be considered when interpreting 1D NMR spectra of oligonucleotides. These include the number and position of peaks in the spectrum, the area under each peak (an integral), splitting patterns of the peaks and their shapes.

5.2.1.1 Chemical Shift

The number of resonance lines in the NMR spectrum usually correlates with the number of chemically non-equivalent nuclei; however, some signals may overlap in the spectrum leading to a reduced number of resonance lines. The positions of the NMR signals indicate the electromagnetic environment of the nuclei, which generate these signals. Each nucleus in a molecule is surrounded by a number of electrons and atoms, which are capable of generating local magnetic fields. Therefore, the resonance frequency of each nucleus in the static magnetic field is influenced by its chemical and magnetic environment. Since the nucleus is shielded by its electronic environment, the magnetic field experienced by each nucleus (B_{eff}) is reduced as compared with the field generated by the applied magnetic field B_0, which is illustrated by equation (5.6) [5]

$$B_{eff} = B_0 - \sigma B_0 = (1 - \sigma)B_0 \tag{5.6}$$

where σ is defined as a shielding factor.

Since chemically non-equivalent nuclei in a molecule have different electronic environments, they experience different shielding effects from surrounding electrons, hence generating separate resonance signals which depend on the values of the shielding factor σ. The variation of resonance frequency ν_0 with chemical environment of the nucleus is known as a *chemical shift*. Differences in resonance frequency ν_0 are much greater than resonance line widths. As a result, for the same type of nuclei, distinct NMR resonance lines can be observed for chemically different nuclei.

Conventionally, chemical shift δ is measured against a standard reference (e.g. tetramethyl silane for organic solvents or 3-trimethylsilyl-propionate (TSP) for aqueous solutions) and quoted in parts per million (ppm) units

$$\delta = \frac{\nu_{sample} - \nu_{reference}}{\nu_{reference}} \times 10^6 \tag{5.7}$$

where δ is the chemical shift (ppm) of a sample peak, ν_{sample} is the resonance frequency (Hz) of a sample peak and $\nu_{reference}$ is the resonance frequency (Hz) of the reference.

5.2.1.2 Relative Integral Intensities

For most nuclei present in nucleic acids (e.g. ^1H or ^{31}P), the relative integral intensities of the NMR signals are directly proportional to the number of nuclei responsible for these signals. However, this is not the case for ^{13}C NMR spectra, owing to the fact that relaxation times may strongly differ for various ^{13}C nuclei.

5.2.1.3 Signal Multiplet Structure and Shape

Fine multiplet structure of proton NMR signals depend on their spin–spin coupling interactions with other nuclei, which are transmitted by electrons through intermediate chemical bonds. The shape of NMR signals often depends on the dynamic properties of the oligonucleotides in solution. Also, some signals arising from exchangeable protons may show considerable line broadening due to exchange with the media.

5.2.2 General Principles of 2D NMR

The most useful information that can be obtained from NMR spectroscopy comes from the interactions between two or more nuclei, either through bonding interactions, which could be detected *via* J-coupling interactions (also termed indirect dipole–dipole coupling), or through space connectivities by way of observation of the nuclear Overhauser effect (NOE). It is possible to follow these interactions one at a time, while irradiating one resonance in the 1D NMR proton spectrum (either during the relaxation delay or during acquisition) and looking at the effect on the integral intensity or coupling pattern of a resonance signal from another proton. However, this can be achieved in a more efficient and informative way using 2D NMR spectroscopy, which allows simultaneous irradiation of all nuclei of the same type in the sample (e.g. protons) in order to generate a 2D map of all affected nuclei. In 2D NMR spectroscopy two dimensions refer to two *frequency* dimensions, although higher-dimensionality NMR approaches (e.g. 3D or four-dimensional) are also available.

Generally, any 2D-NMR technique consists of the following key experimental steps, namely, *preparation*, *evolution*, *mixing* and *detection* [2]. The preparation period consists of an appropriate delay time during which the thermal equilibrium is reached, followed by 1 or more radio-frequency pulses (e.g. 90°, 180° and/or 45° pulses) to flip the net magnetisation M_0 in order to produce transverse magnetisation in the X, Y plane. This step is followed by evolution, which is the time period in a 2D pulse sequence during which a net magnetisation is allowed to precess in the X, Y plane during t_1 time period prior to mixing and detection. (Note that in the case of correlation spectroscopy (COSY) experiments, the evolution and mixing times occur simultaneously [2].) Variation of the t_1 delay in a 2D pulse sequence, which is incremented by a fixed amount, generates the t_1 time domain. During the mixing time, which is usually a combination of radio-frequency pulses and/or delay periods, the phase-encoded spins are allowed to mix with each other *via* exchange of t_1-encoded phase information from one nucleus to another, either through J-coupling (in COSY type NMR experiments) or through dipole–dipole interactions (in nuclear Overhauser effect spectroscopy (NOESY) experiments). The final step is the detection period, during which the NMR spectrometer collects the FID of excited nuclei when they are relaxing back to equilibrium state. The collected matrix data are then subjected to two Fourier transformations to produce the frequency domain 2D spectrum. This 2D NMR spectrum usually contains cross peaks that correlate information on chemical shifts on one axis with those on the second axis. There are a number of 2D NMR techniques currently available in modern NMR spectroscopy; however, most of them can be classified into two different categories. The first category is represented by COSY type

NMR spectroscopy, whereas the second type involves NOESY type experiments; both are briefly outlined in the following sections.

5.2.2.1 COSY

COSY type techniques (e.g. ^{1}H-^{1}H COSY, total coherance transfer spectroscopy (TOCSY), double quantum filtered-COSY (DQF-COSY), heteronuclear single-quantum correlation spectroscopy (HSQC) or heteronuclear multiple-bond correlation spectroscopy (HMBC)) detect correlations between nuclei *via* scalar spin–spin interactions, thus giving information about their connectivity through chemicals bonds. Using this type of spectroscopy, it is possible to map nuclei sharing a mutual scalar coupling and thus establish a net of interacting nuclei involved in integrated spin systems.

Homonuclear Correlation Spectroscopy

A homonuclear 2D COSY NMR spectrum (e.g. ^{1}H-^{1}H COSY) normally shows chemical shifts from the same type of nuclei on both the F1 and F2 axes.

Signals usually appear on the diagonal of the 2D spectrum, where the analogous 1D spectrum contains resonances. Off-diagonal signals in COSY spectra are called cross-peaks and correlate the resonance lines corresponding to spins that are coupled to each other. The intensity of COSY cross-peaks qualitatively reflects the magnitude of the J-coupling between the two nuclei, but does not allow spin-coupling constants to be measured precisely. Accurate measurements of J-coupling constants can be achieved, however, by using other COSY-related techniques (e.g. DQF-COSY). This offers the opportunity to evaluate associated torsion angles and thus extract essential structural information.

In oligodeoxyribonucleotides and oligoribonucleotides H5 and H6 protons of cytidines and uridines, H6 and -CH_3 protons of thymidines as well as H1′, H2′, H2″, H3′, H4″ and H5′/H5″ sugar ring protons (see Figure 5.4 in the colour insert) form unique spin systems and thus can be recognised and assigned using J-correlation spectroscopy (see Figure 5.5). COSY experiments allow detection of relatively strong spin coupling interactions *via* two or three chemical bonds. However, weak spin–spin interactions *via* four or more chemical bonds might not always be detectable in COSY [4] and require running different types of correlation spectroscopy experiments (e.g. TOCSY). TOCSY allows observation of spin–spin interactions between distantly located nuclei, which are separated by up to six chemical bonds, and thus it is possible to observe long-distance correlations, in addition to usual J-coupling interactions seen between closely located spins. Although both COSY and TOCSY spectra are broadly used for signal assignments, they do not provide any structural information since they are recorded in the absolute mode. Recording COSY experiments in a phase-sensitive mode may generate very broad diagonal peaks which hinder the assignment of cross-peaks near the spectral diagonal. To overcome this problem, DQF-COSY can be used to eliminate this effect. DQF-COSY allows the recording of spectra in an anti-phase mode in order to generate cross-peaks showing fine structural elements, from which a coupling constant could be measured. Figure 5.5 shows an expanded region of a DQF-

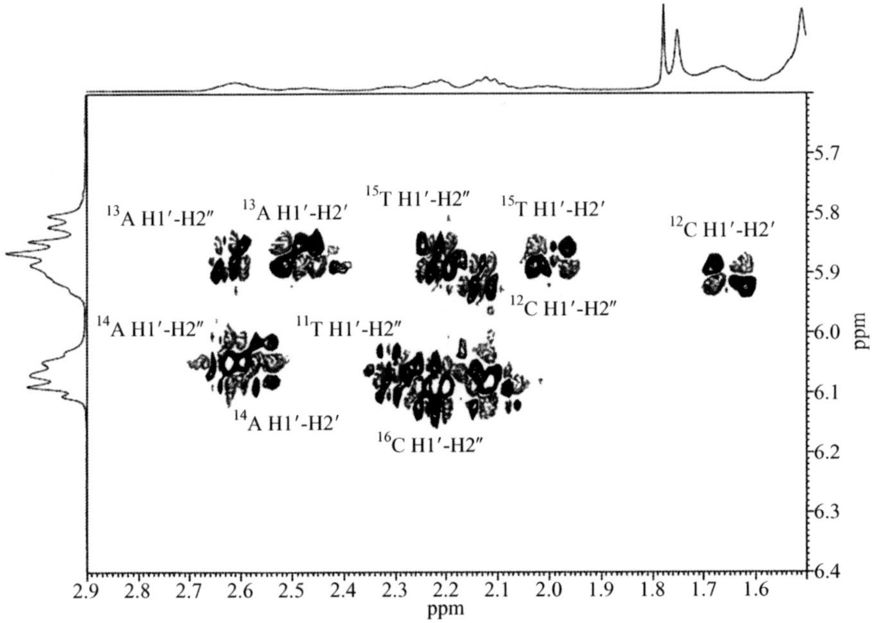

Figure 5.5 The expanded region of the DQF-COSY spectrum of the oligonucleotide analogue showing the H1′-H2′ and H1′-H2″ region along F2 axis. Some cross-peaks show fine splitting patterns, from which J-coupling constant can potentially be measured. The digital resolution of this spectrum in F2 axis is 0.3 Hz/point.

COSY spectrum of an oligonucleotide showing fine structural components of cross-peaks.

Cross-peaks in a DQF-COSY spectrum may reflect the sugar ring conformations and thus provide vital information about the dominant conformation of oligonucleotides. Another advantage of DQF-COSY is the ability to eliminate strong signals of uncoupled protons, such as undesirable signals from solvent.

Heteronuclear Correlation Spectroscopy

Heteronuclear correlated spectroscopy allows the detection of spin–spin coupling interactions between different types of nuclei (e.g. ^{1}H-^{13}C, ^{1}H-^{31}P or ^{1}H-^{15}N) *via* one or even more bonds. This type of spectroscopy (see Figure 5.6 as an example) is particularly important for nucleic acids, as in many cases evaluation of structural information of oligonucleotides and their derivatives requires measuring torsion angles that involve heavy atoms such as ^{31}P or ^{13}C, which constitute vital structural elements of the sugar-phosphate backbone. For example, coupling of ^{31}P nuclei to the attached protons may provide valuable structural data on the nucleic acid backbone conformation that can potentially be used as experimental constraints for subsequent restrained molecular modelling.

The examples of this type of spectroscopy include HMQC and HSQC (Figure 5.6). Cross-peaks in HMQC and HSQC spectra represent 1-bond correlations between

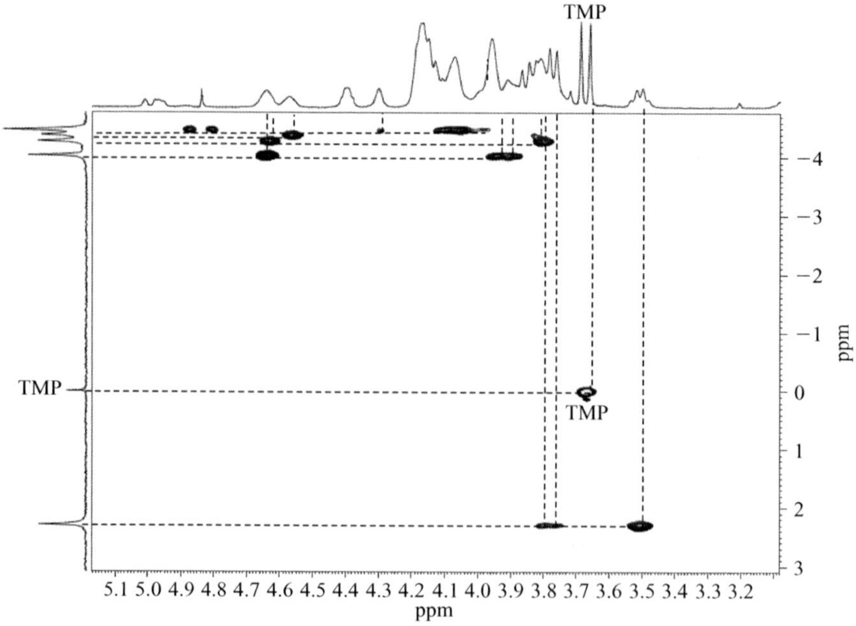

Figure 5.6 ^{1}H -^{31}P HSQC NMR spectrum of the oligonucleotide conjugate recorded at 5°C (400 MHz) showing the coupling between the protons and connected phosphorus atoms of the sugar-phosphate backbone. The *X*-axis corresponds to the ^{1}H dimension (F2), while the *Y*-axis represents the ^{31}P dimension (F1). In this figure, the reference signal from trimethyl phosphate (TMP) in the ^{31}P dimension was referenced to 0.0 ppm.

a heteronucleus (e.g. ^{13}C, ^{31}P or ^{15}N) and the attached proton (demonstrated by dashed lines on Figure 5.6 for selected ^{1}H-^{31}P correlations). In contrast, HMBC allows long-range correlations to be obtained between a heteronucleus and the attached proton via ^{2}J and ^{3}J coupling interactions (in addition to one-bond correlations). Therefore, a cross-peak in a HMBC spectrum represents a long-range bond correlation between the heteronucleus and the attached proton.

5.2.2.2 NOESY

NOESY (e.g. ^{1}H-^{1}H NOESY, ^{1}H-^{1}H rotational frame NOESY (ROESY) is based on dipole–dipole interactions of different nuclei through space and thus allows detection of the close proximity between nuclei that are not necessarily connected by chemical bonds. These techniques allow for the detection of the spatial connectivity between closely located nuclei, which are separated by less than 5 Å. By measuring distances between ^{1}H-^{1}H protons through space, it is possible to extract valuable structural information, which is even more beneficial when combined with high-label computational techniques. Figure 5.7 gives an example of 2D NMR pulse sequences, which are used to generate a NOESY experiment. A typical NOESY experiment requires three 90° pulse pulsations in order to record the cross relaxation intensities [6]. It starts with a preparation period in order to ensure that nuclear spins have attained their

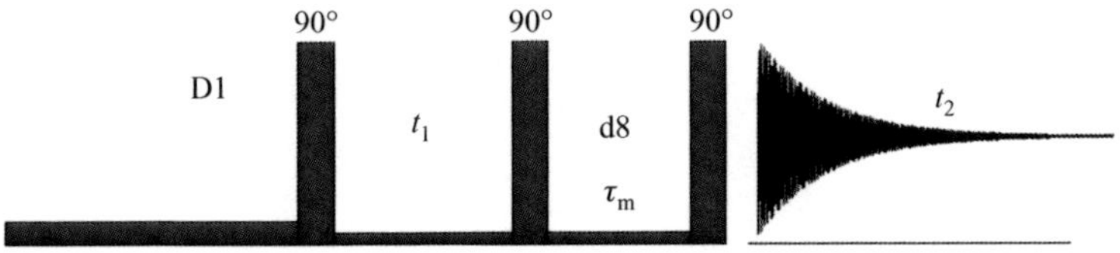

Figure 5.7 Schematic representation of a typical pulse sequence used to carry out a NOESY experiment. During the preparation period with relaxation delay D1, the thermal equilibrium is achieved followed by the first 90° pulse, which produces transverse magnetisation in the *X,Y* plane. During the evolution period with the t_1 duration, a net magnetisation is allowed to precess in the *X,Y* plane followed by the second 90° pulse. The mixing time τ_m allows the magnetisation to be distributed among the various spin states of interacting nuclei. The third 90° pulse returns the magnetisation vector into the *X,Y* plane leading to subsequent detection of the signal in the final period, t_2.

thermal equilibrium. After the preparation period the first 90° pulse is applied in order to flip the longitudinal *Z*-magnetisation into the *X, Y* plane. After the first pulse the evolution time, t_1, is allowed, during which the transverse magnetisation precesses in the *X, Y* plane.

The signal sampling during this period gives the chemical shifts or the first dimension of the spectrum. By applying a second 90° pulse some of the magnetisation is transferred to the *Z* axis to give new longitudinal magnetisation. During the mixing time, τ_m, the cross relaxation occurs between nuclei closely located in space. Each longitudinal polarisation of each spin is labelled with its resonance frequency at which its signal appears in the spectrum. The third 90° pulse creates transverse magnetisation from the unrelaxed longitudinal magnetisation. The detection period, t_2, of the final spectrum starts immediately after the third 90° pulse, and the transverse magnetisation is recorded as a function of time t_2. An example of 2D NOESY spectrum is given in Figure 5.8, which demonstrates the most important resonance regions used for the subsequent assignment of the ^1H signals.

The importance of NOESY experiments in structural analysis of oligonucleotides and their derivatives comes from the fact that the cross-relaxation rate σ_{ij} between the interacting protons is inversely related to the sixth power of the *inter*-nuclear separation [1, 3] (see equation (5.8)).

$$\sigma_{ij} = \frac{\hbar^2\gamma^4}{10}\frac{1}{r_{ij}^6}\left(\frac{6\tau_c}{1+4(\omega\tau_c)^2}\right) - \tau_c \tag{5.8}$$

where σ_{ij} is the cross-relaxation rate between the interacting protons i and j, r_{ij} is the distance between them, ω is the spectrometer frequency in radians and τ_c is the correlation time of the molecular species. The intensity of the NOESY cross-peak formed by the interacting protons i and j is proportional to the cross-relaxation rate σ_{ij} and is related to the interproton distance by the following equation

$$\mathrm{NOE}_{ij} \propto \left\langle \frac{1}{r_{ij}} \right\rangle^{1/6} \times f(\tau_c) \tag{5.9}$$

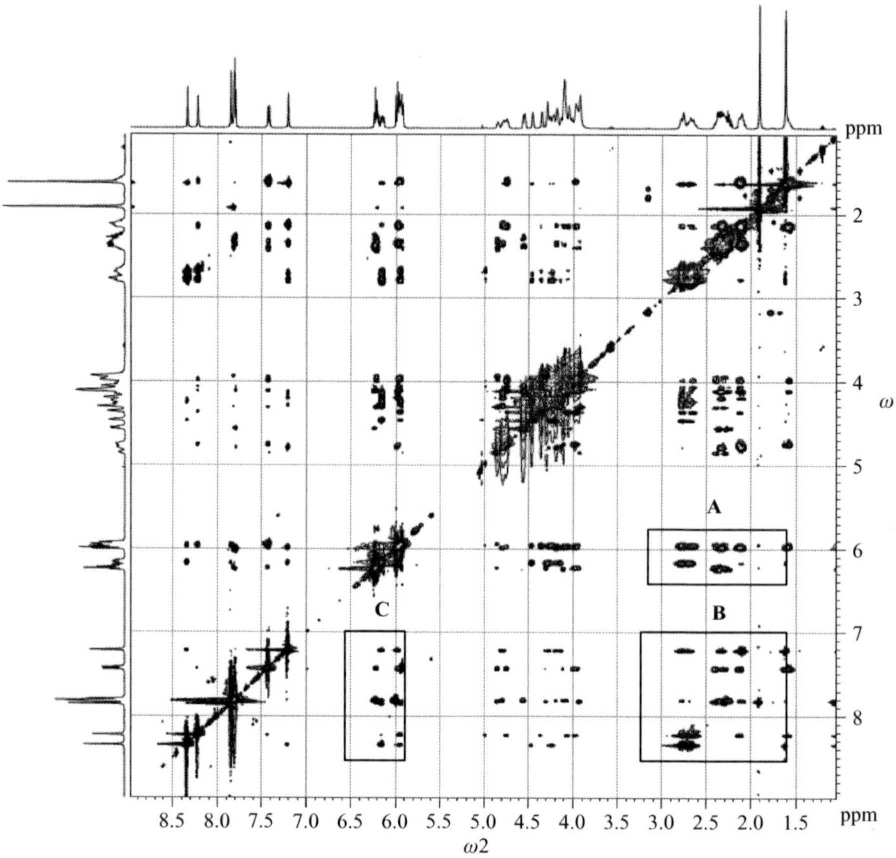

Figure 5.8 An example of 2D NOESY spectrum (500 MHz) of 6mer oligodeoxynucleotide sequence, showing the most important resonance regions used for the subsequent assignment of the ^{1}H signals. Region **A** shows cross-peaks corresponding to the NOE interactions between H1′ and H2′/H2″. Regions **B** and **C** show cross-peaks between aromatic protons (A H8, C H6 and T H6) and sugar protons H2′/H2″ and H1′, respectively.

where $f(\tau_c)$ is the function of the correlation time that takes into account the effect of the molecular motions on the observed NOEs in the NOESY experiment. However, very often experimental evaluations of the correlation times is difficult or impossible; therefore, an alternative approach is to estimate the *relative* ^{1}H-^{1}H distances using some fixed reference distances r_{ref}, which are independent from the conformation of the molecule

$$r_{ij} = r_{ref}\left\langle \frac{a_{ref}}{a_{ij}} \right\rangle^{1/6} \tag{5.10}$$

where r_{ij} is the unknown distance between nuclei i and j, a_{ij} is the intensity of a NOESY cross-peak formed by these two interacting nuclei, r_{ref} is a known distance between two reference protons and a_{ref} is the intensity of a NOESY cross-peak generated by these two nuclei. In the case of oligonucleotides, the distance between aromatic protons H5 and H6 of cytidine can be used as a reference distance, which is constant, independent from the oligonucleotide structure and equal to 2.46 Å [7].

An approach that considers NOE to be the result of interactions between the isolated pair of nuclei *i* and *j* is called an 'isolated spin-pair approximation'. In reality, each proton is usually integrated into a complicated net of interacting protons. This means that the cross-relaxation between the two neighbouring spins *i* and *j* occurs not only through direct magnetisation transfer mechanisms, but also *via* indirect cross-relaxation pathways through the closely located spins (see Figure 5.9 for illustration). This phenomenon is known as multi-spin diffusion [8] and can be influenced by mixing time τ_{mix} used in NOESY experiments. The longer τ_{mix}, the higher the impact of multi-spin diffusion on cross-peak intensities. Substantial interference from multi-spin diffusion may potentially lead to incorrect evaluation of inter-proton distances, and therefore generate less precise 3D structures of oligonucleotides.

The main disadvantage of the 'isolated spin-pair approach' in calculation of interproton distances using equation (5.10) is that the multi-spin diffusion effect is not taken into consideration. One way to minimise the impact of spin-diffusion mechanisms is to record NOESY spectra at relatively short mixing times. However, this may result in losing some valuable long-distance NOE interactions, which could be vital for structure determination. An alternative approach can be based on analysis of the full relaxation matrix to calculate ^1H-^1H distances more precisely. One example of the full relaxation matrix approach is matrix analysis of relaxation for discerning

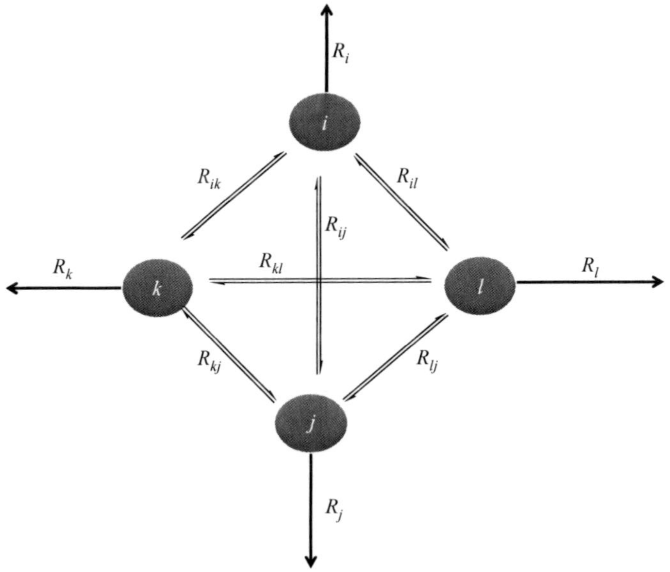

Figure 5.9 Schematic diagram showing a network of through-space NOE interactions between nuclei *i*, *j*, *k* and *l*, which may lead to multi-spin diffusion.

geometry of an aqueous structure (often shortened to MARDIGRAS), which generates a proton–proton distance network taking into consideration all possible multi-spin diffusion pathways [9].

5.2.2.3 Combination of NMR and Molecular Modelling for 3D Structure Calculation of Oligonucleotides

The proton–proton distance constraints and sugar-ring torsion angles obtained from NMR spectroscopy can then be used as input structural parameters for the subsequent structure calculation utilising molecular modelling techniques. Molecular modelling is a mathematical approach that uses the equations of classical physics and often experimentally derived parameters (e.g. obtained from NMR) to represent 3D structures and energetics of molecules numerically [10, 11].

Molecular Mechanics Force Field

Molecular mechanics force field is a modelling approach based on classical physics, which attempts to calculate the total energy (E_{total}) and thus predict the geometry of a chemical structure. To decrease computational time, molecular mechanics treats the molecule as a collection of interacting point masses with harmonic forces [11]. Equation (5.11) represents a simple molecular mechanics equation used to calculate the total energy.

$$E_{\text{total}} = E_{\text{stretching}} + E_{\text{bending}} + E_{\text{torsion}} + E_{\text{vdw}} + E_{\text{elec}} \tag{5.11}$$

The components of this equation, together with the parameters which are necessary to describe the behaviour of atoms and bonds, are called a force field. For example, equation (5.11) represents the general force field used in molecular mechanics. Each term of this equation represents the contribution of the specified energy to the total energy as described below [11]

$$E_{\text{stretching}} = 1/2k_{\text{b}}(b - b_0)^2 \tag{5.12}$$

$$E_{\text{bending}} = 1/2k\theta(\theta - \theta_0)^2 \tag{5.13}$$

$$E_{\text{torsion}} = 1/2k_{\varphi}\left[1 + \cos(n\varphi - \varphi_0)\right] \tag{5.14}$$

$$E_{\text{vdw}} = \sum \frac{A_{ij}}{r_{ij}^{12}} - \frac{B_{ij}}{r_{ij}^{6}} \tag{5.15}$$

$$E_{\text{elec}} = -\sum_{ij} \frac{1}{\varepsilon} \frac{Q_i Q_j}{r_{ij}} \tag{5.16}$$

where $E_{\text{stretching}}$ is the energy of bond stretching, k_{b} is the bond stretching force constant, b_0 (the equilibrium value) is the unstrained bond length and b is the actual bond length in the molecule or molecules (5.12). k_θ is the angle bending force constant, θ is the actual value of the angle and θ_0 is the equilibrium value for the angle in the angle bending force field term (5.13). k_φ is the torsion barrier, φ is the actual torsion angle value, φ_0 is the reference torsion angle, which is 0° in the cosine function, with an energy maximum at 0°, and n is the periodicity (the number of energy minima at the full cycle completion) in the torsion angle force field term (5.14). A_{ij} is the repulsive coefficient, B_{ij} is the attractive coefficient and r_{ij} is the distance between the two atoms i and j in the van der Waals' (denoted vdw) force field term (5.15). ε is the dielectric constant, Q_i and Q_j are the atomic charges of the two atoms i and j, and r_{ij} is the distance between the two atoms in the electrostatic force field term (5.16).

The force field parameters, the force constant and the equilibrium values of the bond lengths and angles, are used to calculate the total energy of the molecule. Deviation from the equilibrium values results in an increase of the total molecular energy, which is a measurement of the intramolecular strain energy relative to that of a hypothetical ideal molecule. Different force field parameters could be used in the equations (5.12) to (5.16). Accordingly, some force fields are specific to small organic molecules, which represent a wide range of molecular classes, while other force fields are targeted at biomolecules (e.g. oligonucleotides) with specifically tuned force field parameters [11].

Molecular Dynamics

To ensure that molecules are able to find their lowest energy conformations and/or overcome energy barriers, molecular dynamics (MD) can be used to monitor molecular motion and thus explore the conformational space of a chemical structure [10]. This method combines the energy derived by the force field with the Newtonian law of motion. MD simulation is carried out by solving Newton's equations of motion over small time steps. Newton's second law is represented in equation (5.17) [12]

$$a_i = F_i / m_i \tag{5.17}$$

where a_i and m_i are the acceleration and the mass of atom i, respectively, and F_i is the force on atom i. Since the force is related to the negative gradient of the potential energy, Newton's equation can relate the derivative of the potential energy to the changes in the position of an atom as a function of time [12]

$$\frac{\mathrm{d}^2 x_i}{\mathrm{d}t^2} = \frac{1}{m_i}\left(-\frac{\mathrm{d}V}{\mathrm{d}x_i}\right) \tag{5.18}$$

where V is the potential energy of the system, x_i and m_i are the position and mass of atom i, respectively, and t is the time. Integration of equation (5.18) obtains x_i as a function of time

$$x_i = a_i t^2 + v_0 t + x_0 \tag{5.19}$$

where t is time, a_i and v_0 are the acceleration and the initial velocity of atom i, x_i is the position of the atom i after time period t and x_0 is the initial position of the atom. Acceleration can be calculated from the gradient of the potential energy function. It is therefore possible to generate a dynamics trajectory of the molecule based on the initial atom velocities, initial atom positions and the acceleration, which could be obtained from the potential energy. The initial velocities are assigned from a

Maxwell–Boltzmann distribution at a given temperature [13, 14]. Applying MD simulations for the calculation of 3D structures of oligonucleotides, their complexes and derivatives are rather expensive since they have a relatively large number of atoms, especially if explicit water is used in solvating these biomolecules. However, new advances in computational power (for example using graphics processing units) enable simulation of biomolecules on the microsecond timescale, and hence facilitate a better understanding of the structure and dynamic properties of oligonucleotides.

Restrained MD

A variant of MD simulation, restrained MD (rMD) searches the conformational space to find a final conformation of the biomolecule (e.g. oligonucleotide) that satisfies the experimental restraints obtained from NMR spectroscopy (e.g. proton–proton distances (from NOESY) and torsion angles (from DQF COSY). Different random starting conformers are needed to test the convergence into one or more molecular ensemble(s), which satisfy the experimental constraints. In rMD, the simulation is carried out at constant temperature using the full empirical force field in the calculations in combination with pseudoenergetic terms, which penalise violation of the experimental restraints [12]. However, performing rMD at a constant temperature can trap the system in a local minimum, which might prevent finding a global conformation minimum of the molecule. Therefore, simulated annealing can be used to assist the system to escape local minima [12].

Simulated Annealing Algorithm

A simulated annealing algorithm can be used to search the conformational space of the biomolecule. To achieve this, the annealing process is simulated by heating the system up to a certain high temperature in order to ensure that all conformers are energetically accessible. This initial stage is followed by cooling the system down to room temperature (or below) to find the natural conformation at that temperature [12]. This algorithm requires a starting structure which can be built in an extended form or in starting biomolecule conformations such as the A- or B-form for nucleic acids and oligonucleotides. Often, an implicit rather than explicit solvent is used in this type of simulation, which substantially decreases the computation time. The force field used in simulated annealing experiments includes the pseudoenergy terms, which incorporate distance and torsion angles constraints obtained from NMR.

5.2.2.4 Diffusion-Ordered Spectroscopy

Diffusion-ordered spectroscopy (DOSY) is a smart 2D NMR technique that allows the evaluation and measurement of diffusion coefficients of various molecular species in an NMR sample. DOSY experiments [15–18] are often used to resolve compounds in a complicated mixture according to their diffusion constants, which depend on the hydrodynamic radii of the corresponding molecular species. These experiments are often used in a non-quantitative fashion (e.g. using DOSY spectra), to resolve components of a mixture qualitatively. However, this type of spectroscopy can also be used to measure diffusion coefficients.

The diffusion constant D of a molecular species is related to its shape and size by the Stokes–Einstein equation (5.20) [17]

$$D = \frac{KT}{6\pi\eta r_s} \tag{5.20}$$

where D is the diffusion coefficient, K is Boltzmann constant, T is the temperature, η is the viscosity of the solution and r_s is the hydrodynamic radius of the particle.

The concept of the DOSY experiment is based on the idea of 'coding' each proton spatially in the NMR sample tube followed by 'decoding' it after allowing the sample to diffuse for a certain period of time. This could be achieved by applying a field gradient pulse, which introduces a gradient magnetic field along a desired axis, mainly along the Z-axis. In the absence of a magnetic field gradient, spins experiencing the same chemical environment evolve with the same frequency, regardless of their relative location within the sample, and thus have the same chemical shift and phase. However, in the presence of a magnetic field gradient, the same spins would evolve with different frequencies that depend on their relative position in the sample. In addition, phases of individual spins would directly depend on their physical location in the sample, and this dependence becomes stronger with the increased field gradient, which can be achieved by using specialised diffusion probes. In other words, in a DOSY experiment each spin becomes 'spatially labelled'.

Generally, a DOSY experiment relies on the effect of a couple of gradient pulses separated by a refocusing pulse (Figure 5.10). During the first gradient pulse, the spins precess in one direction, thus generating zero net signals at the end of this gradient pulse, as they are completely distributed in phase. However, the overall signal has not been lost, and it is possible to refocus it. If the second gradient pulse is applied for the same period of time, which has opposite polarity, but equal strength, the spins will evolve in exactly the opposite direction to become 'refocused'. The NMR signals of those spins that are significantly relocated in the sample due to diffusion processes

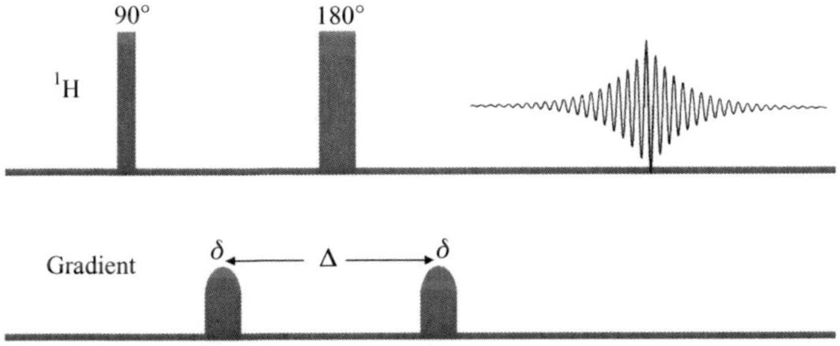

Figure 5.10 The pulse sequence of a typical DOSY experiment, which consists of a 90° pulse, two gradient pulses and a 180° pulse. The pulse sequence starts with the 90° pulse, which produces the net magnetisation in the *xy* plane followed by the first gradient pulse to 'encode' each spin spatially. The applied 180° pulse reverses spins' evolution, which is followed by the second gradient pulse (with the same strength, but with opposite sign) that refocuses spins' signals. The diffused spin will not recover the full (initial) signal intensity, which leads to the signal intensity attenuation.

will not be refocused, since during the second gradient pulse they experience a different strength of the gradient field as compared with the original gradient used during the first gradient pulse. Therefore, the total signal for spins that have diffused during this experiment will be considerably decreased as compared with the signals of those spins that moved to a lesser extent. This means that the attenuation of the NMR signals can be modulated by the diffusion properties of the corresponding molecular species. The specialised diffusion probes use a much higher power than standard probes which is generated by a gradient amplifier. As a result, the diffusion probes can generate a maximum gradient strength of 1200–1500 Gauss/cm, whereas usual probes will produce a much lower gradient field (a maximum of 50 Gauss/cm). Therefore, the linear distortion in the magnetic field is much greater with the specialised diffusion probes.

The ratio between the intensity of the attenuated NMR signal I and the initial signal I_0 is related to the diffusion properties of the molecule by the Stejskal–Tanner equation (5.10) [18]

$$\frac{I}{I_0} = e^{-D\gamma^2 g^2 \delta^2 (\Delta - \delta/3)} \tag{5.21}$$

where D is the diffusion coefficient, γ is the gyromagnetic ratio, g is the gradient field pulse strength, δ is the gradient field pulse length and Δ is the diffusion time.

The data obtained in DOSY experiments can be used to generate a 2D spectrum (see Figure 5.11) with a chemical shift axis in the F2 dimension (as in the usual 2D experiment), but with diffusion coefficients in the indirect dimension. Alternatively, numerical values for the diffusion constant at a given chemical shift can be obtained using a different way of processing the generated data. Indeed, plotting the Stejskal–Tanner equation at different gradient strength values results in a correlation, which allows calculation of diffusion coefficients for different molecular species [17].

In some cases, diffusion coefficients of the components of the sample mixture have similar values, which results in producing one broad peak due to a continuous distribution of particle sizes, thereby reflecting contributions from various diffusion coefficients. In these cases an inverse Laplace transform known as a CONTIN algorithm can be used to resolve the individual diffusion coefficients [19]. CONTIN is a program, which allows an inversion of data represented by linear algebraic or integral equations through an inverse Laplace transform. This mathematical algorithm uses the regularisation method by utilising least-squares analysis in order to approximate the solutions, and then the simplest solution that is consistent with the experimental data can be selected [19]. Analysis of the signal attenuation in the DOSY experiment by the CONTIN algorithm results in evaluation of the distribution of the diffusion coefficients in the complex sample mixture [17].

5.2.2.5 Modern NMR Instrumentation

Modern NMR spectrometers are based on a pulsed field Fourier transform NMR spectroscopy technique (FT-NMR), which was first introduced in the 1960s. Similar to the old continuous wave (CW) NMR spectrometers, pulsed field NMR spectrometers

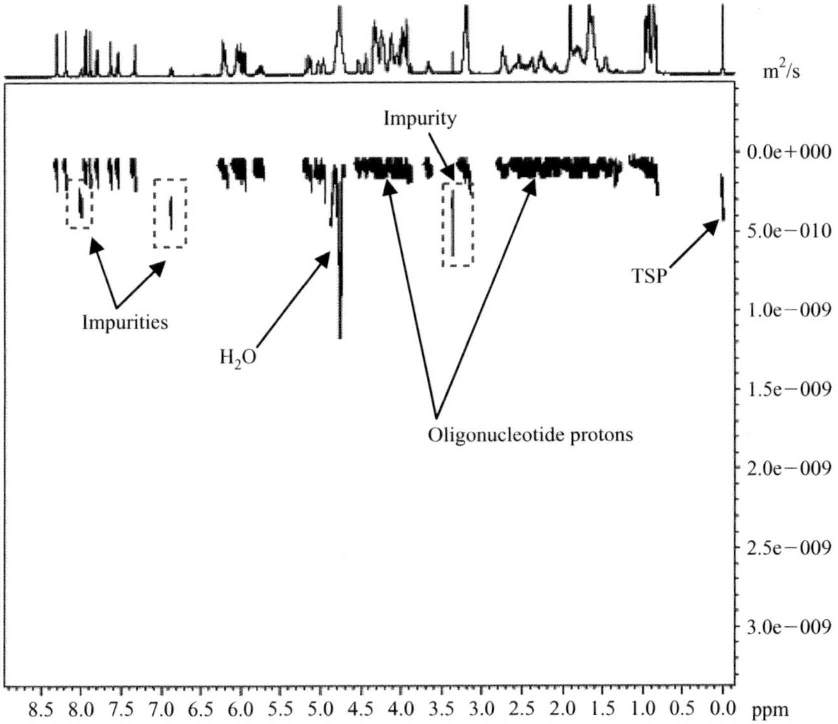

Figure 5.11 An example of a ^1H DOSY spectrum of the oligonucleotide recorded in D_2O (400 MHz) with internal reference (TSP). Impurities are highlighted by the dotted line boxes.

use a magnetic field to align the nuclei. However, instead of applying the external magnetic field followed by scanning through the frequencies, modern NMR spectrometers excite all the resonant nuclei by utilising a single pulse of radio-frequency energy followed by collecting the time-domain response, which is then converted to the frequency-domain spectrum using Fourier transformation.

The application of Fourier transformation to NMR spectroscopy allowed considerable enhancement of the sensitivity of this method by increasing the signal-to-noise ratio, which is a major concern of any analytical technique. Moreover, the use of FT-NMR offered the opportunity to develop correlational multi-dimensional NMR spectroscopy.

5.3 Application of ^1H NMR for Oligonucleotides

5.3.1 Use of 1D and 2D ^1H NMR Techniques

The main problem in application of 1D NMR spectroscopy for identification and characterisation of oligonucleotides and their analogues is associated with the fact that they often produce highly overlapping resonance regions in the ^1H NMR spectra due to a large number of protons, some of which display a close similarity in chemical shifts. Spectra interpretation can also be complicated by the fact that the assignment of specific nuclei to particular nucleotide residues might be a challenging task, especially without implementing 2D (or even higher dimensionality) NMR techniques. However, even 1D NMR spectra may provide invaluable information on the composition of oligonucleotide sequences and on the overall quality of samples.

5.3.1.1 Main Types of Oligonucleotide Protons Detectable by NMR

Figure 5.4 shows the chemical structures of four DNA (A, G, C and T) and RNA (A, G, C and U) nucleotides bases along with the deoxyribose and ribose sugar moieties. Two types of protons could be recognised in the DNA and RNA, namely, non-exchangeable and exchangeable protons (the latter are labelled in red).

Non-exchangeable Protons from Nucleotide Bases

In the case of DNA, cytidine (cytosine attached to ribose sugar) and thymidine (thymine attached to ribose sugar) (Figure 5.4) constitute two unique spin systems (CH5 ↔ CH6 for cytidine and TH6 ↔ T(CH_3) for thymidine), which are not J-coupled to the sugar ring protons. Cytosine is the only nucleotide base in DNA containing vicinal protons (H5 and H6) on adjacent carbons, which are usually split into doublets and expected to produce an intensive cross-peak in COSY and TOCSY. In addition, these two protons are located in close proximity as they are separated by only 0.25 nm (2.5 Å) (Figure 5.12, see colour insert). Therefore these protons can also be recognised *via* corresponding cross-peak in the NOESY spectrum. TH6 and TCH_3 protons within the same thymidine nucleotide residue usually display a very weak J-coupling constant (ca. 1.2–1.5 Hz), and thus are seen in the NMR spectrum as broad singlets. Despite this poor spin–spin coupling interaction, TH6 and TCH_3 often generate a low-intensity cross-peak in J-correlated spectra (e.g. COSY and especially in TOCSY). These protons are closely located in space with the distance of 2.5 Å between them (Figure 5.12), and thus can be detected *via* a NOESY spectrum in a form of a medium intensity cross-peak.

In RNA the two pyrimidine bases (uridine, which replaces thymidine, and cytidine) form similar spin systems (CH5 ↔ CH6 and UH5 ↔ UH6). In a 1D NMR spectrum CH5, CH6, UH5 and UH6 protons are seen as doublets and could be detected in both COSY (or TOCSY) and NOESY. The remaining non-exchangeable aromatic protons (i.e. AH8, GH8 and AH2) usually appear as singlets and are not involved in any spin–spin connectivity.

Non-exchangeable Sugar Ring Protons

Sugar ring protons of both deoxyribose and ribose moieties usually form isolated spin systems for each nucleotide residue, which are involved in a net of J-coupling interactions indicated by arrows in Figure 5.13 (see colour insert). In the case of DNA, H1′ protons of the deoxyribose moiety are expected to produce doublets of doublets due to their J-coupling interactions with H2′ and H2″ protons of the same sugar ring. Their resonance region (5.2–6.4 ppm) is usually isolated from the spectral areas of other sugar ring protons. Geminal protons H2′ and H2″, which are attached to the same C2′ carbon, are normally seen as multiplets due to their additional spin–spin coupling with H1′ and H3′ protons. The resonance area for H2′ and H2″ protons of deoxyribose is located between 1.7 and 3.0 ppm. H3′ are also seen as multiplets due to their J-coupling interactions with H4′, H2′ and H2″ from the same nucleotide residue as well as with ^{31}P atom of the 3′-phosphodiester group. The resonance areas for H3′ protons range between 4.6 and 5.1 ppm, which partially overlap with the HOD signal (note that a NMR water signal in aprotic solvents is seen as an H_2O signal, while in protic solvents it is seen as a HOD signal from a water molecule in which one of the protons has been exchanged for a deuterium of the solvent) of the residual water signal. H4′, H5′ and H5″ usually form a strongly overlapped region between 3.6 and 4.6 ppm. H5′ and H5″ display $^{1}H-^{31}P$ coupling interactions with the ^{31}P atom of the 5′-phosphate group.

In RNA, H2″ is absent, whereas H2′ is shifted considerably downfield (4.4–5.0 ppm) due to the presence of the electron withdrawing OH group at the 2′ position. As a result, the resonance area of the H2′ protons substantially overlaps with that observed for H3′, H4′, H5′ and H5″ protons, which are usually seen between 3.7 and 5.2 ppm (see Table 5.2).

Exchangeable Amino and Imino Protons

Amino and imino protons are exchangeable protons, as they usually exist in a fast exchange with the solvent. Exchangeable protons are usually not observable in D_2O,

Table 5.2 Summary of proton chemical shifts in DNA and RNA.

Protons	DNA, δ (ppm)	RNA, δ (ppm)
GH8, AH8, AH2	7.3–8.4	7.3–8.4
CH6, TH6 and/or UH6	6.9–7.6	6.9–7.6
CH5 and/or UH5	5.3–6.0	5.3–6.0
T-CH_3	1.2–1.6	[a]
AH1′, GH1′, CH1′, TH1′ and/or UH1′	5.2–6.4	5.0–6.0
AH2′ and GH2′	2.3–2.9	4.4–5.0
CH2′, TH2′ or UH2′	1.7–2.3	4.4–5.0
AH2″ and GH2″	2.4–3.1	[a]
CH2″, TH2″	2.1–2.7	[a]
AH3′, GH3′, CH3′, TH3′ and/or UH3′	4.3–5.2	4.3–5.2
AH4′, GH4′, CH4′, TH4′ and/or UH4′	3.8–4.3	3.8–4.3
AH5′/5″, GH5′/H5″, CH5′/H5″, TH5′/H5″ and/or UH5′/H5″	3.8–4.3	3.8–4.3
G(N1H)[b]	11.5–13.6	11.5–13.6
T(N3H) and U(N3H)[b]	13.0–14.0	13.0–14.0
A(NH_2), C(NH_2) and C(NH_2)[c]	6.6–9.0	6.6–9.0

[a]Absent in RNA.

[b]Exchangeable imino protons, which are involved in formation of hydrogen bonding and can only be detected in stable duplexes in H_2O-based solutions.

[c]Exchangeable amino protons which can be detected only in H_2O-based solutions.

but could be detected in H_2O-based buffers (typically 90–95% H_2O and 10–5% D_2O to ensure deuterium field lock) using experiments with selective non-excitation of the water resonance (e.g. water suppression by gradient-taylored excitation experiments).

There are a number of new pulse sequences currently available for the efficient and selective suppression of a water signal without considerable distortion of other spectral regions. Figure 5.14 shows expanded regions of two NMR spectra of synthetic oligonucleotides after HPLC purification, recorded with and without suppression of the water signal ((a) and (b), respectively). This example clearly demonstrates that the undesirable HOD signal dominating the NMR spectrum can successfully be eliminated by selecting an appropriate pulse program.

Although amino protons can in principle be observed even in the single-stranded form of nucleic acids, imino protons can only be detected by NMR in a double-stranded (duplex) form. In Figure 5.15 (see colour insert) imino and amino protons involved in the hydrogen bond formation are highlighted in blue. The presence of imino proton signals in the ^{1}H NMR spectrum of oligonucleotides usually signifies the

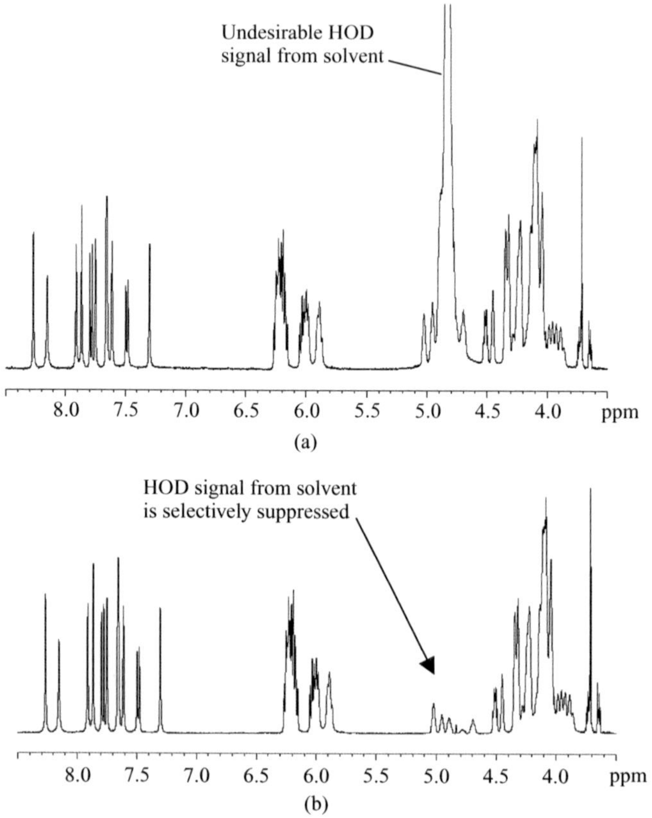

Figure 5.14 Expanded regions of the NMR spectra of synthetic oligonucleotide recorded after HPLC purification (a) without and (b) with suppression of residual water signal.

formation of specific hydrogen bonding, which may confirm the presence of folded conformations for single-stranded nucleic acids and/or the formation of complementary double-stranded complexes.

The number of imino proton resonances corresponds to the number of hydrogen-bonded base pairs formed during the hybridisation process of two complementary regions, whereas the line widths indicate the relative stability of the corresponding base-pairs and their accessibility to water molecules, which may facilitate the exchange. The G and T imino protons in canonical Watson–Crick G·C and A·T pairs are found at 11.5–13.6 ppm and 13.0–14.0 ppm, respectively, while the imino proton signals in protonated cytidines and unpaired thymidines are located around 14–16 ppm and 10–11 ppm, respectively (Table 5.2). Figure 5.16 shows an example of the ^{1}H NMR spectrum of the 6mer duplex recorded in phosphate buffer prepared using a H_2O:D_2O mixture (90%:10%), which shows the region of the exchangeable imino protons involved in the formation of Watson–Crick hydrogen bonds. The exchangeable imino protons are visible only when they are involved in the hydrogen bonding [20, 21] and start broadening when the exchange rate with the bulk solvent

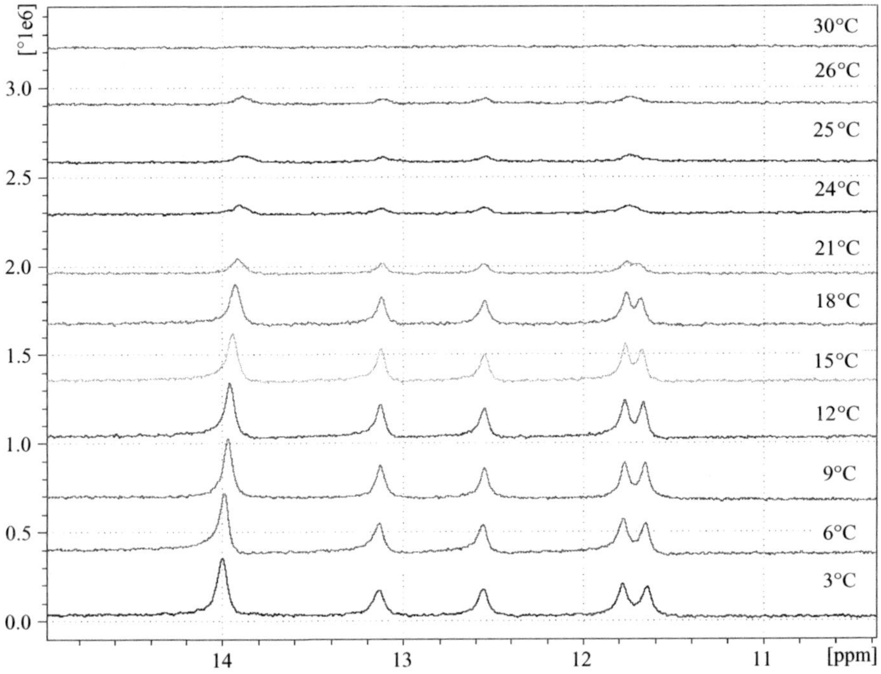

Figure 5.16 ^{1}H NMR spectra of the hybrid complex between 6mer $p^1T^2C^3A^4A^5T^6C$ and RNA target $^{5'}$GC**<u>GAUUGA</u>**AAACG$^{3'}$ recorded at different temperatures (indicated above each spectrum), which show the resonance region of the exchangeable imino protons. The spectra were recorded in 90%H_2O:10%D_2O phosphate buffer (10 mM phosphate, 200 mM NaCl, 0.1 mM ethylenediaminetetraacetic acid, pH 7.0) at 500 MHz. (The complementary region within the RNA target is underlined and shown in bold.)

exceeds 200/s [20]. It is seen from Figure 5.16 that increase of the temperature of the NMR sample induces thermal denaturation of the oligonucleotide duplex, leading to a decrease of the imino proton signals with their eventual disappearance. Therefore, the presence of any secondary structures within oligonucleotide sequences and their relative stability can easily be monitored using NMR spectroscopy.

Exchangeable amino protons involved in Watson–Crick hydrogen bonding can also be detected in H_2O-based solutions. However, their resonance signals, which are usually seen between 6.6 and 9.0 ppm, often overlap with the aromatic protons of the nucleotide bases, making it difficult to monitor them by ^1H NMR during thermal denaturation experiments.

Another approach for detecting internal structures within oligonucleotide chains and/or monitored hybridisation events between different oligonucleotide species could be melting temperature profiles obtained using variable temperature (VT) NMR experiments (see Chapter 6 for fuller details). This can be achieved by following the chemical shifts of the non-exchangeable protons of the oligonucleotide residues against temperature. An example of such T_m (melting temperature) profiles is shown later in Figure 5.17(a) for the hybrid complex (1:1) between 6mer pdTCAATC and RNA target $^{5'}$rGC**<u>GAUUGA</u>**AAACG$^{3'}$ obtained by plotting the chemical shift of the $^1T(CH_3)$ protons of the nucleotide residue as a function of temperature. A sigmoidal-like character of the melting temperature profile usually indicates a co-operative transition within the double-stranded regions of nucleic acids. In contrast, the NMR-based T_m profile recorded for the equimolar mixture of the same 6mer oligodeoxynucleotide pdTCAATC and the non-complementary oligoribonucleotide ($^{5'}$rCACUGGGAAGUUU$^{3'}$) shows nearly linear character of the temperature-induced change in chemical shift (Figure 5.17(b)), thus demonstrating the absence of any stable secondary structures formed by these two oligonucleotides. The advantage of performing VT NMR experiments for these purposes is that this

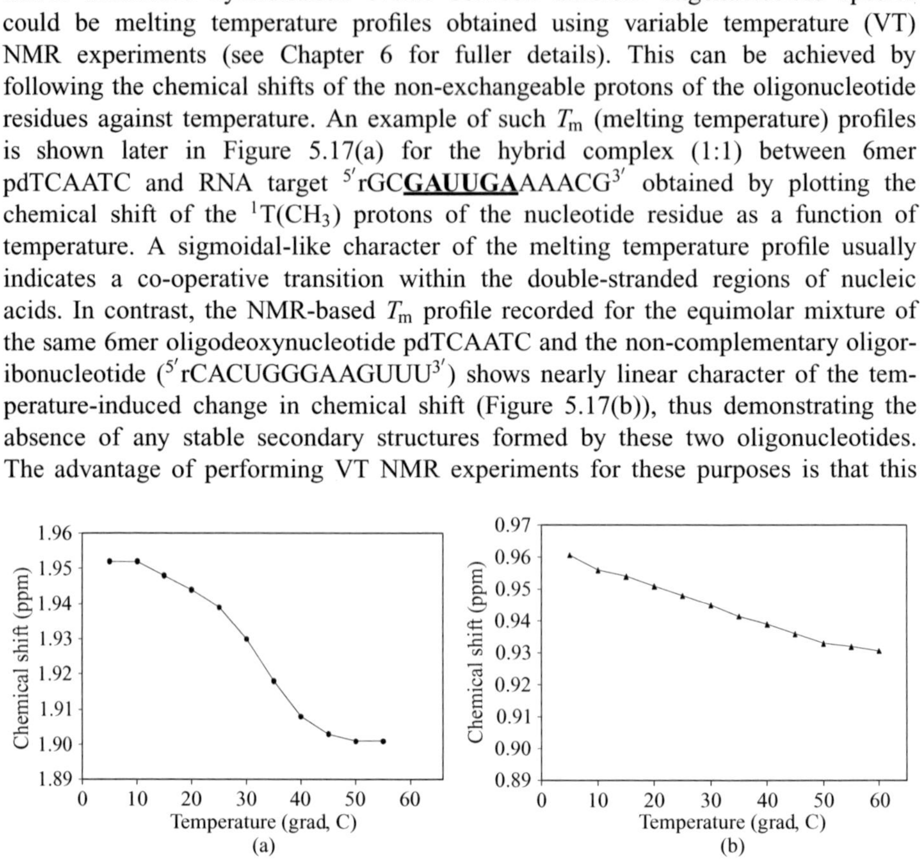

Figure 5.17 NMR-based melting temperature profiles obtained for the equimolar mixtures of the 6mer oligodeoxynucleotide pTCAATC with (a) complementary RNA target $^{5'}$GC**<u>GAUUGA</u>**AAACG$^{3'}$ and (b) non-complementary RNA sequence $^{5'}$CACUGGGAAGUUU$^{3'}$ in phosphate buffer. The melting profiles were obtained by plotting the chemical shifts of the selected protons of the 6mer oligodeoxynucleotide against temperature.

method allows monitoring of the participation of individual nucleotide residues in the formation of double-stranded structures by following signals of corresponding proton signals.

VT ^{1}H and ^{31}P NMR-based experiments have been used by Jaroszewski *et al.* [22] to investigate duplex stability of phosphorothioate and phosphorodithioate DNA analogues, in which all phosphodiester linkages were replaced by phosphorothioate or phosphorodithioate backbones. They also used NMR spectra of exchangeable imino proton resonances recorded at different temperature regimes to look at thermal stabilities of individual Watson–Crick base pairs of the mono- and the dithioate duplexes. Temperature dependence of the imino proton resonances and ^{31}P signals of the phosphorodithioate allowed detection of the heterogeneity of the base-pairing, compatible with the presence of a hairpin loop. The phosphorothioate analogue showed strongly reduced base-pair lifetimes evident from NMR experiments.

Chemical shifts for the most important oligonucleotide protons are given in Table 5.2, but these can serve only as guidance as they strongly depend on oligonucleotide sequences and their interactions with complementary counterparts. Media (e.g. ionic strength, pH, concentration, presence of metal ions) may also strongly affect proton chemical shifts.

5.3.1.2 Typical 1D NMR Spectra of Oligonucleotide Samples

Figure 5.18 shows a typical ^{1}H NMR spectra of two complementary oligonucleotide sequences, (a) dp^{1}T^{2}T^{3}T^{4}T^{5}C^{6}A^{7}A^{8}T^{9}C and (b) dp^{1}G^{2}A^{3}T^{4}T^{5}G^{6}A^{7}A^{8}A^{9}A, in their single-stranded forms, demonstrating the characteristic resonance areas for each type of oligonucleotide proton. It can be seen from this example, that even 1D ^{1}H NMR spectra of oligonucleotides can provide essential information on nucleotide composition and purity. For example, within the aromatic region of ^{1}H NMR spectrum of dp^{1}T^{2}T^{3}T^{4}T^{5}C^{6}A^{7}A^{8}T^{9}C contains eleven protons, namely, five T(H6) broad singlets (due to ^{4}J-coupling interactions with corresponding T(CH_3) protons), two doublets from two C(H6), two A(H8) singlets and two sharp singlets from A(H2) protons. The upfield region of the spectrum between 1.2 and 1.6 ppm contains three intensive signals from five T(CH_3) protons, two of which have an integral intensity of 3H, whereas the most intensive signal at 1.9 ppm has an integral intensity of 9H due to overlapping T(CH_3) signals.

The second oligonucleotide, dp^{1}G^{2}A^{3}T^{4}T^{5}G^{6}A^{7}A^{8}A^{9}A, produces a substantially different ^{1}H NMR spectrum, and facilitates the distinguishing of two separate oligonucleotides samples. The aromatic area of the ^{1}H NMR spectrum of dp^{1}G^{2}A^{3}T^{4}T^{5}G^{6}A^{7}A^{8}A^{9}A contains eleven separate signals including two T(H6) broad singlets, two singlets from G(H8), five A(H8) singlets and five sharp singlets from A(H2) protons. The aliphatic region of the spectrum between 1.2 and 1.8 ppm contains only two intensive signals from $-CH_3$ groups of two thymidines.

This quick overview of ^{1}H NMR spectra also allows some important assessments regarding the purity of these oligonucleotides samples. For example, sharp peaks at around 3.4 ppm seen in both spectra (labelled by asterisk) signify the presence of an impurity in each sample, presumably from triethylamine (see Chapter 2). The other,

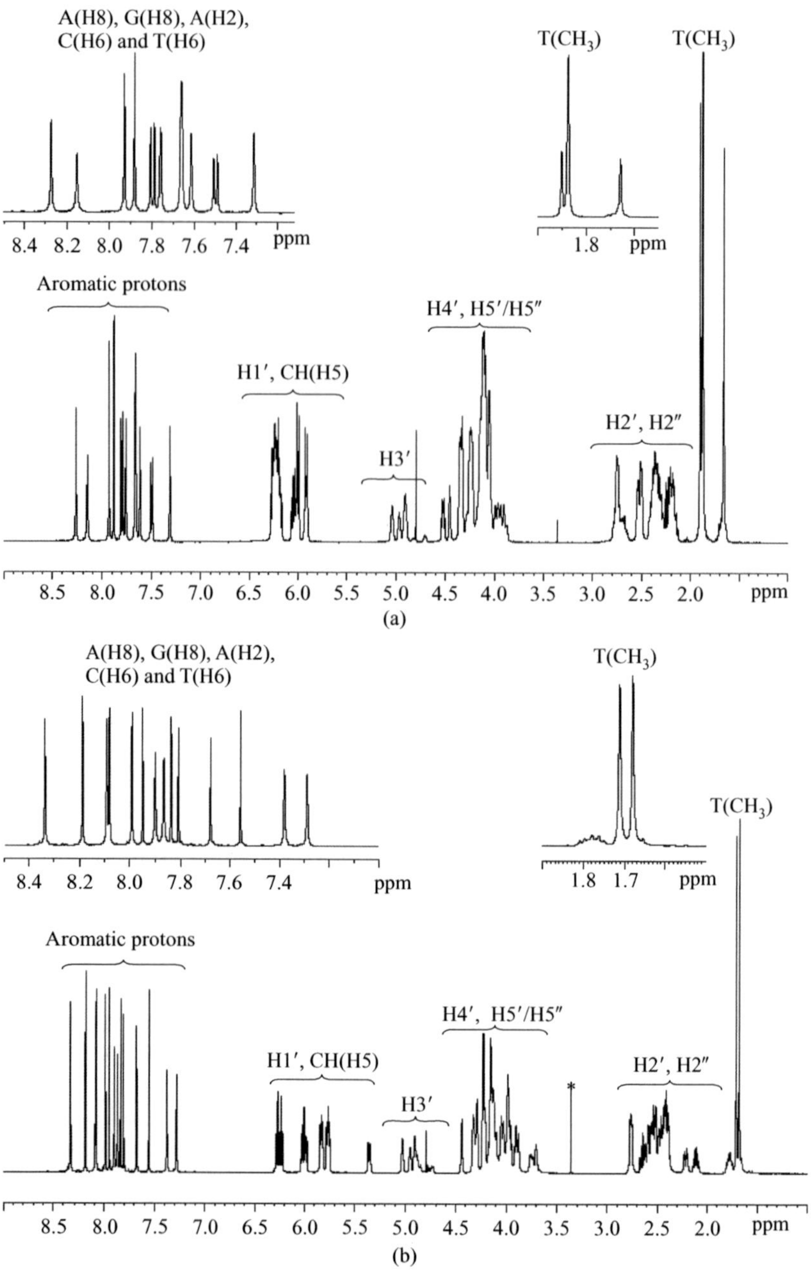

Figure 5.18 ^{1}H NMR spectra of the 9mer oligonucleotides (a) $dp^1T^2T^3T^4T^5C^6A^7A^8T^9C$ and (b) $dp^1G^2A^3T^4T^5G^6A^7A^8A^9A$ showing the characteristic regions of the oligonucleotide protons. The spectra were recorded in D_2O at 20°C using 500 MHz NMR (Bruker). Insertions on the top of each spectra show expanded areas of aromatic and T(CH_3) protons. Sharp peaks at ~3.4 ppm in both spectra labelled by * indicate an impurity.

very minor impurities seen, for example in the aromatic region of $dp^1G^2A^3T^4T^5G^6A^7A^8A^9A$ (Figure 5.18(b)) do not exceed 1%.

However, 1D NMR spectra do not always confidently confirm the exact oligonucleotide sequence of the oligonucleotide sample since this method does not provide direct information on how the individual nucleotide blocks are connected to each other within the oligonucleotide chain. There are two main reasons of this. First, regardless of the quality of the spectra produced, full identification of oligonucleotide protons from 1D NMR spectra alone remains challenging, especially for relatively long oligonucleotide sequences. Although some information about proton spin–spin interactions can be gained from the fine multiplet structures of signals as well as from measurements of J-coupling constants, complete identification of individual spin systems entirely from 1D NMR spectra is usually problematic. The second reason is that 1D NMR spectra provide only limited information on long-distance interactions between different nuclei, and, thus, it is difficult (if possible at all) to make any confident judgement on the mutual orientation of individual nucleotide residues within the oligonucleotide strand.

This essential information can be obtained from 2D NMR experiments by using both J-correlated-based spectroscopy techniques (e.g. COSY, TOCSY, DQF-COSY or their analogues) and NOE-based methods (e.g. NOESY, ROESY or alternative variants of these techniques), which constitutes the main focus of the following sections.

5.3.1.3 Sequence Specific Assignment of Proton Signals in Nucleic Acids

The first step of the assignment strategy for oligonucleotides involves the identification of each signal in the spectrum to a specific type of spin system followed by the assignment of these systems to their sequential positions in the oligonucleotide chain. In most cases this is done using a combination of COSY (see Figure 5.19 as an example), TOCSY and DQF-COSY.

Sequence-specific assignment of oligonucleotide protons using NOESY is based on a very distinctive interaction network of aromatic bases and sugar ring protons within the double-stranded regions of nucleic acids. Figure 5.20 (see colour insert) represents the 3D structure of the short fragment of the DNA duplex, showing only one oligonucleotide strand for reasons of simplicity. The structure maintains typical π–π stacking interactions between nucleotide bases, which, together with Watson–Crick hydrogen bonding (not shown for clarity), maintain a relatively stable helical conformation. Figure 5.20 also highlights the most important protons, which are normally involved in through-space interactions and usually generate cross-peaks in ^{1}H NOESY spectra.

The main strategy for sequential assignment of oligonucleotide protons is outlined in Figure 5.21 (see colour insert), which shows some typical intra- and inter-residue NOE-connectivities (indicated by red and blue arrows, respectively) seen in the NOESY spectra of oligonucleotides. The main network of the key interactions involves through-space contacts between H1′, H2′, H2″ and H3″ sugar ring protons of the residue *n* with its own aromatic H6/H8 protons, as well as with the H6/H8 protons of the next ($n + 1$) residue

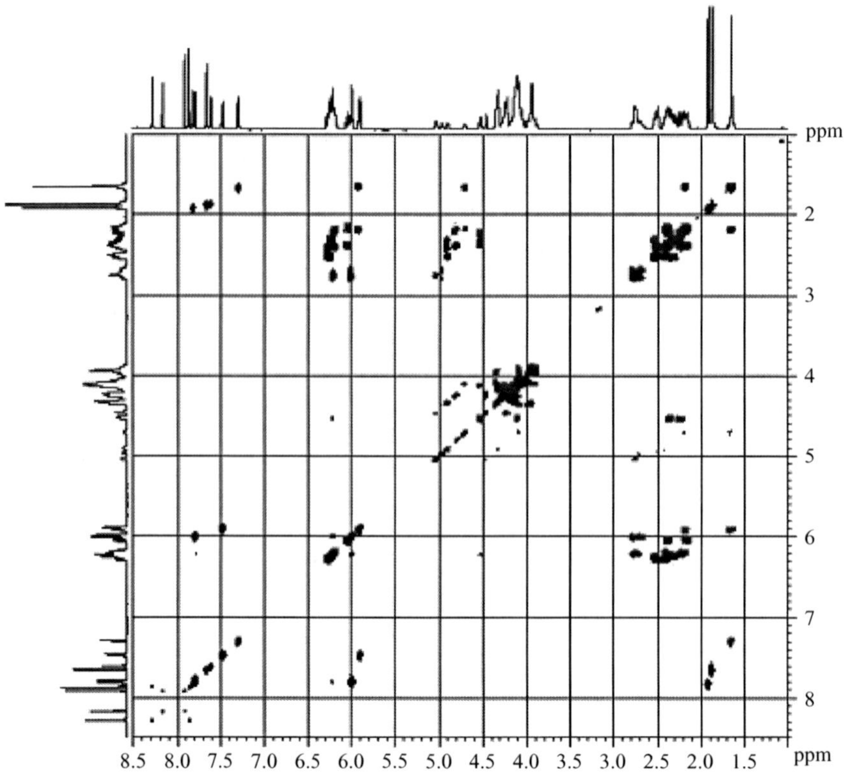

Figure 5.19 COSY spectrum of single-stranded oligonucleotide $dp^1T^2T^3T^4T^5C^6A^7A^8T^9C$ recorded in D_2O at 20°C using 500 MHz NMR (Bruker).

$$\leftrightarrow H6/H8_n \leftrightarrow (H1', \; H2'/H2'', H3')_n \leftrightarrow H6/H8_{n+1}$$

$$\leftrightarrow (H1', H2'/H2'', H3')_{n+1} \leftrightarrow \ldots$$

In addition, aromatic H8/H6/H5 protons and CH_3 protons of thymidine can form the additional network of intra- and inter-residue interactions

$$\ldots \leftrightarrow (H5/H6/H8/CH_3)_n \leftrightarrow \; H5/CH_3H6/H8_{n+1} \leftrightarrow \ldots$$

The starting point of this sequential assignment could be, for example, the 5′ terminal residue, as it will have only intra-residue H6/H8-H1′, H6/H8-H2′, H6/H8-H2″ and H6/H8-H3′ NOESY cross-peaks. However, each sugar ring proton of the 5′ terminal residue (H1′/H2′/H2″ and H3′) will have an additional (inter-residue) cross-peak with the aromatic protons of the 3′-adjacent residue. Taking this as a starting point, the other aromatic and sugar ring protons can be assigned in a sequential manner by following this strategy.

There are usually several independent assignment pathways of sequential assignment, which provides a resolution to spectral overlap. Any over-looked connectivity (as a result of spectral overlap) in one region can be compensated by results from the other regions.

Recently, it was demonstrated that this strategy, previously used predominantly for double-stranded nucleic acids, could also be applicable for single-stranded oligonucleotide sequences, which in aqueous solutions often maintain relatively rigid conformations similar to those found in analogous duplex forms.

An example of the sequence-specific assignment of aromatic and H1′ protons as well as -CH_3, H2′ and H2′′ protons of the single-stranded $dp^1G^2A^3T^4T^5G^6A^7A^8A^9A$ sequence is demonstrated in Figure 5.22, which shows the expanded regions of the ^{1}H NOESY spectrum of this oligonucleotide with indicated H8/H6–H1′ (a) and H8/H6–H2′/H2″/CH_3 connectivities (b).

By following the dashed lines in Figure 5.22, it is possible to track the most important NOE-interactions in this oligonucleotide (e.g. $H6/H8_n \leftrightarrow (H1', H2'/H2'', H3')_n \leftrightarrow H6/H8_{n+1} \leftrightarrow (H1', H2'/H2'', H3')_{n+1}$).

5.4 Application of ^{31}P NMR Spectroscopy for Oligonucleotide Analogues

^{31}P NMR spectroscopy [23,24] is broadly used for the characterisation of oligonucleotides and their derivatives, as phosphorus atoms are the key components in the sugar-phosphate backbone of nucleic acids. ^{31}P has a 100% natural abundance and nuclear spin quantum number of $\frac{1}{2}$, which makes it detectable by NMR spectroscopy [24]. The relative sensitivity of the phosphorus nucleus is considerably lower than that of the proton (i.e. 6.63% of the proton sensitivity see Table 5.1). However, the ^{31}P NMR spectra of oligonucleotides are relatively simple and may provide vital information about oligonucleotide purity, identity and conformation of the sugar-phosphate backbone. The concepts that have been discussed above in application to proton NMR spectroscopy in Sections 5.1.1–5.1.3 can also be applicable to ^{31}P NMR spectroscopy. The integral intensity of the peak is indicative of the number of phosphorus atoms responsible for this peak. Furthermore, each phosphorus functional group has a distinct region in the ^{31}P NMR spectrum, which extends over 500 ppm (from −180 to +220 ppm) [23]. A slight change in the chemical structure around the phosphorus nucleus results in a significant chemical shift difference. Furthermore, the torsion angles around the phosphorus atom strongly affect its chemical shift [23]. A proton-decoupled ^{31}P NMR spectrum of the 6mer oligonucleotide with R modification group attached to 5′-terminal phosphate group via the phosphoramidate bond is shown on Figure 5.23 as an example.

As demonstrated in Figure 5.23, phosphodiester ^{31}P signals of oligonucleotides and their analogues are usually seen in the upfield area of the NMR spectrum as compared with the TMP signal and especially with that observed for phosphoramidate ^{31}P nucleus. Therefore, chemical modification of oligonucleotides via covalent attachment of various functional groups through terminal 5′- or 3′-phosphate groups can easily be tracked using ^{31}P NMR spectroscopy. Phosphorothioate oligonucleotide derivatives, which are broadly used as therapeutic oligonucleotide analogues, can also be analysed and quantified by following distinctive ^{31}P NMR signals of phosphorothioate groups at around 55 ppm.

This method also allows the evaluation and quantification of some degradation

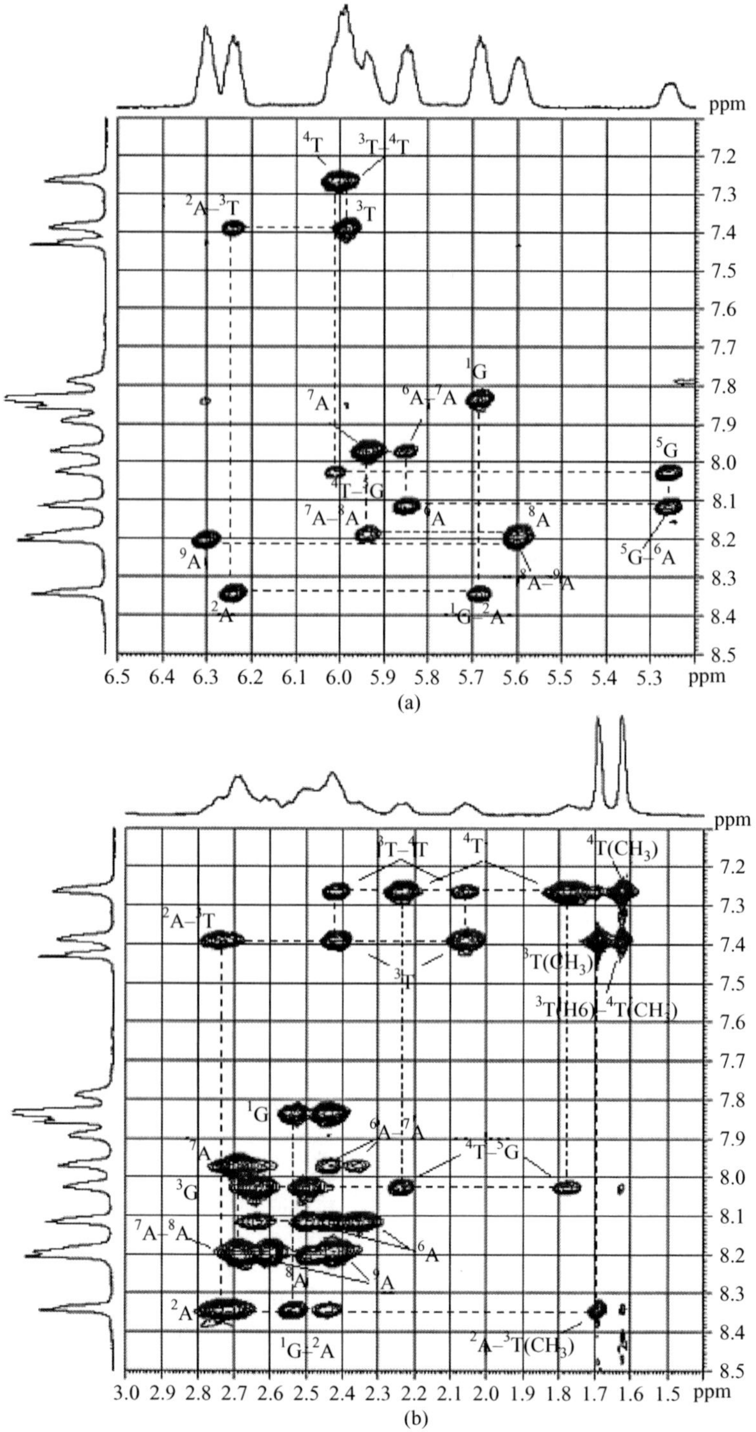
ppm
6.5 6.4 6.3 6.2 6.1 6.0 5.9 5.8 5.7 5.6 5.5 5.4 5.3 ppm
7.2 7.3 7.4 7.5 7.6 7.7 7.8 7.9 8.0 8.1 8.2 8.3 8.4 8.5
3.0 2.9 2.8 2.7 2.6 2.5 2.4 2.3 2.2 2.1 2.0 1.9 1.8 1.7 1.6 1.5 ppm

(a)

(b)

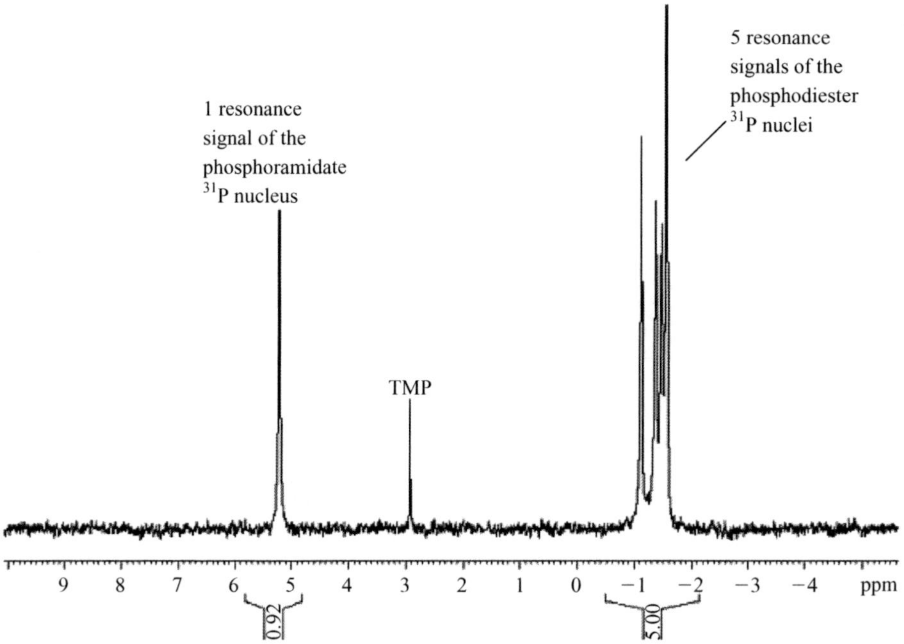

Figure 5.23 1D (proton decoupled) ^{31}P NMR spectrum of the oligonucleotide synthetic analogue R—NH-pTCAATC with covalently attached modification group R via phosphoramidate linker. TMP was used as internal reference with chemical shift set to 2.92 ppm. The spectrum shows the integration of the ^{31}P resonance signal corresponding to six phosphorus atoms for the oligonucleotide part.

products and/or impurities of oligonucleotide nature *via* integration of isolated ^{31}P NMR signals. Cieslak *et al.* [25] used ^{31}P NMR spectroscopy to investigate the desulfurisation process of oligonucleoside phosphorothioates induced by 'aged' trichloroacetic acid solutions in the solid-phase synthesis of oligonucleoside phosphorothioates. It was demonstrated that the ^{31}P NMR signal of phosphorothioate diesters appeared as a broad signal at 56 ppm, whereas the resonance corresponding to desulfurised phosphodiesters appeared at around 0 ppm, thus allowing quick and unambiguous detection of the degradation products. Based on the findings of this research it was recommended that the deblocking solutions should be prepared from solid TCA prior to use in the solid-phase synthesis to ensure the integrity of phosphorothioate diesters.

The phosphorus atom can also be involved in spin–spin coupling interactions with nuclei (e.g. protons) connected to it *via* chemical bonds. These coupling

---

Figure 5.22 *(page 230)* Expanded (a) H8/H6–H1′ and (b) H8/H6–H2′/H2′′/CH_3 regions of the ^{1}H NOESY spectrum of the single-stranded dp^{1}G^{2}A^{3}T^{4}T^{5}G^{6}A^{7}A^{8}A^{9}A sequence recorded in D_2O (500MHz). This figure shows the sequential assignment of inter-nucleotide and intra-nucleotide NOE interactions within the oligonucleotide part involving the H6/H8 aromatic protons and sugar ring protons.

interactions and associated coupling constants can potentially be exploited to determine torsion angles of the sugar-phosphate backbones in nucleic acids. The application of the 2D ^{1}H-^{31}P 2D NMR techniques provides further details about structural features of oligonucleotides and their structural analogues. Therefore, ^{31}P NMR, together with the 1D and 2D ^{1}H NMR spectroscopy, is often used to extract some important structural parameters, which might be essential for conformational analysis of nucleic acid.

5.5 Diffusion-Ordered Spectroscopy for Oligonucleotide Characterisation

In the area of nucleic acids, DOSY techniques [15–18] can be used, for example to analyse the purity of the synthesised oligonucleotide products and their analogues. Very often post-synthetic reaction mixtures contain many different starting materials, intermediates and products and thus produce complicated NMR spectra with strongly overlapping signals. This makes it difficult (if possible at all) to characterise the reaction products. This issue might potentially be resolved by applying DOSY, as NMR signals from each individual molecular species can be easily identified using this type of spectroscopy. This technique can also be used for detection of self-assembly and/or aggregation of nucleic acids and their analogues [17]. If particulate matter exists in equilibrium between aggregated forms and individual unbound species, they may have different diffusion properties, which could be measured by DOSY. Finally, DOSY can also be used to detect the interaction of oligonucleotides with bigger biomolecules (i.e. proteins), since bound forms might have diffusion properties close to those of the large biomolecules and very different from those of the unbound oligonucleotide species.

5.6 Application of NMR for Structural Analysis of Oligonucleotides with Therapeutic or Diagnostic Potentials

This section provides only a limited number of examples to illustrate powerful and versatile application of NMR spectroscopy (both alone and in its combination with other analytical and computational techniques) to oligonucleotide chemistry, biochemistry and structural biology. Although it was impossible to review here all recently published advances in NMR characterisation and structural analysis of oligonucleotide analogues, these examples make it possible to demonstrate the opportunities that this type of spectroscopy could offer in the area of oligonucleotides with potential therapeutic and diagnostic properties.

It has been demonstrated in recent research that NMR spectroscopy could be extremely useful to examine a structural basis of the increased affinity of some oligonucleotide derivatives towards target DNA and/or RNA sequences. This information could be crucial to elucidate *structure–activity* relationships and potentially improve the therapeutic potential of antisense oligonucleotides. Using ^{1}H and ^{31}P NMR as well as 2D ^{1}H NMR spectroscopy (e.g. NOESY, COSY, TOCSY, ^{1}H-^{31}P HMBC) Thiviyanathan *et al.* [26] studied the effect of R and S stereochemistry on structural properties of chirally pure methyl phosphonate oligonucleotides that could potentially be used as antisense or antigene agents with high resistance towards cellular nucleases. Figure 5.24 represents 1D ^{31}P NMR spectra of two heteroduplexes containing Sp and Rp isomers of methyl phosphonate oligonucleotides ((a) and (b), respectively) showing only the phosphonate region. It is seen that the methyl-

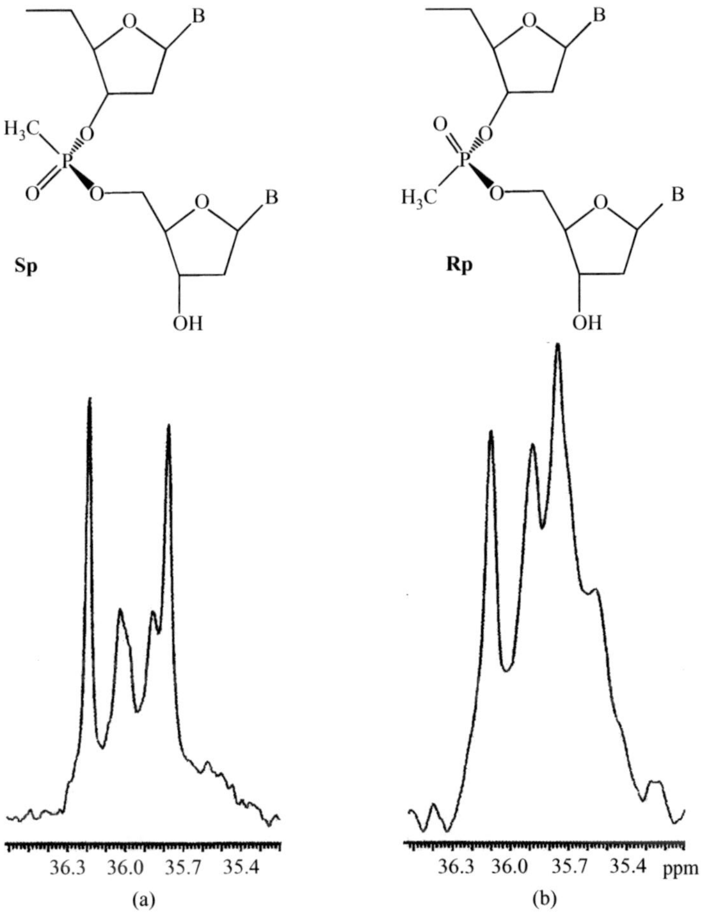

Figure 5.24 1D ^{31}P NMR spectra of (a) Sp and (b) Rp isomer heteroduplexes. Only the phosphonate region of the spectra is shown. The ^{31}P NMR spectra were collected at 162 MHz and at 25°C. Chemical shifts were referenced to 85% H_3PO_4 at 0.0 ppm. (Source: Reprinted with permission from Thiviyanathan, V., Vyazovkina, K.V., Gozansky, E.K., Bichenkova, E., Abramova, T.V., Luxon, B.A., Lebedev, A.V., Gorenstein, D.G., Structure of hybrid backbone methylphosphonate DNA heteroduplexes: Effect of R and S stereochemistry, *Biochemistry*, 41, 827–838. Copyright (2002) American Chemical Society.)

phosphonate ^{31}P nuclei resonates downfield between 35 and 37 ppm with the Sp phosphonate centre resonating further downfield compared to the phosphorus atoms in the corresponding Rp isomer.

^{1}H-^{1}H distance restraints obtained from integrating the NOESY cross-peaks were used in the rMD structure calculations using the iterative hybrid-matrix MORASS program to calculate the most probable conformations of the heteroduplexes incorporating chirally pure Sp and Rp isomers of methyl phosphonate oligonucleotides (Figure 5.25, see colour insert). NMR provided experimental evidence that the methylphosphonate strands in the heteroduplexes showed increased dynamics compared to the DNA strand. This research revealed that substitution of one chiral centre from R_p to S_p exhibited a profound effect on the hybridisation ability of the methylphosphonate strand. The significant differences in physical properties and hybridisation abilities observed between the R_p and S_p diastereomers have been correlated with the structural properties of these oligonucleotide analogues within their hybrid complexes with DNA. This research provided the first NMR analysis demonstrating structural differences between the diastereomeric methylphosphonate duplexes, thus providing a conformational basis for improved design of antisense and/or antigene agents (Figure 5.25).

Seth *et al.* [27] investigated the effect of R and S stereoisomerism on biophysical and biological properties of the R- and S-5′-Me-LNA modified oligonucleotides. While both of these modifications showed an increased resistance toward exonucleases, only the S-5′-Me-LNA analogue showed hybridisation properties similar to those of LNA. Moreover, the R-5′-Me-LNA analogue showed a destabilising effect compared to the parent LNA derivative. NMR studies (i.e. ^{1}H, ^{13}C, ^{31}P 1D NMR and ^{1}H-^{1}H NOESY) indicated that the R-5′-Me group changes the orientation around torsion angle γ, and this might be responsible for the poor hybridisation abilities seen for this type of modification. S-5′-Me-LNA modified antisense oligonucleotides showed slightly reduced potency relative to the parent LNA analogues, but improved the therapeutic profile. The structural data obtained from NMR contributed to further evaluation of the S-5′-Me-LNA modification for potential diagnostic and therapeutic applications.

Nauwelaerts *et al.* [28] investigated cyclohexenyl nucleic acids, which represent a new class of nucleic acid based biomimetics with a potential for antisense therapeutic activity. The conformational flexibility of cyclohexenyl nucleosides was assessed by NMR spectroscopy (e.g. NOESY, TOCSY, DQF-COSY and ^{1}H-^{31}P heteronuclear correlation spectroscopy (often abbreviated to HETCOR)) in combination with the HexRot program in order to calculate possible conformations using scalar coupling constants of cyclohexenyl moieties. It was demonstrated that the conformational equilibria and the thermodynamic parameters of the cyclohexenyl nucleoside were very similar to those observed for the natural ribose nucleosides. The NMR experiments carried out in this research demonstrated for the first time that a synthetic nucleoside was capable of adopting different conformations when incorporated in different double-stranded DNA sequences.

High-field NMR was also used to investigate the influence of the incorporation of highly fluorescent tricyclic cytosine base analogue (tC) on DNA duplex conformation after hybridisation [29]. This fluorescence tag can potentially be used for preparation of highly selective DNA and RNA detection probes for application in molecular diagnostics and single nucleotide polymorphism detection. Some of the most important properties required for fluorescent-based analogues include: the ability to form specific base pairs with the naturally occurring complementary base, having a minimal effect on the overall DNA structure and also to give sufficient fluorescent quantum yield, both for unbound oligonucleotide probe and after the hybridisation event with the biological target. These challenges inspired structural analysis of the selected oligonucleotide model system containing fluorescent tricyclic cytosine base analogue. The NMR structure determination of the self-complementary sequence 5′-CTC(tC)ACGTGGAG showed that the DNA double-stranded structure was consistent with the B-form for the entire duplex. The NMR experiments confirmed the formation of the correct Watson–Crick hydrogen bonding for tC by the observation of NOE interactions for both the intra-strand and interstrand iminoproton. The collected NMR data provided experimental evidence that that tC behaves as a cytosine analogue by forming hydrogen bonds with guanosine residues in the complementary strand without considerable distortion of the sugar-phosphate backbone conformation, thus providing structural justifications of this fluorescent tag in hybridisation bio-assays.

Another example of a fruitful application of NMR spectroscopy (1D ^{1}H NMR, COSY TOCSY, NOESY and ^{1}H-^{13}C HSQC) to oligonucleotide structural analysis was provided by Quang Do and co-workers [30], who studied inter-residue interactions within G-quadruplexes seen for G-rich oligonucleotides. The importance of this research was underpinned by the recently recognised therapeutic potential of this type of oligonucleotide structure against HIV and cancer, although the structural data remained to be conflicting. The oligonucleotides chosen for this study are shown in Table 5.3.

Interestingly, the analysis of the imino proton regions of the ^{1}H NMR spectra of T40214 (Figure 5.26(a)) and T30695 (Figure 5.26(b)) showed very similar shift patterns although there had been a cytidine-to-thymidine substitution. Also, ^{1}H resonances of both non-exchangeable and exchangeable imino protons of these two oligonucleotides were strongly overlapped, presumably due to the quasi-symmetry of the structure resulting from the repetitive nature of the sequences. This could potentially complicate the NMR structure determination. However, the authors demonstrated that a single substitution of the guanosine residue at the 2 position to inosine (to produce the J19 oligonucleotide) resulted in improved spectral resolution (Figure 5.26(c)), presumably due to a disruption of the symmetry of the structure. This single nucleotide substitution considerably enhanced spectral quality and facilitated

Table 5.3 DNA sequences used in high-field 2D NMR study (NOESY) [30].

Name	Sequence (5′–3′)
T30695	GGG T GGG T GGG T GGG T
J19	GIG T GGG T GGG T GGG T
T40214	GGG C GGG C GGG C GGG C

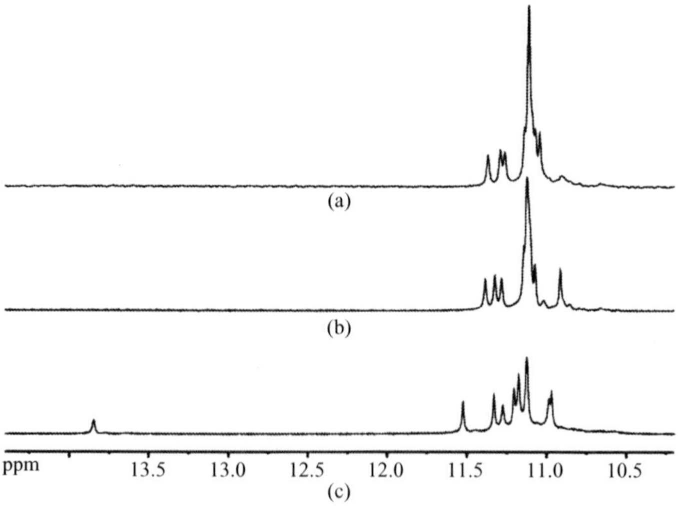

Figure 5.26 Imino proton region of the 1D NMR spectra of (a) T40214, (b) T30695 and (c) J19 in K^+ solution at 25°C. (Source: Do, N.Q., Lim, K.W., Teo, M.H., Heddi, B., Phan, A.T., Stacking of G-quadruplexes: NMR structure of a G-rich oligonucleotide with potential anti-HIV and anticancer activity, *Nucleic Acids Res.*, 2011, 39, 9448–9457, by permission of Oxford University Press.)

detailed NMR structural analysis of the G-quadruplexes by the example of J19 oligonucleotide.

The combination of high-field NMR spectroscopy (e.g. NOESY, TOCSY, COSY, ^{1}H-^{13}C HSQC) and rMD demonstrated that a DNA sequence derived from the G-rich oligonucleotides (Table 5.3) adopts a dimeric G-quadruplex, formed by the stacking of two identical propeller-type parallel stranded G-quadruplex subunits at their 5′-terminals. The structure of the stacking interface that formed stacking interaction between the two subunits was disfavoured by extension of non-G residues from the terminals (Figure 5.27, see colour insert).

5.7 Future Perspectives for NMR Characterisation of Oligonucleotides

The future direction of applying NMR to characterising oligonucleotides is focused mainly on improvements in magnetic fields strengths (currently, up to 23.5 Tesla) using supercooled superconducting magnets. The consequence of this innovation, especially when combined with Fourier transformation, was further improvement in spectral resolution and sensitivity, which allowed observation of nuclei with low natural abundance (e.g. ^{13}C or ^{15}N). The development of cryogenically cooled probes (e.g. CryoProbes) in the last decade offered a further considerable jump in NMR sensitivity, which provided the opportunity to bring down NMR detection limits and work with significantly lower sample concentrations.

The world's first 1 GHz NMR spectrometer (1000 MHz, 23.5 Tesla) with ultra-high resolution capabilities and sensitivity (Figure 5.28) was installed by Bruker BioSpin in 2009 at the Centre de RMN à Très Hauts Champs in Lyon (France). The ultra-shielded superconducting NMR magnet incorporated into the AVANCE 1000 MHz NMR spectrometer brought unique 1 GHz technology to the NMR field. The AVANCE 1000 system incorporated a record-breaking 23.5 Tesla UltraStabilized superconducting magnet. The high field strength and high field stability in combination with the triple resonance CryoProbe enabled unique application of 1 GHz NMR to study structure, function and dynamics of complex biological molecules, including proteins and nucleic acids.

The development of numerous numbers of new pulse NMR techniques as well as pulsed field gradient selected spectroscopy has offered further improvements in the quality and sensitivity of NMR spectra. These advancements have fundamentally changed the way of data collection *via* mapping different types of interactions between different nuclei.

Hyphenation of NMR with other analytical techniques such as HPLC and MS offers another promising strategic direction to enhance power, speed and reliability of analysis and characterisation of nucleic acids alone and with other biomolecules. As discussed earlier, HPLC chromatography, mass spectrometry and NMR spectroscopy represent the key analytical techniques for characterisation of purity and identity of oligonucleotides and their synthetic analogues. The emergence of hyphenated techniques, such as HPLC-NMR (Figure 5.29) and/or MS-NMR, opened doors to application of coupled analytical methods in order to eliminate the rate-determining, time-consuming isolation step and gain as much as possible information in a single

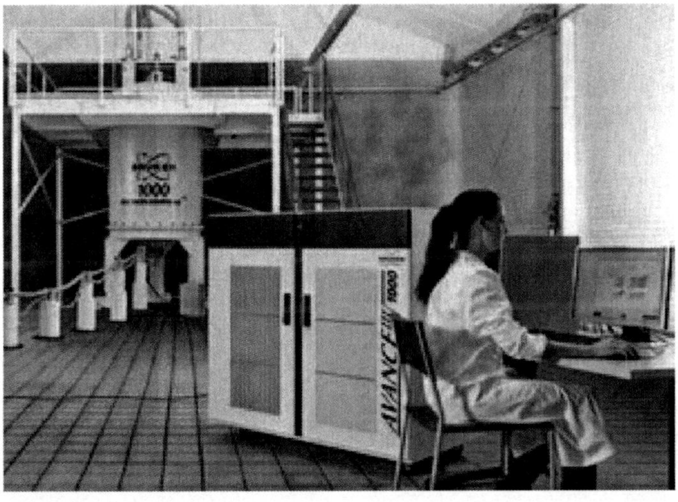

Figure 5.28 AVANCE 1000 MHz NMR (23.5 T) spectrometer with the first 1 GHz CryoProbe installed. (This figure was kindly provided by Bruker BioSpin specifically for this book.)

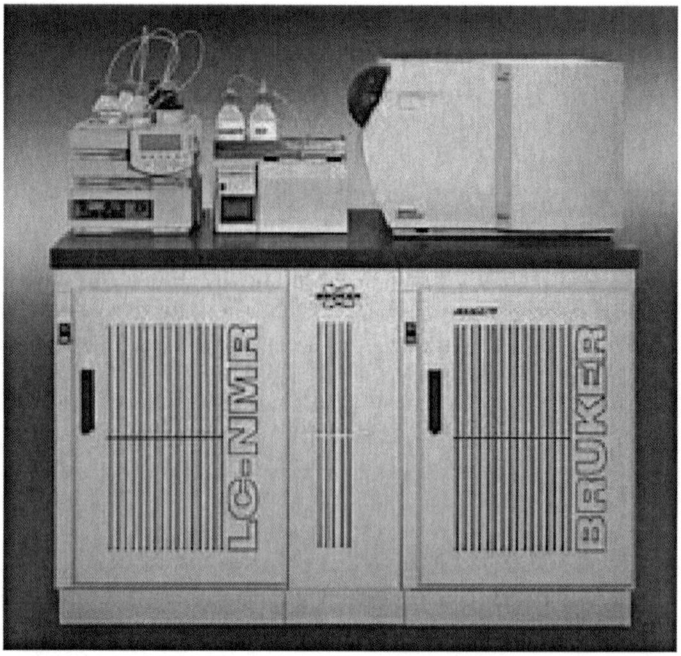

Figure 5.29 HPLC-NMR system. The liquid chromatography system is coupled to the NMR spectrometer. The sample is transferred into the NMR during the chromatographic separation followed by NMR spectra acquisition. (This figure was kindly provided by Bruker BioSpin specifically for this book.)

analytical run. This synergetic combination of vital analytical techniques can significantly accelerate the characterisation process of valuable samples available only in small quantities in complicated reaction mixtures.

HPLC-NMR, which can be applied either in continuous-flow mode or in stop-flow mode, represents a powerful, well-established laboratory tool. In the latest development of the HPLC-NMR system, solid-phase extraction (SPE) cartridges can be used to immobilise selected fractions after the chromatographic separation. The solvents used during chromatography can then be removed followed by elution of the samples from the cartridges using fully deuterated solvents for subsequent NMR analysis. The most recent development of hyphenated methods includes a synergetic combination of three analytical techniques (i.e. liquid chromatography, magnetic resonance and MS) into an HPLC-NMR/NMR-MS system (Figure 5.30), which was first introduced by Bruker BioSpin in 1999. An HPLC-NMR system was initially coupled with a Bruker Daltonics esquire series ion trap mass spectrometer *via* a Bruker NMR-MS interface. However, now the more advanced Bruker Daltonics microTOF-HPLC time-of-flight mass spectrometer can be integrated into an HPLC-NMR system to provide the mass spectra of the eluted peaks, thus facilitating full characterisation of all components of the complicated analyte sample. Suitable configuration of the system allows the mass spectrometer to be used as an experi-

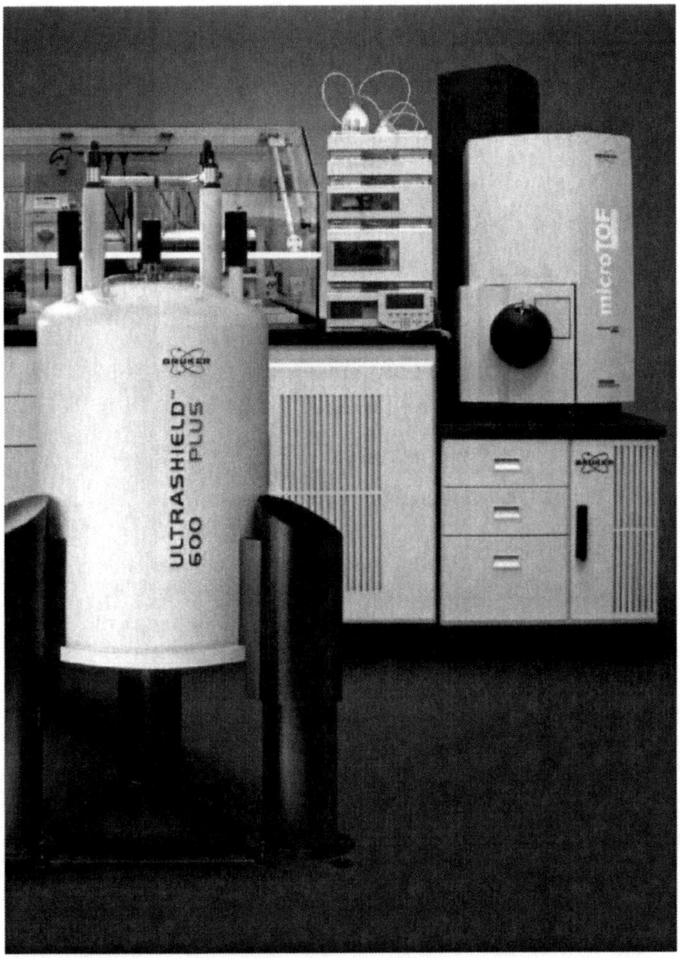

Figure 5.30 The HPLC-NMR/NMR-MS system (Bruker BioSpin). (This figure was kindly provided by Bruker BioSpin specifically for this book.)

mental control device to trigger the time-consuming NMR experiment. During chromatographic separation microTOF-HPLC generates the mass spectra of the eluted peaks, thus initiating the collection of chromatographic peaks into loops (in the case of HPLC-NMR), or SPE cartridges (in the case of HPLC-SPE NMR) based on molecular masses of the eluted components.

A broad variety of versatile NMR techniques available nowadays together with a powerful arsenal of sophisticated NMR instrumentation not only makes routine analysis of therapeutic oligonucleotides quick, reliable and cost-efficient, but also offers a great opportunity to explore their structural and dynamic properties for future applications in therapy and molecular diagnostics.

Acknowledgements

The author is grateful to Bruker BioSpin Ltd, and particularly to Dr Andrew Gibbs and Mr Andy McAlister for providing images of the Bruker NMR instrumentation for this book. The author also thanks Dr Waleed A. Zalloum for very useful technical support and assistance in preparation of figures. The author is also grateful to Dr Richard Bryce for invaluable discussions on molecular modelling aspects.

References

1. Claridge, T.D.W., Introduction. In *High-Resolution NMR Technologies in Organic Chemistry*, Backvall, J.-E., Baldwin, J.E., Williams, R.M. (Eds), 2009, Vol. 19. Elsevier, Oxford.
2. Simpson, J.H., *Organic Structure Determination Using 2-D NMR Spectroscopy*, 2008, Elsevier, Oxford.
3. Wüthrich, K., *NMR of Proteins and Nucleic Acids*, 1996, Wiley-Interscience, New York.
4. Jacobsen, N.E., Fundamentals of NMR spectroscopy in liquids. In *NMR Spectroscopy Explained*, 2007, Wiley-Interscience, New York.
5. Friebolin, H., *Basic One- and Two- Dimension NMR Spectroscopy*, 1999, VCH, Weinheim/New York.,
6. Gray, G.A., Two-dimensional NMR spectroscopy. In *Applications for Chemists and Biochemists*, 2nd edn, Croasmun, W.R. (Ed.), 1994, John Wiley and Sons, New York.
7. Davis, D.R., Stillman, D.J. Altered structure of the DNA duplex recognized by yeast transcription factor Reb1p, *Nucleic Acids Res.*, 1997, 25, 668–674.
8. Marion, D., Genest, M., Ptak, M., Reconstruction of NOESY maps: A requirement for a reliable conformational analysis of biomolecules using 2D NMR, *Biophys. Chem.*, 1987, 28, 235–244.
9. Borgias, B.A., James, T.L., MARDIGRAS-A procedure for matrix analysis of relaxation for discerning geometry of an aqueous structure, *J. Mag. Res.* 1990, 87, 475–487.
10. Richon, A.B., An introduction to molecular modelling, *Mathematech*, 1994, 1, 83.
11. Hans-Dieter Holtje, G.F., Small molecules. In *Molecular Modeling Basic Principles and Applications*, 3rd, revised, expanded edn, Mannhold, R., Kubinyi, H., Timmerman, H. (Eds), 1996, pp. 9–64, VCH.
12. Güntert, P., Structure calculation using automated techniques. In *BioNMR in Drug Research*, Zerbe, O. (Ed.), 2002, pp. 39–66, Wiley-VCH.
13. Case, D.A., Darden, T.A., Cheatham, T.E. III, Simmerling, C.L., Wang, J., Duke, R.E., Luo, R., Crowley, Crowle, M., Walker, R.C., Zhang, K.M., Merz, B., Wang, S., Hayik, A., Roitberg, G., Seabra, I., Kolossváry, Won, K.F., Paesani, J., Vanicek, X., Wu, S.R., Brozell, T., Steinbrecher, H., Gohlke, L., Yang, C.T., Mongan, J., Hornak,V., Cui, G., Mathews, D.H., Seetin, M.G., Sagui, C., Babin, V., Kollman, P.A., *AMBER 10*, 2008, University of California, San Francisco.
14. Allen, M.P., In *Introduction to Molecular Dynamics Simulation*, *Computational Soft Matter: From Synthetic Polymers to Proteins*, Attig, N., Binder, K., Grubmüller, H., Kremer, K. (Eds), 2004, NIC Series, Vol. 23, pp. 1–28, NIC, Germany.
15. Khajeh, M., Botana, A., Bernstein, M.A., Nilsson, M., Morris, G.A., Reaction kinetics studied using diffusion-ordered spectroscopy and multiway chemometrics, *Anal. Chem.*, 2010, 82, 2102–2108.
16. Morris, G.A., Diffusion-ordered spectroscopy. In *Encyclopedia of Magnetic Resonance: Supplementary Volume*, Grant, D.M., Harris, R.K. (Eds), 2002, pp. 35–44, J. Wiley and Sons, New York.
17. Valentini, M., Vaccaro, A., Rehor, A., Napoli, A., Hubbell, J.A., Tirelli, N., Diffusion NMR spectroscopy for the characterization of the size and interactions of colloidal matter: the case of vesicles and nanoparticles. *J. Am. Chem. Soc.*, 2004, 126, 2142–2147.

18. Kerssebaum, R., Salnikov, G., DOSY and diffusion by NMR. In *A Tutorial for TopSpin 2.0, Version 2.0.0*, 2006, pp. 1–32, Bruker BioSpin, Rheinstetten.
19. Provencher, S.W., CONTIN: A general purpose constrained regularization program for inverting noisy linear algebraic and integral equations, *Comput. Phys. Communs*, 1982, 27, 229–242.
20. Patel, D.J., Kozlowski, S.A., Ikuta, S., Itakura, K., Bhatt, R., Hare, D.R., NMR studies of DNA conformation and dynamics in solution, *Cold Spring Harbor Symposia on Quantitative Biology*, 1983, 47, 197–206.
21. Bichenkova, E., Marks, D., Dobrikov, M.I., Vlassov, V.V., Morris, G.A., Douglas, K.T., Refined high-field NMR solution structure of a binary-addressed pyrene/perfluoro azide complementary DNA oligonucleotide system shows extensive distortion in the central nick region, *J. Biomol. Struct. Dyn.*, 1999, 17, 193–211.
22. Jaroszewski, J.W., Clausen, V., Cohen, J.S., Dahl, O., NMR investigations of duplex stability of phosphorothioate and phosphorodithioate DNA analogues modified in both strands, *Nucl. Acids Res.*, 1996, 24, 829–834.
23. Lois D.Q. and Antony J.W., Introduction to P-31 NMR spectroscopy. In *Practical Interpretation of P-31 NMR Spectra and Computer Assisted Structure Verification*, 2004, Advanced Chemistry Development Inc., Toronto.
24. Kühl, O., Short review of NMR theory. In *Phosphorus-31 NMR Spectroscopy A Concise Introduction for the Synthetic Organic and Organometallic Chemist*, 2008, Springer-Verlag, Berlin/Heidelberg.
25. Cieslak, J., Ausın, C., Chmielewski, M.K., Kauffman, J.S., Snyder, J., Del-Grosso, A., Beaucage, S.L., ^{31}P NMR study of the desulfurization of oligonucleoside phosphorothioates effected by 'aged' trichloroacetic acid solutions, *J. Org. Chem.*, 2005, 70, 8.
26. Thiviyanathan, V., Vyazovkina, K.V., Gozansky, E.K., Bichenkova, E., Abramova, T.V., Luxon, B.A., Lebedev, A.V. and Gorenstein, D.G., Structure of hybrid backbone methylphosphonate DNA heteroduplexes: Effect of R and S stereochemistry, *Biochemistry*, 2002, 41, 827–838.
27. Seth, P.P., Allerson, C.R., Siwkowski, A., Vasquez, G., Berdeja, A., Migawa, M.T., Gaus, H., Prakash, T.P., Bhat, B., Swayze, E.E., Configuration of the 5′-methyl group modulates the biophysical and biological properties of locked nucleic acid (LNA) oligonucleotides, *J. Med. Chem.*, 2010, 53, 8309–8318.
28. Nauwelaerts, K., Lescrinier, E., Sclep. G. and Herdewijn, P., Cyclohexenyl nucleic acids: conformationally flexible oligonucleotides, *Nucleic Acids Res.*, 2005, 33, 2452–2463.
29. Engman, K.C., Sandin, P., Osborne, S., Brown, T., Billeter, M., Lincoln, P., Norden, B., Albinsson, B., Wilhelmsson, L.M., DNA adopts normal B-form upon incorporation of highly fluorescent DNA base analogue tC: NMR structure and UV-Vis spectroscopy characterization, *Nucleic Acids Res.*, 2004, 32, 17, 5087–5095.
30. Do, N.Q., Lim, K.W., Teo, M.H., Heddi, B., Phan, A.T., Stacking of G-quadruplexes: NMR structure of a G-rich oligonucleotide with potential anti-HIV and anticancer activity, *Nucleic Acids Res.*, 2011, 39, 9448–9457.

6 Analytical Characterisation of Oligonucleotide using Thermal Melting Curves

George Okafo, David P. Elder and Mike Webb

6.1 Introduction

The thermodynamic stability of an oligonucleotide duplex (self-association of two complementary single strands) is a function of the H-bonding between the complementary nucleotide base pairs and the stacking efficiency of the aromatic rings. Certain fundamental thermodynamic parameters, for example, enthalpy (ΔH) and entropy (ΔS) can be measured utilising thermal denaturing (often called melting and designated as T_m, the melting temperature) using a variety of different measurement techniques, for example, ultra-violet spectroscopy (UV), differential scanning calorimetry (DSC), nuclear magnetic resonance spectroscopy (NMR), etc., where T_m is defined as the temperature at which half of the oligonucleotide strands are in the double-helical state and half are in the 'random-coil' state. The T_m depends on the length of the oligonucleotide as well as the specific nucleotide sequence, but can also be affected by several intrinsic variables of the measurement technique [1].

Historically, the vast majority of these polynucleotide molecules that have been studied have 2-dimensional (2D) structures that are either monomolecular (single-stranded hairpins) or bimolecular (duplexes). Importantly, the supporting theoretical equations necessary to translate the measured T_m data into the derived thermodynamic parameters, for example, ΔG (Gibb's free energy) and ΔH, are well known [2–4].

The performance and outcome of several molecular biology techniques are dependent on the accurate prediction of T_m. This is particularly true for those techniques involving simultaneous use of several oligonucleotides. For example, deoxyribonucleic acid (DNA) microarrays of fixed-length, short sequences, typically termed 'affymetrix chips' [5], quantitative polymerase chain reaction (PCR) [6] and

Analysis of Oligonucleotides and their Related Substances, edited by George Okafo, David Elder and Mike multiplex PCR all rely on accurate prediction of T_m, as errors can lead either to inappropriate hybridisation performance or to increases in non-specific products [7]. In this chapter different methods for determining T_m will be discussed, including *in-silico* approaches, different spectroscopic detection modes and thermal techniques. A brief overview of the analytical approaches, applications and advantages/disadvantages is provided in Table 6.1.

6.2 *In-Silico* Modelling Approaches

Historically, there has been a wide variety of different methods employed to estimate T_m theoretically. Early approaches based theoretical T_m determinations on the relative cytosine/guanine (C/G) contents [8], see equation (6.1).

$$T_m = 64.9 + 41.0x(y\mathrm{G} + z\mathrm{C}–16.4)/(w\mathrm{A} + x\mathrm{T} + y\mathrm{G} + z\mathrm{C}) \qquad (6.1)$$

where x, y, w and z are the number of T, G, A and C Watson–Crick bases, respectively. Equation (6.1) assumes annealing occurs under standard conditions in a buffered solution of 50 mM Na+ and 50 mM oligonucleotide, with pH close to 7.0, and that T_m is unaffected by pH in this region due to lack of titratable groups.

This equation was later modified to account for the impact of salt concentrations [9], see equation (6.2).

$$\begin{aligned} T_m &= 64.9 + 41.0x(y\mathrm{G} + z\mathrm{C}–16.4)/(w\mathrm{A} + x\mathrm{T} + y\mathrm{G} + z\mathrm{C}) \\ &\quad -(820.0)/(w\mathrm{A} + x\mathrm{T} + y\mathrm{G} + z\mathrm{C}) + 16.6 \log\{[\mathrm{Na}]^+\} \end{aligned} \qquad (6.2)$$

The second term in equation (6.2) adjusts for G/C base content; whereas the third term adjusts for the oligonucleotide sequence length. However, it has become apparent that the sequence order of nucleotide bases has a significant impact on calculated T_m values [10] (see equation (6.3)).

This led to the adoption of the nearest neighbour (NN) model [11]. This model postulates that the ΔG of duplex formation is dependent on (i) the impact of reduction in entropy (ΔS) as a result of decrease in translational freedom as a consequence of the first nucleotide base pair formation and (ii) the summation of complementary base pair formation between the two single oligonucleotide sequences [11]. In addition, there is a further ΔS constraint which accounts for the maintenance of symmetry in self-complementary sequences.

$$T_m = [\Sigma(\Delta H_d) + \Delta H_i]/[\Sigma(\Delta S_d) + \Delta S_i + \Delta S_{self} + Rx \ln Ct/b] \qquad (6.3)$$

where sums of enthalpy (ΔH_d) and entropy (ΔS_d) are calculated over all internal NN doublets, ΔS_{self} is the entropy impact of self-complementary sequences, and ΔH_i and ΔS_i are the sums of initiation enthalpies and entropies, respectively. R is the gas constant (1.987 cal/K mol), Ct is the total strand concentration in molar units (typically fixed at 50 mM), b is a constant with value of 4 for non self-complementary sequences or 1 for duplexes. Two-state transition is assumed.

Information obtained from naturally occurring DNA and ribonucleic acid (RNA) molecules, as well as model synthetic oligonucleotides have allowed thermodynamic databases to be established that have helped to characterise all ten Watson–Crick NNs for both RNA [2, 12–14] and DNA [2, 15].

An in-depth comparison of different tabulated data is difficult because several independent variables are utilised in each derivation, but some efforts have been made [16, 17]. The findings of SantaLucia [17] seemed to show significant consensus amongst the different approaches. However, in contrast, Panjkovich and Melo [10] appeared to show the opposite and that large differences in the predicted T_m were likely to be based on inherent differences in the input thermodynamic parameters. Panjkovich and Melo [10] summarised the potential biases that may be encountered, as follows.

1. Oligonucleotide length is important, that is short-length 10mers are over-represented in the published data sets.
2. C/G content is between 40 and 60%, with few examples of the extremes.
3. A large percentage of oligonucleotides represented in the database have used their own experimental T_m data to validate these predictions, thus biasing the datasets towards oligonucleotides of similar properties.
4. Some large oligonucleotides are known to not follow the 2-state melting requirement of the NN model.

Panjkovich and Melo [10] indicated that these limitations were not disclosed to potential users of web-based software packages. Only 4 out of 17 web-based software packages made potential users aware of these limitations. They suggested that users should:

1. be aware of intrinsic limitations
2. use oligonucleotide sequences that fall in the middle of CG content with chain length less than 20–22mers
3. avoid oligonucleotide sequences inadequately covered by existing datasets
4. where a 2-state transition might reasonably be expected not to apply, use a consensus T_m method.

Experimental approaches, for example, crystallography, NMR, etc., are not yet capable of direct study of the 3-dimensional (3D) structure of oligonucleotides which are capable of forming a prototype double helix, that is ca. 30mers in length. Recently, *ab initio* molecular dynamics (MD) simulations of DNA oligonucleotides of intermediate length in aqueous solution have become possible allowing the modelling of helix-phasing sequences of different compositions allowing the determination of sequence-dependent DNA curvature and flexibility as a function of temperature [18].

Beveridge *et al.* [18] performed MD simulations of a 25mer oligionucleotide in solution, at eight different temperatures between 273 and 338 K. They were able to show that there is a temperature discontinuity between 310 and 330 K, where the overall curvature of the DNA sequence decreases as a function of temperature. This phenomenon occurs very near the B′ to B pre-melting transition, which is between

Table 6.1 Overview of analytical approaches to determine T_m.

Analytical approach to determine T_m	Applications	Comments	Advantages/Disadvantages
UV spectroscopy	Standard approach to T_m determination	Identical heating/cooling curves indicates association/dissociation is rapid compared to heating gradient. In contrast, hysteresis indicates either slow kinetics or presence of secondary (or higher structures)	Rapid, cost effective, easy to perform. Disadvantages are (i) the absorbance changes are relatively small, (ii) low throughput, (iii) requires large volumes (1–3 ml) with absorbances of about 0.2, (iv) higher-order oligonucleotide structures typically show complex transition, (v) some triplexes are not accompanied by changes in absorption, (vi) model-dependent approach
UV Resonsance Raman spectroscopy (UVRR)	Utilised for higher-order structure determination	Yields both vibrational and rotational information about oligonucleotide. Can investigate effect of cations on higher-order structure	Sensitive and applicable for higher-order structure determination. Disadvantages are (i) complex hyphenated technique, (ii) low throughput, (iii) expensive, (iv) model-dependent approach
Fluorescence spectroscopy	Utilised for higher-order structure determination	Can use fluorescence probes, for example 2-aminopyridine, to assess local/global conformational changes	Can be used in tandem with UV spectroscopy. Provides more insight into conformational changes. Disadvantages are (i) complex technique requiring fluorescent probe, (ii) low throughput, (iii) expensive, (iv) model-dependent approach
NMR spectroscopy	Utilised for higher-order structure determination	Can use Nuclear Overhauser Effect (NOE) enhancements to examine local/global conformational changes	Provides more insight into conformational changes. Disadvantages are (i) complex technique, (ii) low throughput, (iii) expensive, (iv) slow, (v) model-dependent approach

(*continued*)

Table 6.1 (*continued*)

Analytical approach to determine T_m	**Applications**	**Comments**	**Advantages/Disadvantages**
Circular dichroism (CD)	Utilised for concentration-dependent measurements, for higher-order structure determination	Extremely sensitive (useful for low-concentration/low-solubility samples). Useful to assess impact of other factors, such as salts, acids, bases, etc.	Simple, rapid, inexpensive insight into conformational changes. Can be used in tandem with UV spectroscopy. Disadvantage is that it is a model-dependent approach
Differential scanning calorimetry (DSC)	Utilised for model-independent assessment of thermodynamic parameters	If $\Delta H_{UV} < \Delta H_{cal}$ then the transition involves a high proportion of intermediate states; whereas, in contrast, if $\Delta H_{UV} = \Delta H_{cal}$ then the transition proceeds in a 2-state manner	Simple, rapid, inexpensive approach. DSC (model independent) can be used in tandem with UV spectroscopy (model dependent) to provide indirect evidence of higher order structure. Disadvantages are (i) larger sample sizes, (ii) cannot assess higher-order structures directly
Isothermal titration calorimetry (ITC)	Utilised for low-temperature ($<T_m$) structural transitions and higher-order structure determination	Enthalpy and entropy contributions to higher-order structure can be assessed	Provides more insight into conformational changes. Disadvantages are (i) complex technique, (ii) low throughput, (iii) expensive, (iv) slow

228 and 230 K. The MD simulation also predicts a transition in the key YpR steps (pyrimidine–purine) from a high roll, open hinge to a low roll, closed hinge substrate.

6.3 Spectroscopic Methods for Determining T_m

6.3.1 UV Spectroscopy

UV spectroscopy (sometimes termed UV hyperchromicity) has been used to determine T_m for over 50 years and it is still the simplest and often the most accurate approach. The aromatic nucleotide bases, for example purine or pyrimidine, absorb strongly in the UV region of the spectra with absorbance maxima at about 260 nm. Typically, the UV absorption of oligonucleotides is lower than the corresponding nucleotides (attributed to π–π interactions due to base-pair stacking, which affect the transition dipoles of the bases) and this is called the hypochromic effect or hypochromicity. Conversely, when moving from a double-stranded duplex to a single strand the opposite effect is observed, that is hyperchromicity [19]. Absorbance readings of the oligonucleotide are taken as the temperature of the solution is incrementally increased. The T_m is determined based on the first derivative of the resultant temperature versus absorbance plot (melting curve). The inflection point (T_m) is derived using the first derivative sign change [1] (see Figure 6.1). In addition, 2-point averages or non-linear fit criteria can be applied. The advantages/disadvantages of all three approaches have been reviewed by Simonian [20].

Identical heating and cooling curves indicate that the processes of association and dissociation of the single and double strands are rapid relative to the heating gradient employed. In contrast, where this is not the case and heating and cooling curves show hysteresis, this may be attributable to either slow kinetics or to the presence of other secondary (or higher) structures. In the latter case, the assessment of the melting temperature from the heating or cooling curves should be approached with

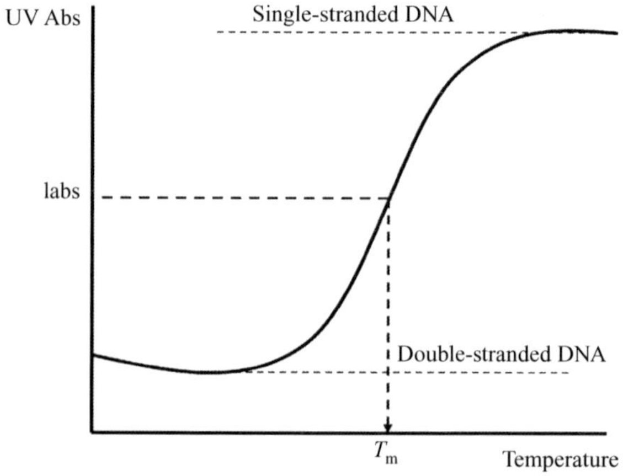

Figure 6.1 A typical absorbance versus temperature melting curve for an oligonucleotide [1].

caution. Mergny and Lacroix [21] used the heating and cooling curves to determine the kinetic constants for association and dissociation (K_{on} and K_{off}, respectively) from which the corresponding activation energies could be ascertained, which can then be used to determine the theoretical T_m.

If the fraction of single strands in the duplex is also defined as α, where $\alpha = 0.50$ at T_m, then any typical absorbance versus temperature melting curve can be converted into the corresponding α versus temperature curve (see Figure 6.2).

Thus the enthalpy of transition (ΔH_t) can then be determined as follows (see equation (6.4)) [2]

$$\Delta H = RT^2 \times \mathrm{d} \ln K/\mathrm{d}T \tag{6.4}$$

where K is expressed in terms of α, and it is assumed that the transition proceeds in a 2-state fashion

$$K = \alpha/[(Ct/n)^{n-1} \times (1 - \alpha)^n] \tag{6.5}$$

For the special case of $\alpha = 0.5$ at the transition temperature, substitution can be carried out to obtain

$$KT_m = 0.5/[(Ct/n)^{n-1} \times (0.5)^n] = 1/[(Ct/n)^{n-1} \times (0.5)^{n-1}] = 1/(Ct/2n)^{n-1} \tag{6.6}$$

where n is a constant with value of 1 for monomolecular, for example, single strand of hairpin, or 2 for duplexes, and Ct is the total strand concentration in molar units (typically fixed at 50 mM).

KT_m can now be expressed in terms of α, by substituting equation (6.5) into equation (6.4)

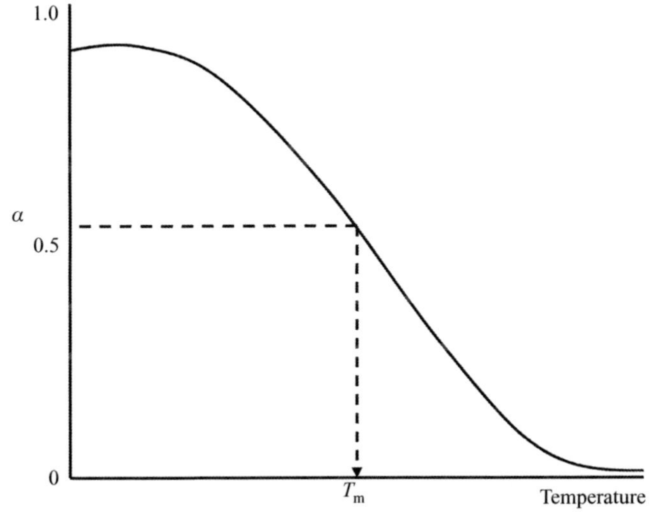

Figure 6.2 A typical fraction of single strands in the duplex (α) versus temperature melting curve for an oligonucleotide [2].

$$\Delta H = (2 + 2n) \times RT_{\mathrm{m}}^2 \times \mathrm{d}\alpha/\mathrm{d}T \tag{6.7}$$

For single strands or hairpins ($n = 1$) this simplifies to

$$\Delta H = 4RT_{\mathrm{m}}^2 \times \mathrm{d}\alpha/\mathrm{d}T \tag{6.8}$$

And for duplexes ($n = 2$) this simplifies to

$$\Delta H = 6RT_{\mathrm{m}}^2 \times \mathrm{d}\alpha/\mathrm{d}T \tag{6.9}$$

Some of the possible factors that can affect accurate T_{m} measurements were delineated and investigated by Carmody [1] including variability in both buffer and sample preparations, buffer and sample concentration and data sampling rates. Carmody [1] used a wavelength of 260 nm, a starting temperature of 25°C and an end temperature of 95°C, with a heating rate of 1°C/min and a reading rate of 1°C. Ten 20-mer oligonucleotides were assessed in buffer concentrations ranging from 342.5 to 343.3 mM sodium phosphate and sample concentrations ranging from 0.0195 to 0.0201 mg/ml. Then a further six 20mer oligonucleotides were evaluated in buffer concentrations ranging from 175 to 1750 mM sodium phosphate and sample concentrations ranging from 0.01 to 0.06 mg/ml. She found that sample/buffer preparation had a minimal impact on T_{m}. In contrast, there was a 4–6°C increase in T_{m} when the buffer concentration was increased from 2- to 5-fold. There was less of an impact of sample concentration on T_{m}. The T_{m} increased 1–2°C with a 2–5-fold increase in sample concentration. Finally, the read rate was assessed over the range 0.3–1.0°C and found to have limited impact on the absolute value, but lower read rates gave better precision.

Carmody [1] also showed the profound impact of changes in nucleotide base sequences on T_{m}. She assessed two 16mer oligonucleotides with nucleotide base complements differing at one location (5′-****GGAGC****-3 versus 5′-****GGCGC****-3). All other factors were constant. The former gave a T_{m} of 56.4°C, whereas the latter gave a T_{m} of 65.3°C, that is an 8.9°C difference.

Nelson *et al.* [22] assessed DNA and RNA thermodynamic stabilities, together with the effect of mismatched bases on double-helix stability, using UV spectrophotometric-derived melting curves. They showed that deoxyoligonucleotides are more stable than the corresponding RNA oligomers, owing to more favoured enthalpy considerations. Double helices can be formed with mismatched base pairs, and can form several atypical 2D structures. They found that structures with a dangling mismatched base were more favoured than a helix with a bulging base, the energy differential being about 1.6 kcal/mol at 10°C.

6.3.2 UV/Resonance Raman Spectroscopy

UV/resonance Raman spectroscopy (UVRR) has been used to facilitate better understanding of the 2D structure before and after the melting transition [23]. Raman yields information about both vibrational and rotational changes within an oligonucleotide; therefore, the individual contribution of the various oligonucleotide components (e.g.

nucleotide base, sugar or phosphate) to the overall stability of the duplex can be assessed. Thus, the change of the oligonucleotide conformation can be studied by assessing the temperature dependency of a specific vibrational modality [19]. For example, it is well known that both poly(G) and poly(rI), can form 4-stranded structures, containing guanine or inosine tetrads, see Figure 6.3.

In addition, it is known that the stability of these tetrads is cation-dependent [24]. Mukerji *et al.* [23] investigated the structure of poly(rI) in the presence of K or Na ions and assessed the relative stabilities with T_m measurements using UVRR. The UVRR studies revealed that the K-stabilised-poly(rI) is more stable with a T_m of 47°C versus 30°C for the Na-stabilised-poly(rI) (see Figure 6.4).

They demonstrated using Raman that the ribosyl conformation and the coordination of the carbonyl groups are cation-dependent. The $C_6{=}O$ stretching frequency demonstrates that the Na cation binds more strongly than K (1672 cm^{-1} for Na versus 1684 cm^{-1} for K) (see Figure 6.5). In addition, it appears that the Na-stabilised-poly(rI) exhibits a $C_{3'}$-endo ribose conformation; whereas, in contrast, the K-stabilised-poly(rI) exhibits a $C_{2'}$-endo ribose conformation; possibly arising from the larger ionic radius of the K cation.

Chan *et al.* [25] demonstrated that poly(dA).poly(dT) and DNA duplexes with four or more adenine bases in a row exhibit a broad, solid-state structural transition at 35°C. They correlated this transition with 'bent-DNA' (the base-pairs are twisted and form a third hydrogen bond, a balance between the gain in electronic energy arising

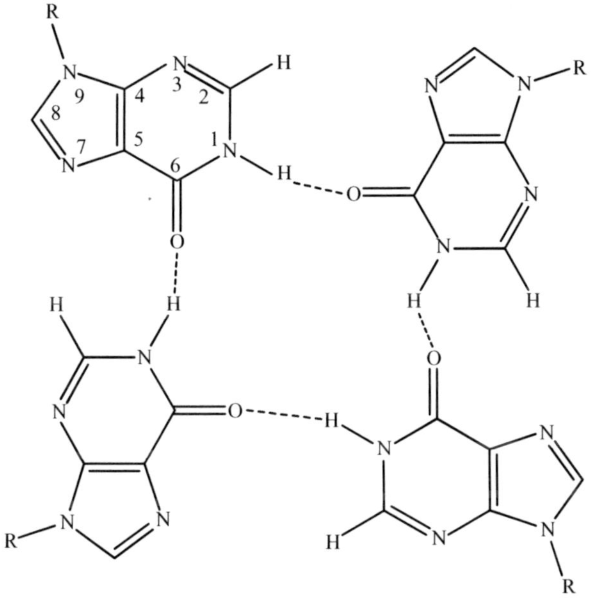

Figure 6.3 Expected H-bonding pattern for inosine tetrad. (Source: Mukerji, I., Sokolov, L., Mihailescu, M-N., A UV resonance Raman investigation of poly(rI): Evidence of cation-dependent structural perturbations, *Biopolymers*, 1998, 46, 475–487, by permission of John Wiley and Sons.)

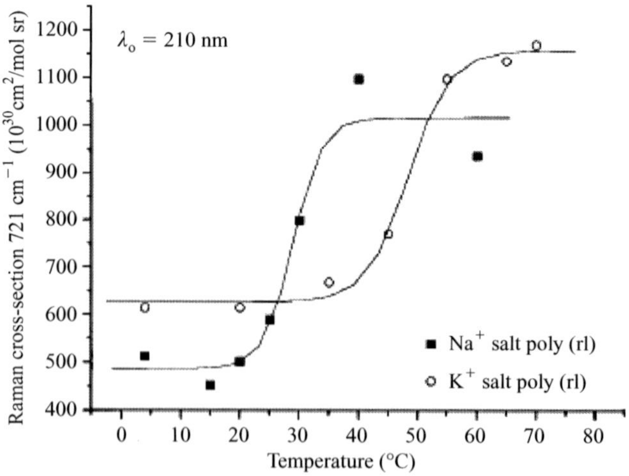

Figure 6.4 Raman cross-section at 721 cm^{-1} plotted as a function of temperature for (Na-stabilised-poly(rI) exhibits a C3′-endo ribose conformation 13 mM poly(rI) in 0.3 M Na_2SO_4, 0.01 M cacodylate, pH 7.0, and 6 mM poly(rI) in 0.1 M KCl, 0.2 M cacodylate, pH 7.0. T_m of 30±2°C for Na-poly(rI) and T_m of 48±2°C for K-poly(rI). (Source: Mukerji, I., Sokolov, L., Mihailescu, M-N., A UV resonance Raman investigation of poly(rI): Evidence of cation-dependent structural perturbations, *Biopolymers*, 1998, 46, 475–487, by permission of John Wiley and Sons.)

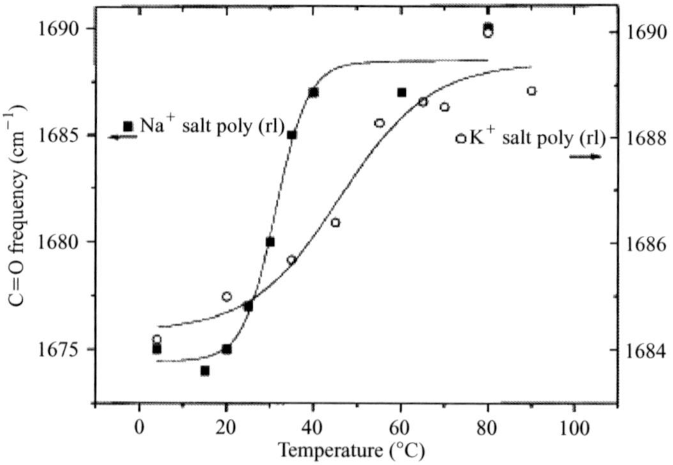

Figure 6.5 Frequency of the localised Raman C_6=O stretching frequency as a function of temperature for (Na-stabilised-poly(rI) exhibits a C3′-endo ribose conformation 13 mM poly(rI) in 0.3 M Na_2SO_4, 0.01 M cacodylate, pH 7.0, and 6 mM poly(rI) in 0.1 M KCl, 0.2 M cacodylate, pH 7.0. T_m of 30 ± 1°C for Na-poly(rI) and T_m of 50 ± 2°C for K-poly(rI). (Source: Mukerji, I., Sokolov, L., Mihailescu, M-N., A UV resonance Raman investigation of poly(rI): Evidence of cation-dependent structural perturbations, *Biopolymers*, 1998, 46, 475–487, by permission of John Wiley and Sons.)

from H-bond stabilisation versus an increase in mechanical strain energy arising from base pair twist). They performed temperature-dependent UVRR measurements utilising poly(dA).poly(dT) under physiological conditions and identified changes associated with the C4 carbonyl stretching mode of the thymine ring and the *N*6 scissors stretching vibration of the amine in the adenine ring of the duplex. These observations supported the 'bent-DNA' model involving a set of cross-strand bifurcated H-bonds between consecutive dA/dT pairs.

6.3.3 Fluorescence Spectroscopy

The fluorescence method for measuring T_m relies on the change in fluorescence of an intercalating dye, for example fluorescein, as the oligonucleotide moves from duplex to single-stranded or from higher state, for example quadraplex, triplex to intermediate states to single-stranded DNA [26]. The standard UV spectrophotometric approach for T_m determination suffers several shortcomings.

1. The absorbance changes are relatively small (ca. 25%) and the approach has low throughput.
2. The approach requires large volumes (1–3 ml) with absorbances of about 0.2 (i.e. ca. 20 nmol of nucleotide bases).
3. Higher order oligonucleotide structures, for example triplexes and quadruplexes, typically show complex transition phenomena, such as triplex → duplex → single strands, which often overlap and are difficult to resolve.
4. The formation of some triplexes are not accompanied by changes in absorption [27].

Several researchers have investigated using oligonucleotides labelled with acridine [28], malachite green [29], fluorescein [30, 31], 5-carboxyfluorescin [31], tetramethyrhodamine [31] at the 5′ end and tetramethylrhodamine [31], ethidium [28] at the 3′ end, to study the formation of triplexes in solution and in real time (Figure 6.6).

Yang *et al.* [31] described the melting behaviour of a DNA triple helix using both fluorescence spectroscopy and UV spectrophotometric approaches. The triplex → duplex and duplex → single-strand transitions are monitored as temperature increases. For a triplex containing a fluorescent label (fluorescein or tetramethylrhodamine) the efficiency of the non-radiative energy transfer decreases as the first transition occurs, resulting in an increase in the fluorescence intensity of the donor group. The fluorescence melting curve was measured by measuring the fluorescence intensity of fluorescein at 518 nm with an excitation wavelength of 485 nm. The UV melting curve was obtained by subtracting the absorbance at 260 nm of the DNA duplex (F1.2; flurescein labelled) at 1.0 μM from that of mixture of F1.2 duplex and the R5 oligonucleotide rhodamine labelled. The melting temperature of the fluorescein labelled duplex (66.3°C) and the normal duplex (66.6°C) were essentially similar, indicating that the fluorescent probe does not affect the stability of duplex. The melting temperatures of the F1.2*R5 oligomer show good agreement by both UV and fluorescence measurements and indicate that this is the triplex → duplex transition (see Figure 6.7).

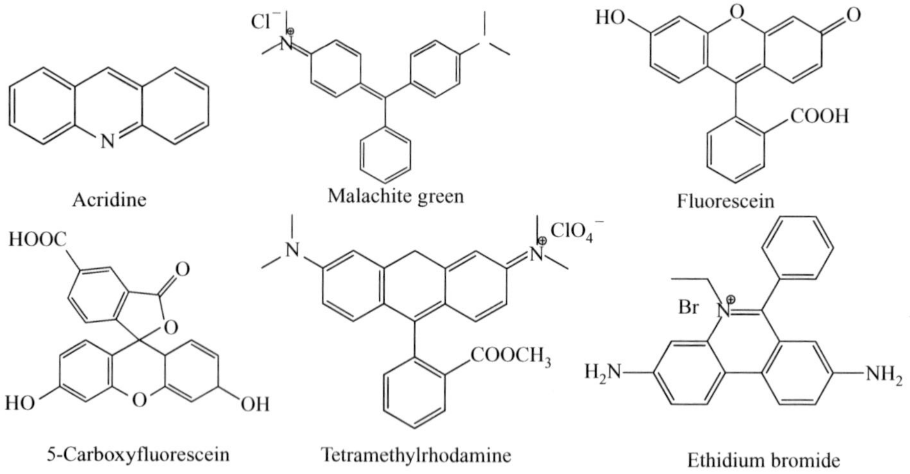

Figure 6.6 Structures of molecular probes used to study formation of triplexes in solution and real time.

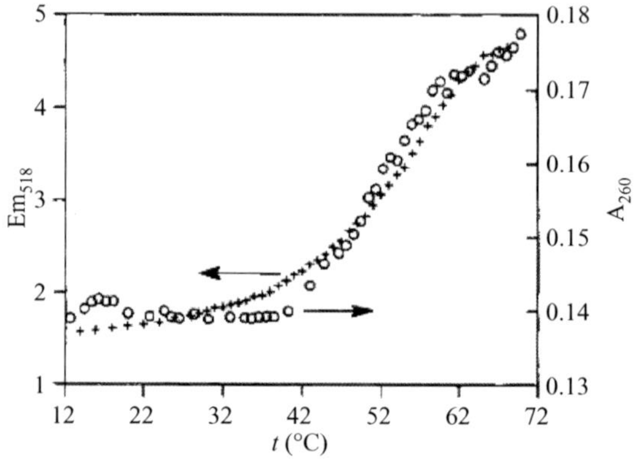

Figure 6.7 Dissociation profiles of the DNA triplex F1.2*R5 (1.0 μM) in buffer by fluorescence spectroscopy (x) and UV absorbance (o). Where F indicates a fluorescein label and R indicates that the label was rhodamine. (Source: Reprinted with permission from Xu, D., Evans, K.O., Nordlund, T.M., Melting and premelting transitions of an oligomer measured by DNA base fluorescence and absorption, *Biochemistry*, 33, 9592–9599. Copyright (1994) American Chemical Society.)

Xu *et al.* [32] used a fluorescence probe (2-aminopurine (2-AP)) to provide an optical probe of local and global DNA conformational changes. DNA with 2-AP in place of adenine exhibits fluorescence which is easily detected at ambient temperature. This substitution does not inhibit the classical 3D β-helical structure of these oligonucleotides. Thus 2-AP can be placed at any position in the duplex sequence to measure local conformational changes. The method is very sensitive, the fluorescence can be determined in the presence of large amounts of non-fluorescent natural DNA.

Figure 6.8 shows the melting curves for the decamer (21 μM) measured at 260 nm for the UV melting curve and at 330 nm for 2-AP absorption.

A plot of fluorescence-quenched fraction versus temperature is shown in Figure 6.9. The melting temperatures measured by UV at 260 nm (28.3 ± 0.5°C), UV at 330 nm (28.6 ± 1.0°C) and fluorescence quenching (28°C) are all very similar.

6.3.4 NMR

The temperature dependence of specified chemical shifts can be used to assess the T_m. The melting behaviour of aptamers can be monitored by assessing the impact of H-bonding on the imino protons involved in base-pairing of the duplex versus those same protons in the single-stranded oligonucleotide. Resonances in the region of 12–16 ppm using one-dimensional ^{1}H-NMR were used to demonstrate formation of a secondary duplex structure. NMR (like DSC) has the advantage that highly concentrated oligonucleotide solutions, often in the formulation matrix, can be directly measured [33]. Patel *et al.* [34] used NMR and DSC to assess the helix-to-coil transition (T_m) of the duplex formed by a self-complementary 12mer containing the standard Watson–Crick base pairs (dG-dC) at the 3-position (12mer GC duplex) versus the synthetic 12mer oligionucleotide with non-standard Watson–Crick base pairs (dG-dT) at the 3-position (12mer GT duplex). The three broad thymidine H6 singlets in the 12mer GT duplex were well resolved in a 500 MHz NMR spectrum between 7.1 and 7.4 ppm at ambient temperature and could be independently

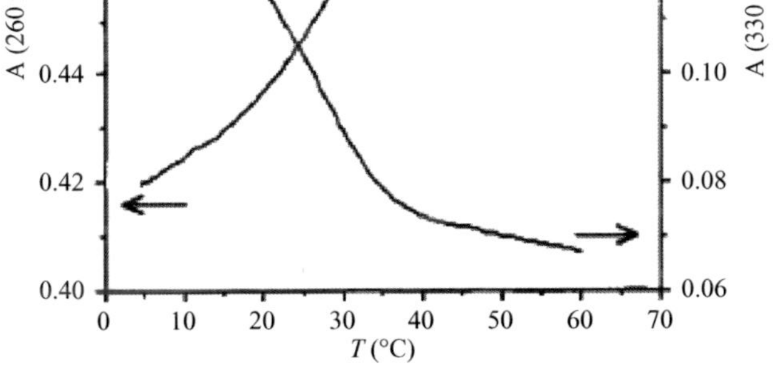

Figure 6.8 Melting curves for decamer at 260 nm (normal base pair absorption) and 330 nm (2-AP absorption). (Source: Reprinted with permission from Yang, M., Ghosh, S.S., Millar, D.P., Direct measurement of thermodynamic and kinetic parameters of DNA triple helix formation by fluorescence spectroscopy, *Biochemistry*, 33, 15329–15337. Copyright (1994) American Chemical Society.)

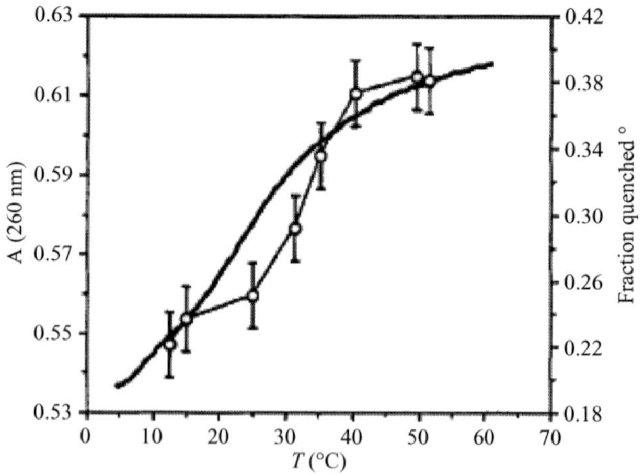

Figure 6.9 Melting curves for decamer measured by temperature dependence of 260 nm absorption and of quenched fraction (o). (Source: Reprinted with permission from Yang, M., Ghosh, S.S., Millar, D.P., Direct measurement of thermodynamic and kinetic parameters of DNA triple helix formation by fluorescence spectroscopy, *Biochemistry*, 33, 15329–15337. Copyright (1994) American Chemical Society.)

monitored through the melting transition. The T_m data from NMR and DSC showed essentially equivalent data. Both techniques revealed that a replacement of dG-dC by dG-dT gave rise to an approximately 20°C difference in T_m (12mer GC $T_m = 71.3$°C versus 12mer GT $T_m = 51.3$°C). The authors concluded that, as the enthalpies were similar in the two oligonucleotides, the destabilisation was entropic in nature. The authors showed that, for both oligonucleotides, the derived enthalpies were much smaller than the calorimetrically measured enthalpies ($\Delta H_{\text{derived}} < \Delta H_{\text{cal}}$) and concluded that the transitions involved partial opening of the helix prior to the cooperative constituent of the transition rather than in an all-or-none fashion.

Temperature-dependent conformational changes of deoxyoligonucleotides can also be monitored by measuring the ^{31}P chemical shifts, spin-lattice relaxation times (T_1) and ^{31}P[H] nuclear Overhauser enhancements (NOEs), see Figure 6.10 [35]. In parallel the T_m (32°C) was assessed using UV spectrophotometric techniques.

NMR spin echo techniques have also been used to assess T_m. Lubas *et al.* [36] assessed both the longitudinal spin–lattice (T_1) and the transverse spin–spin (T_2) relaxation times and showed similar T_m to the UV spectrophotometric approach. Additional information on the non-rotational hydration of the DNA was obtained using the NMR approach.

6.3.5 Circular Dichromism

Circular dichromism (CD) is based on the differential absorption of left/right circularly polarised light caused by chiral molecules. CD has seen extensive applications in the evaluation of secondary structure of proteins. In addition to its use for

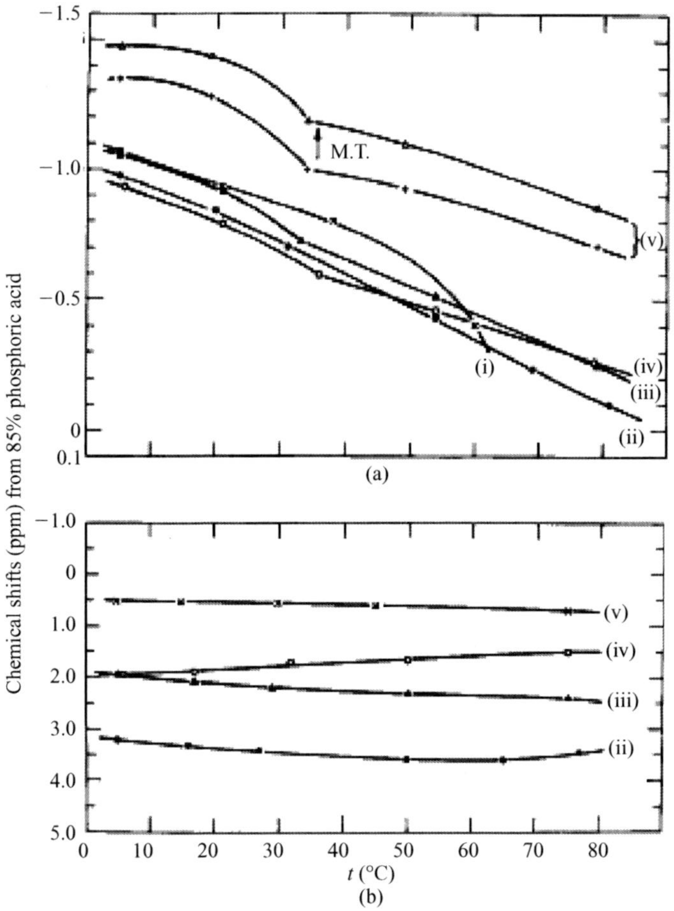

Figure 6.10 Temperature-dependent ^{31}P chemical shifts of deoxyoligonucleotides in 8 mM EDTA, D_2O. (a) Temperature dependence of the 3′-5′ phosphate diester chemical shifts. (b) Temperature dependence of the terminal phosphate chemical shifts. (i) $d(pA)_8$, 2.1 mM in strands; (ii) $d(pA)_8$, 37.7 mM, pH 7.28; (iii) $d(pT)_8$, 23.3 mM, pH 7.29; (iv) $d(pT)_8$, 16.98 mM, pH 6.95; (v) $d(pA)_3pGpC(pT)_3$, 8.6 mM, pH 7.86 in strands. T_m = melting temperature. Source: Davanloo, P., Armitage, I.M., Crothers, D.M., 31P NMR spin lattice relaxation studies of deoxyoligonucleotides, *Biopolymers*, 1979, 18, 663–680, by permission of John Wiley and Sons.

determining the T_m of oligonucleotides [37], it can be used in the empirical measurement of secondary structure, particularly conformational analysis. CD is extremely sensitive, allowing low-concentration (20 μg/ml) analysis, which is particularly useful for low-solubility samples, for example long, G-rich DNA fragments. It is also chain-length independent (short-chain oligonucleotides and long-chain RNA/DNA). The impact of agents causing conformational changes (e.g. salts, acids, bases, alcohols, drugs) can easily be assessed. Measurements are fast and inexpensive, unlike other comparative techniques, such as X-ray diffraction, NMR.

Kamiya *et al.* [38] evaluated the melting characteristics of both triple-stranded and single-stranded oligonucleotides. They assessed the melting temperature of four 15mer/23mer triplexes: T.AT, T.TA, A.TA and A.AT. The triplexes with a base-pair mismatch were less stable than the T.AT triplexes. The melting temperatures were T.AT ($T_m = 70°C$) > A.AT ($T_m = 58°C$) > A.TA ($T_m = 53°C$) > T.TA ($T_m = 48°C$), see Figure 6.11.

They also assessed the melting temperatures of the single-stranded oligonucleotides to see if there were any temperature-dependent conformational changes. They concluded that alternating T/C sequences were much more stable (T_m 27–30°C) than mixed T/C sequences (T_m 5–10°C).

In contrast, Plum *et al.* [39] assessed the relative stability of a family of 13mer duplexes (with and without 8-oxydeoxyguanosine (G*)) and used CD to assess the 3D structure of the resultant duplexes. Despite quantitative differences in the CD spectra, the authors found that the four duplexes formed with dG* (G*C, G*A, G*G and G*T), together with the corresponding dG base (GC, GA, GG and GT), all formed a β-like 3D structure. The authors showed that, despite these structural similarities, the dG to dG* modification does somewhat influence the duplex thermal and thermodynamic properties and some of this is dependent on the nature of the base opposite to dG*.

The different conformations adopted by both r(purine).d(pyrimidine) and d(purine).r(pyrimidine) hybrids in β-DNA and α-DNA have been assessed using CD [40]. Similarly, CD was used to ascertain whether individual strands could form self-

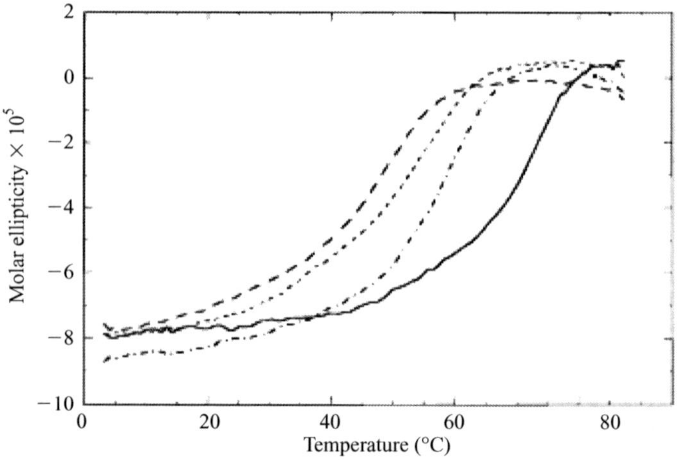

Figure 6.11 Melting profiles of the 4 15mer/23mer triplexes: T.AT (solid line), T.TA (dash line), A.TA (dotted line) and A.AT (dashed-dotted line), measured by CD. The solution contains 10 μM of each strand in 10 mM sodium acetate/10 mM cacodylic acid, pH 4.8, 0.2 M NaCl and 20 mM mgCl$_2$. Source: Kamiya, M., Torigoe, H., Shindo, H., Sarai, A., Temperature dependence and sequence specificity of DNA triplex formation: an analysis using isothermal titration calorimetry, *J. Am. Chem. Soc.*, 1996, 118, 4532–4538, by permission of the American Chemical Society.

complexes that could in turn compete with duplex or triplex formation [41]. CD has also been used to understand concentration dependence of higher structure formation. Davis *et al.* [42] used CD to follow a duplex → hairpin → single-strand transition (via two well-defined transitions by CD); in comparison UV melting studies only showed a single transition.

6.3.6 Dynamic Light Scattering

Dynamic light scattering (DLS) techniques allow for the assessment of gross changes in particle size during DNA denaturation. It is well established that large deviations from a spherical shape are required to invalidate the Stokes–Einstein relationship; hence this equation can be used to provide a good estimate of DNA size during the denaturation process. Although DLS has been used to determine melting temperatures [43], it appears to have more general usage in determining higher-order structures, for example supercoiled and nicked DNA [44], or in a complementary fashion with other techniques, such as CD and UV melting experiments [45].

Wilson *et al.* [45] used DLS measurements of single-stranded DNA as a function of sodium chloride concentration to reveal the presence of two dynamic decaying components. This was correlated with the CD and UV melting data, which indicated a decrease in flexibility and internal motions due to increases in both base stacking and the stability of stacked regions with increasing salt concentrations.

6.4 Thermal Method for Determining T_m Values

6.4.1 DSC

DSC can also be utilised to determine the T_m. In this case it is the heat capacity (ΔCp) that is measured rather than the absorbance, and enthalpy can be determined using a modified version of the van't Hoff equation [2] (see equation (6.10))

$$\Delta H = \int \Delta Cp \, dT \tag{6.10}$$

The enthalpy can thus be directly obtained from the area under the melting endotherm.

Thus, in contrast to the model-dependent nature (typically two states) of UV determination of enthalpy, the calorimetrically derived enthalpy is model-independent and enthalpy (Figure 6.12) and entropy (Figure 6.13) can be directly obtained from the primary measurements [2]. The van't Hoff transition enthalpy can also be calculated from the calorimetric data. Values derived in this manner do not agree with either the directly measured calorimetric values [34] or the corresponding measured values derived by UV measurements [46, 47].

Thus comparison of the UV (model-dependent) and calorimetric (model-independent) enthalpies gives an insight into the nature of the transition [2]. If $\Delta H_{UV} < \Delta H_{cal}$ then the transition involves a high proportion of intermediate states;

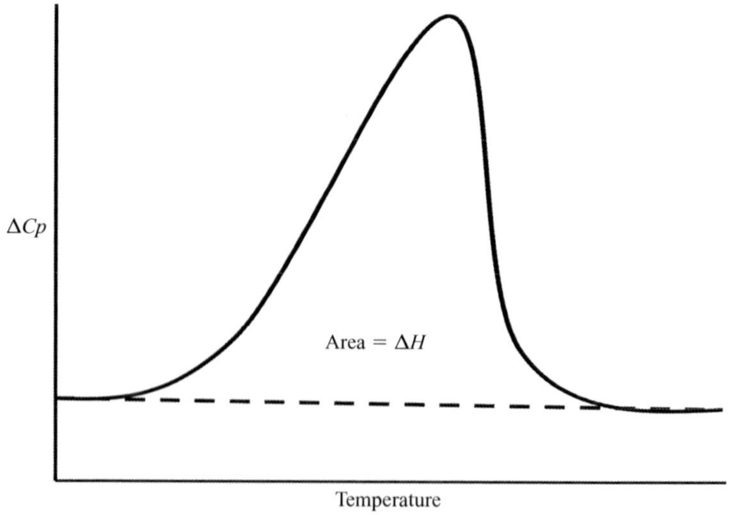

Figure 6.12 A typical heat capacity versus temperature melting curve for an oligonucleotide. (Source: Chakrabarti, M.C., Schwartz, F.P., Thermal stability of PNA/DNA and DNA/DNA duplexes by differential scanning calorimetry, *Nucleic Acid Res.*, 1999, 27, 4801–4806, by permission of Oxford University Press.)

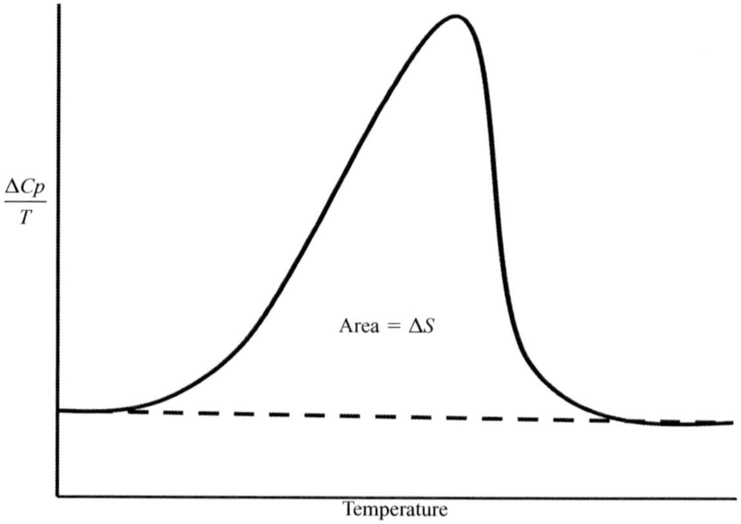

Figure 6.13 A typical heat capacity/temperature versus temperature melting curve for an oligonucleotide. (Source: Chakrabarti, M.C., Schwartz, F.P., Thermal stability of PNA/DNA and DNA/DNA duplexes by differential scanning calorimetry, *Nucleic Acid Res.*, 1999, 27, 4801–4806, by permission of Oxford University Press.)

whereas, in contrast, if $\Delta H_{UV} = \Delta H_{cal}$ then the transition proceeds in a 2-state manner.

Chakrabarti and Schwartz [48] evaluated the impact of peptide nucleic acids (PNAs) on duplex stability, where the four nucleotide bases are attached via a *N*-(2-aminoethyl)glycine backbone in PNAs rather than the negatively charged deoxyribose phosphate backbone in naturally occurring DNA. They used DSC to measure the T_m of some 10mer oligonucleotides and hence were able to directly determine thermodynamic parameters (see Figure 6.14).

The PNA/DNA T_m of the 10mer oligonucleotides ranged from 329 to 343 K and the calorimetrically derived transition enthalpies ranged from 209 ± 6 to 283 ± 37 KJ/mol. In contrast, the corresponding DNA/DNA melting temperatures were 7–20 K lower and the transition enthalpies were also lower, ranging from 72 ± 29 to 236 ± 24 KJ/mol [34]. Figure 6.15 shows DSC heating curves of the PNA(TG)/DNA duplex and the DNA(TG)/DNA duplex showing the intrinsically higher enthalpy values of the PNA/DNA duplex.

The authors [48] showed good agreement between the DSC- and UV-derived transition enthalpies, supporting the conjecture that a 2-state transition model was applicable. In contrast, Plum *et al.* [39] assessed the relative stability of a family of 13mer duplexes (with and without 8-oxydeoxyguanosine) and demonstrated that transition enthalpies from DSC data (model independent) were higher than UV spectrophotometrically derived enthalpies (model dependent), which indicated that the underlying 2-state transition model was inaccurate.

Although the requirement for larger sample sizes for DSC (versus UV spectroscopy) is usually considered a disadvantage, it can be a significant advantage when analysing highly concentrated oligonucleotide solutions (1–200 mg/ml) [19].

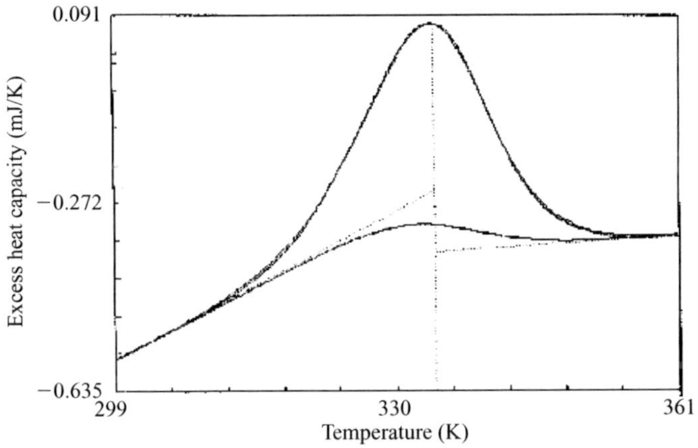

Figure 6.14 DSC heating curve of a 0.06 mM 10mer PNA(TC)/DNA duplex solution at a scan rate of 60 K/h, with a cell volume of 0.51 mL. The fit of a 2-state (AB = A + B) is shown by the broken lines. (Source: Davanloo, P., Armitage, I.M., Crothers, D.M., ^{31}P NMR spin lattice relaxation studies of deoxyoligonucleotides, *Biopolymers*, 1979, 18, 663–680, by permission of John Wiley and Sons.)

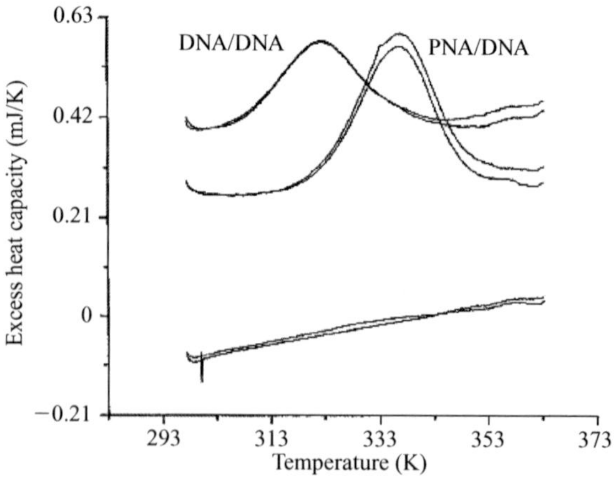

Figure 6.15 DSC heating curves of 0.0344 mM of 10mer PNA (TG)/DNA duplex solution (right-hand side) and a 0.0303 mM of DNA (TG)/DNA duplex solution (left-hand side) and buffer (pH 7.0, 10 mM sodium phosphate containing 100 mM sodium chloride and 0.1 mM EDTA), together with repeat scans. Scan rate of 60 K/h, with a cell volume of 0.51 mL. (Source: Reprinted with permission from Kamiya, M., Torigoe, H., Shindo, H., Sarai, A., Temperature dependence and sequence specificity of DNA triplex formation: an analysis using isothermal titration calorimetry, *J. Am. Chem. Soc.*, 118, 4532–4538. Copyright (1996) American Chemical Society.)

6.4.2 Isothermal Titration Calorimetry

The real value of isothermal titration calorimetry (ITC) lies in the assessment and understanding of intermediate structures and classical melting temperature assessment is better studied using UV hyperchromicity or DSC assessment [49]. ITC offers further insight into the low-temperature ($<T_m$) structural transitions of the single-stranded states, which have been classically difficult to observe [49]. In addition several groups have applied ITC to higher order structures [50, 51]. Diamond *et al.* [50] used ITC to circumvent the non-classical non-2-state behaviour that was evident from the UV melting data. Mikulecky *et al.* [51] used ITC to interrogate the enthalpic and entropic contributions to folding.

There is emerging evidence that simple 2-state models may not be adequate to describe triple helix formation. Kamiya *et al.* [38] studied triple helix formulation using low-temperature ITC. Use of ITC allows direct determination of thermodynamic parameters, for example heat capacity and enthalpy; in addition, the binding free energy and entropy change can be assessed simultaneously. Kamiya *et al.* [38] evaluated triplex formation between a 23mer double-stranded homopurine–homopyrimidine oligonucleotide and a single-stranded 15mer homopyrimidine oligonucleotide, which formed T.AT and C.GC triads at pH 4.8. They found that the reaction was driven by a large negative enthalpy change accompanied by a large negative entropy change. The authors were able to postulate that the triplex formation is a coupled process involving conformational transitions in the single-stranded oligonucleotide and then binding to the double-stranded oligonucleotide.

6.5 Future Directions

One of the more interesting trends from the literature is immobilisation of oligonucleotides onto surfaces (rather than in solution), allowing high-throughput assessments of melting temperature, to support genotyping and other high-volume activities.

Russom *et al.* [52] assessed the use of rapid solid-phase melting curve analysis for accelerated genotyping of single nucleotide polymorphism (SNP). The authors indicated that the melting curve analysis is based on the approach of dynamic allele-specific hybridisation. The DNA duplexes are attached to a bead-support and then immobilised onto the surface of a micro-heater chip with integrated temperature/heater sensors; this allows in situ melting temperatures of the duplexes to be performed. This further allowed ultra-fast heating times (60°C/min), so that analyses could be performed in less than 1 min. The authors commented that 'the dramatic volume reduction and rapid analysis time is promising for the development of highly cost-effective and ultrahigh throughput method for SNP analysis based on arrays of single beads.'

Biswal *et al.* [53] observed that surface bound oligonucleotide complexes show reduced melting temperatures versus their bulk melting temperature and that this difference becomes more pronounced at higher ionic strengths and with longer oligonucleotide chain lengths. Immobilisation of oligonucleotides to surfaces obviously has energetic and entropic effects, but the authors sought to minimise these effects by ensuring that the surface coverage was about 33 nm^2, with an effective radius of 3.3 nm. These absolute changes in melting temperature for immobilised oligonucleotides are probably not an issue, as the technique will be used as part of a high-throughput screen looking at relative differences in melting temperature. The authors felt that this approach (nanomechanical detection of DNA melting on cantilever surfaces) was a complementary alternative to calorimetric or fluorescence approaches.

Oligonucleotide melting analysis is traditionally a time-based approach. However, a recent approach (spatial DNA melting analysis) monitors fluorescence as a function of position, rather than time. This is achievable by use of a characteristic gradient (chemical or temperature) within the sample and fluorescent imaging of the output via a detector array. Liedl and Simmel [54] reported on the fast, reproducible determination of oligonucleotide melting temperature, with single-nucleotide resolution using microfluidics. They utilised stable, linear continuous diffusion gradients of a chemical denaturing agent, formamide. The denaturing agent lowers the T_m and a given formamide concentration can then be correlated to a virtual T_m along the diffusion gradient. The authors applied this approach to the determination of T_m of several sets of dye-labelled and quencher-labelled oligonucleotides of differing chain lengths. The method demonstrated good specificity and could detect single nucleotide mismatches in a complementary sequence. Importantly, comparative assessment of this technique with the standard melting curve approach yielded good results for oligonucleotides of varying chain length (11mer to 17mer), and deviations between the 2 approaches were always $<2\%$. Both approaches overestimated the melting temperature compared with the calculated results obtained by NN determinations.

The low sample consumption and the relatively short time span of the experiments (<2 h), without any loss in performance, demonstrated the promise of this approach. The authors highlighted the need to circumvent costly and time-consuming labelling of the oligonucleotides and suggested the use of UV absorption measurements along the gradient, the use of double-strand specific dyes or the utilisation of molecular beacons as areas for potential future improvement of the method. Spatial DNA melting using a temperature gradient was first utilised by Mao *et al.* [55], who utilised a glass substrate with multiple micro-channels that were each independently heated to a unique temperature. These channels were filled with both double-stranded oligonucleotides (30mer) and a DNA dye. They found the method capable of differentiating a single base mismatch near the centre of the oligonucleotide strand via the resultant melting curve. Similarly, Baaske *et al.* [56] generated a radial temperature gradient of 80°C within a microfluidic device using an infra-red laser. The resultant thermal denaturating of the dye-labelled DNA-hairpin (33mer) was then measured.

This approach was also utilised by Crews *et al.* [57], who developed a continuous-flow, temperature gradient microfluidic device and applied it to spatial oligonucleotide melting analysis for both genotyping and SNP variant screening. The authors used a steady-state temperature gradient over a 10°C sample range (20–30°C) and melting curves were constructed from changes in the fluorescent output (amplicon approach). This approach uses unlabelled probes (in contrast to Liedl and Simmel [54]) to fully genotype single base changes with good signal-to-noise ratios and a temperature resolution as fine as 0.04°C/pixel. The authors validated the approach by genotyping 12 blinded DNA samples at 3 warfarin related sites with 100% accuracy.

Lastly, Fiche *et al.* [58] assessed the opportunities offered by plasmon imaging coupled with temperature scanning for the assessment of single point mutations from the analysis of their melting curves under non-equilibrium conditions. Hybridisation of oligonucleotides or DNA/protein interactions can be detected by simply measuring the change in reflectivity. In addition, the imaging option enables parallel analysis using a DNA chip format. Thus this approach is a low-cost technique allowing for high throughput and detection of hybridisation in real time. The principle of detection lies in the differences in melting behaviour between perfectly matched and mismatched duplexes. All results were in good accord with NN modelling.

6.6 Conclusions

Thermal assessment of secondary, tertiary or higher structures of oligonucleotides (particularly naturally occurring DNA and RNA) have played a key role in our current understanding of base pair and strand–strand interactions. The classical melting temperature (T_m) of duplexes (defined as when half of the oligonucleotide strands are in the double-helical state and half are in the 'random-coil' state) can be determined using a number of different analytical techniques.

UV hyperchromicity or optical melting studies appear to be the method of choice for determining the melting temperature. Other key thermodynamic properties, such as ΔH or ΔS, can then be derived using a model-dependent assumption, which is that there is a 2-state model (duplex → single strands) in operation. Other thermal techniques, such as DSC and ITC, can be used to generate these thermodynamic parameters directly; that is the comparison of the model-dependent and model-independent data can either substantiate or repudiate the underlying hypothesis. ITC offers further insight into the low-temperature ($<T_m$) structural transitions of the single-stranded states, which have been classically difficult to observe. In addition several groups have applied ITC to higher order structural elucidation.

There are several other constraints to the universal applicability of the UV hyperchromicity method, as listed below.

1. The absorbance changes are relatively small and the approach has low throughput.
2. The method requires large volumes (1–3 ml) with absorbances of about 0.2.
3. Higher order oligonucleotide structures, such as triplexes and quadruplexes, typically show complex transition phenomena, for example triplex → duplex → single strands, which often overlap and are difficult to resolve.
4. The formation of some triplexes are not accompanied by changes in absorption.

This often leads to the use of the UV hyperchromicity method in conjunction with other methods (allowing for complementary synergistic output). UVRR has been useful to facilitate better understanding of the 2D and 3D structure before and after the melting transition. Mukerji *et al.* [23] investigated the structure of poly(rI) in presence of cations and showed that the K-stabilised-poly(rI) is more stable than the Na-stabilised-poly(rI). CD can also be used in the empirical measurement of secondary structure, particularly conformational analysis. CD is extremely sensitive allowing low-concentration analysis. Additionally, it is oligonucleotide chain-length independent and the impact of agents causing conformational changes (e.g. salts, acids, bases, alcohols, drugs) can be easily assessed. Measurements are fast and inexpensive, unlike other comparative techniques, such as X-ray diffraction or NMR.

The fluorescence method for measuring T_m is often used in conjunction with UV. This method relies on the change in fluorescence of an intercalating dye, for example fluorescein, as the oligonucleotide moves from duplex to single stranded or from higher state (e.g. quadraplex, triplex) to intermediate states to single-stranded DNA. In addition, fluorescent base pairs, such as 2-AP, can be used to understand the binding forces within a duplex. This substitution does not inhibit the classical 3D β-helical structure of oligonucleotides and 2-AP can be placed at any position in the duplex sequence to measure local conformational changes. The method is very sensitive and the fluorescence can be determined in the presence of large amounts of non-fluorescent natural DNA.

The temperature dependence of specified NMR chemical shifts can be used to assess the T_m. Temperature-dependent conformational changes of deoxyoligonucleotides can also be monitored by measuring the ^{31}P chemical shifts, spin–lattice relaxation times (T_1) and $^{31}P[H]$ NOEs.

DLS techniques allow for the assessment of gross changes in particle size during DNA denaturation. Although DLS has been used to determine melting temperatures, it appears to have more general utility in determining higher-order structures (e.g.

supercoiled and nicked DNA [35]) or in a complementary fashion with other techniques (e.g. DLS, CD and UV melting experiments [36]).

Finally, experimental approaches such as crystallography, NMR and so on, are not yet capable of direct study of the 3D structure of oligonucleotides, which are capable of forming a prototype double helix, that is ca. 30mers in length. Recently, *ab initio* MD simulations of DNA oligonucleotides of intermediate length in aqueous solution have become possible, allowing the modelling of helix-phasing sequences of different compositions and thereby allowing the determination of sequence-dependent DNA curvature and flexibility as a function of temperature [18].

Some of the most interesting future directions are centred on immobilisation of oligonucleotides on a chip, allowing high-throughput microfluidics applications of melting temperature assessments. These approaches are aligned with the need to support genotyping, for example SNP and other high-volume activities.

References

1. Carmody, J.L., Evaluation of the impact of the variables related to thermal melt temperature determination, *Proceedings of Informa EuroTIDES Meeting*, Dusseldorf, Germany, 2008.
2. Breslauer, K., Extracting thermodynamic data from equilibrium melting curves for oligonucleotide order-disorder transitions. In *Methods in Molecular Biology, Vol. 26: Protocols for Oligonucleotide Conjugates*, Agrawal, S. (Ed.), 1994, pp. 347–372, Humana Press Inc., Totowa, NJ, USA.
3. Poland, D., Scheraga, H.A., Phase transitions in one dimension and the helix—coil transition in polyamino acids, *J. Chem. Phys.*, 1966, 45, 1456, 1464.
4. Kallenbach, N.R., Stability of helical nucleic acids. In *Quantum Statistical Mechanics in the Natural Sciences*, Kursonoglu, B., Mintz, S.L., Widmayer, S.M. (Eds), 1974, pp. 95–118, Plenum, New York.
5. Pease, A.C., Solas, D., Sullivan, E.J., Cronin, M.T., Holmes, C.P., Fodor, S.P.A., Light generated oligonucleotide arrays for rapid DNA sequence analysis, *Proc. Natl Acad. Sci.*, 1994, 91, 5022–5026.
6. Buck, K.J., Harris, R.A., Sikela, J.M., A general method for quantitative PCR analysis of mRNA levels for members of gene families; Application to GABAA receptor subunits, *Biotechniques*, 1991, 11, 636–641.
7. Steger, G., Thermal denaturation of double stranded nucleic acids: Prediction of temperatures critical for gradient gel electrophoresis and polymerase chain reaction, *Nucleic Acids Res.*, 1994, 22, 2760–2768.
8. Marmur, J., Doty, P., Determination of the base composition of deoxyribonucleic acid from its thermal denaturation temperature, *J. Mol. Biol.*, 1962, 5, 109–118.
9. Wetmur, J.G., DNA probes: applications of the principal of DNA hybridization, *Crit. Rev. Biochem. Mol. Biol.*, 1991, 26, 227–259.
10. Panjkovich, A., Melo, F., Comparison of different melting temperature calculation methods for short DNA sequences, *Bioinformatics*, 2005, 21, 711–722.
11. Borer, P.N., Dengler, B., Tinoco, I., Uhlenbeck, O., Stability of ribonucleic acid double-stranded helices, *J. Mol. Biol.*, 1974, 86, 843–853.
12. Tinoco, I., Uhlenbeck, O., Levine, M.D., Estimation of secondary structure in ribonucleic acids, *Nature*, 1971, 230, 363–367.
13. Tinoco, I., Borer, P.N., Dengler, B., Levine, M.D., Uhlenbeck, O., Crothers, D.M., Gralla, J., Improved estimate of secondary structure of ribonucleic acids, *Nature New Biol.*, 1973, 246, 40–41.

14. Freier, S.M., Kierzek, R., Jaeger, J.A., Sugimoto, N., Caruthers, M.H., Neilson, T., Turner, D.H., Improved free energy parameters for prediction of RNA duplex stability, *Proc. Natl Acad. Sci.*, 1986, 83, 9373–9376.
15. Breslauer, K.J., Frank, R., Blocker, H., Marky, L.A., Predicting DNA duplex stability from the base sequence, *Proc. Natl Acad. Sci.*, 1986, 83, 3746–3750.
16. Doktyez, M.J., Goldstein, R.F., Paner, T.M., Gallo, F.J., Benight, A.S., Studies of DNA dumbbells. I. Melting curves of 17 DNA dumbbells with different duplex stem sequences linked by T4 endloops: evaluation of nearest neighbour stacking interaction in DNA, *Biopolymers*, 1992, 32, 849–864.
17. SantaLucia, J., A unified view of polymer, dumbbell and oligonucleotide DNA nearest neighbour thermodynamics, *Proc. Natl. Acad. Sci.*, 1998, 95, 1460–1465.
18. Beveridge, D.L., Dixit, S.B., Barreiro, G., Thayer, K.M., Molecular dynamics simulations of DNA curvature and flexibility: Helix phasing and premelting, *Biopolymers*, 2004, 73, 380–403.
19. Zhu, H., Srivatsa, G.S., T_m Analysis of oligonucleotides. In *Handbook of Analysis of Oligonucleotides and Related Products*, Bonilla, J.V., Srivasta, G.S. (Eds), 2011, pp. 219–241, CRC Press, Boca Raton, FL.
20. Simonian, S.H., *Experimental Micro T_m analysis with DU series UV/Visible Spectrophotometers*, 2009, Beckman Coulter Technical Report.
21. Mergny, J.L., Lacroix, L., Kinetics and thermodynamics of i-DNA formation: Phosphodiester versus modified oligonucleotides, *Nucleic Acids Res.*, 1998, 26, 4797–4803.
22. Nelson, J.W., Martin, F.H., Tinoco, I., DNA and RNA oligomer thermodynamics: the effect of mis-matched bases on double-helix stability, *Biopolymers*, 1981, 20, 2509–2531.
23. Mukerji, I., Sokolov, L., Mihailescu, M-N., A UV resonance Raman investigation of poly(rI): Evidence of cation-dependent structural perturbations, *Biopolymers*, 1998, 46, 475–487.
24. Howard, F.B., Miles, H.T., Poly(inosinic acid) helices: essential chelation of alkali metal ions in the axial channel, *Biochemistry*, 1982, 21, 6736–6745.
25. Chan, S.S., Austin, R.H., Mukerji, I., Spiro, T.G., Temperature dependent ultraviolet resonance raman spectroscopy of the premelting states of dA.dT DNA, *Biophysical J.* 1997, 72, 1512–1520.
26. Darby, R.A.J., Sollogoub, M., McKeen, C., Brown, L., Risitano, A., Brown, N., Barton, C., Brown, T., Fox, K.R., High throuput measurement of duplex, triplex and quadruplex melting curves using molecular beacons and a lightcycler, *Nucleic Acid Res.*, 2002, 30, e39.
27. Jetter, M.C., Hobbs, F.W., 7,8-Dihydro-8-adenine as a replacement for cytosine in the third strand of triple helices. Triplex formation without hypochromicity, *Biochemistry*, 1993, 32, 3249–3253.
28. Mergny, J-L., Garastier, T., Rougée, M., Lebedev, A.V., Chassignol, M., Thuong, N.T., Hélène, C., Fluorescence energy transfer between two triple helix-forming oligonucleotides bound to duplex DNA, *Biochemistry*, 1994, 33, 15321–15328.
29. Scaria, P.V., Will, S., Levenson, C., Schafer, R.H., Physicochemical studies of d(G3T4G3)*d(G3A4G3)*d(C3T4C3) triple helix, *J. Biol. Chem.*, 1995, 270, 7295–7303.
30. Ellouze, C., Piot, F., Takahashi, M., Use of fluorescein labelled oligonucleotides for analysis of formation and dissociation kinetics of T:A:T triple-stranded DNA: Effect of divalent cations, *J. Biochem.*, 1997, 121, 521–526.
31. Yang, M., Ghosh, S.S., Millar, D.P., Direct measurement of thermodynamic and kinetic parameters of DNA triple helix formation by fluorescence spectroscopy, *Biochemistry*, 1994, 33, 15329–15337.
32. Xu, D., Evans, K.O., Nordlund, T.M., Melting and premelting transitions of an oligomer measured by DNA base fluorescence and absorption, *Biochemistry*, 1994, 33, 9592–9599.
33. DeRider, M.L., Brooks, D., Burt, G., Structural determination by NMR. In *Handbook of Analysis of Oligonucleotides and Related Products*, Bonilla, J.V., Srivasta, G.S. (Eds), 2011, pp. 361–384, CRC Press, Boca Raton, FL.
34. Patel, D.J., Kozlowski, S.A., Marky, L.A., Rice, J.A., Broka, C., Dallas, J., Itakura, K., Breslauer, K.J., Structure, dynamics, and energetic of deoxyguanosine-thymidine wobble base pair forma-

tion in self complementary d(CGTGAATTCGCG) duplex in solution, *Biochemistry*, 1982, 21, 437–444.

35. Davanloo, P., Armitage, I.M., Crothers, D.M., ^{31}P NMR spin lattice relaxation studies of deoxyoligonucleotides, *Biopolymers*, 1979, 18, 663–680.
36. Lubas, B., Wilczok, T., Daskiewicz, O.K., Thermal transition of DNA measured by NMR spin-echo technique, *Biopolymers*, 1967, 5, 967–974.
37. Kypr, J., Kejnovská, I., Renčiuk, D., Vorličková, M., Survey and summary: Circular dichroism and conformational polymorphism of DNA, *Nucleic Acids Res.*, 2009, 37, 1713–1725.
38. Kamiya, M., Torigoe, H., Shindo, H., Sarai, A., Temperature dependence and sequence specificity of DNA triplex formation: an analysis using isothermal titration calorimetry, *J. Am. Chem. Soc.*, 1996, 118, 4532–4538.
39. Plum, G.E., Grollman, A.P., Johnson, F., Breslauer, K.J., Influence of the oxidatively damaged addut 8-oxodeoxyguanosine on the conformation, energetic, and thermodynamic stability of a DNA duplex, *Biochemistry*, 1995, 34, 16148–16160.
40. Hung, S.-H., Yu, Q., Gray, D.M., Ratliff, R.L., Evidence from CD spectra that d(purine).r(pyrimidine) and r(purine).d(pyrimidine) hybrids are in different structural classes, *Nucleic Acids Res.*, 1994, 22, 4326–4334.
41. Gray, D.M., Hung, S-H., Johnson, K.H., Absorption and circular dichromism spectroscopyof nucleic acid duplexes and triplexes, *Methods Enzymol.*, 1995, 246, 19–34.
42. Davis, T.M., NcFail-Isom, L., Keane, E., Williams, L.D., Melting of a DNA hairpin without hyperchromism, *Biochemistry*, 1998, 37, 6975–6978.
43. Soda, K., Wada, A., Dynamic light scattering studies on thermal motions of native DNAs in solution, *Biophys. Chem.*, 1984, 20, 185–200.
44. Fishman, D.M., Patterson, G.D., Light scattering studies of supercoiled or nicked DNA, *Biopolymers*, 1995, 38, 535–552.
45. Wilson, D.H., Price, H.L., Henderson, J., Hanlon, S., Benight, A.S., Structure and dynamics of M13mp19 circular single-strand DNA: Effects of ionic strength, *Biopolymers*, 1990, 29, 357–376.
46. Breslauer, K., Methods for determining thermodynamic data on oligonucleotide transitions. In *Thermodynamic Data for Biochemistry and Biotechnology*, 1985, pp. 377–394, Academic Press, New York.
47. Marky, L.A., Kallenbach, N., McDonough, K.A., Seeman, N.C., Breslauer, K.J., The melting behaviour of a DNA junction structure: A calorimetric and spectroscopic study, *Biopolymers*, 1987, 26, 1621–1634.
48. Chakrabarti, M.C., Schwartz, F.P., Thermal stability of PNA/DNA and DNA/DNA duplexes by differential scanning calorimetry, *Nucleic Acid Res.*, 1999, 27, 4801–4806.
49. Feig, A.L., Review: Applications of isothermal titration calorimetry in RNA biochemistry and biophysics, *Biopolymers*, 2007, 87, 293–301.
50. Diamond, J.M., Turner, D.H., Matthews, D.H., Thermodynamics of three-way multibranch loops in DNA, *Biochemistry*, 2001, 40, 6971–6981.
51. Mikulecky, P.J., Takach, J.C., Feig, A.L., Entropy driven folding of an RNA helical junction: An isothermal calorimetric titration analysis of the hammerhead ribozyme, *Biochemistry*, 2004, 43, 5870–5881.
52. Russom, A., Haasi, S., Brookes, A.J., Andersson, H., Stemme, G., Rapid melting curve analysis of monolayered beads for high-throughput genotyping of single-nucleotide polymorphisms, *Anal. Chem.*, 2006, 78, 2220–2225.
53. Biswal, S.L., Raorane, D., Chaiken, A., Birecki, H., Majumdar, A., Nanomechanical detection of DNA melting on cantilever surfaces, *Anal. Chem.*, 2006, 78, 7104–7109.
54. Liedl, T., Simmel, F.C., Determination of DNA melting temperatures in diffusion generated chemical gradients, *Anal. Chem.*, 2007, 79, 5212–5216.
55. Mao, H.B., Holden, M.A., You, M., Cremer, P.S., Reusable platforms for high-throughput on-chip temperature gradient assays, *Anal. Chem.*, 2002, 74, 5071–5075.
56. Baaske, P., Duhr, S., Braun, D., Melting curve analysis in a snapshot, *Appl. Phys. Lett.*, 2007, 91, 133901.

57. Crews, N., Wittwer, C.T., Montgomery, J., Pryor, R., Gale, B., Spatial DNA melting analysis for genotyping and variant scanning, *Anal. Chem.*, 2009, 81, 2053–2058.
58. Fiche, J.B., Fuchs, J., Buhot, A., Livache, T., Point mutation detection by surface plasmon imaging coupled with a temperature scan method in a model system, *Anal. Chem.*, 2008, 80, 1049–1057.

7 Oligonucleotide Stability and Degradation

Daren Levin

7.1 Introduction

While a discussion on stability of oligonucleotides can refer to many different types of stability, such as in-process manufacturing, storage and metabolic, this chapter will focus solely on storage stability as it applies to oligonucleotide therapeutic development. Oligonucleotide storage stability needs to be controlled at both the active pharmaceutical ingredient (API) and drug product stages. The International Conference on Harmonization (ICH) of technical requirements for registration of pharmaceuticals for human use provides a stability guidance document, Q1A(R2), which is directly applicable to oligonucleotide therapeutics [1]. While this guidance lays out the framework for exposure conditions and duration of storage for formal registration stability studies, it does not provide any guidance around physicochemical changes on stability and stability indicating methods which are specific to oligonucleotides. The ICH guidance document Q5C, *Stability Testing of Biotechnological Products*, provides some valuable guidance which can be applicable to synthetic oligonucleotides, making this document a good reference [2]. It is important to note that synthetic oligonucleotides are specifically not identified within the scope of this ICH guidance; however, topics covered such as secondary/tertiary structure, biological activity and use of multiple methods to determine purity can provide beneficial guidance for oligonucleotide therapeutics. This chapter will attempt to fill in the gaps of the ICH stability guidance documents as they pertain to oligonucleotide therapeutics and provide an overview of the physicochemical changes which can take place during the storage of oligonucleotides and the methods which can be used to detect these changes.

The primary concern for oligonucleotide stability is whether or not changes to purity and secondary structure during storage can impact the safety and/or efficacy of

Analysis of Oligonucleotides and their Related Substances, edited by George Okafo, David Elder and Mike Webb.

the drug product. Similar to protein therapeutics, oligonucleotide degradation can be separated into chemical degradation and the formation of high molecular weight species. High molecular weight species are typically referred to as aggregates and are the result of multiple oligonucleotide strands which have associated into a stable structure larger than that of the original. The term chemical degradation is used to encompass all degradants formed through chemical reactions resulting in the breaking and/or forming of covalent bonds. Figure 7.1 provides an example of several oligonucleotide functionalities which are labile to chemical degradation under certain environmental conditions. Evaluation of oligonucleotide secondary structure and its impact, if any, on safety and efficacy, is not as straightforward as discussing degradation and changes to purity. The next section will explore this topic as it relates to stability and the different classes of oligonucleotide therapeutics.

7.2 Secondary Structure Considerations

There are many different classes of oligonucleotide therapeutics and each class has its own unique stability considerations. In particular, the secondary structure of the oligonucleotide can be an extremely important stability indicator for some classes and not others. Just as is the case with protein therapeutics, secondary structure (i.e. the proper 3-dimensional arrangement) can play a role in the biological activity of oligonucleotide therapeutics. The secondary structure of an aptamer is its foundation

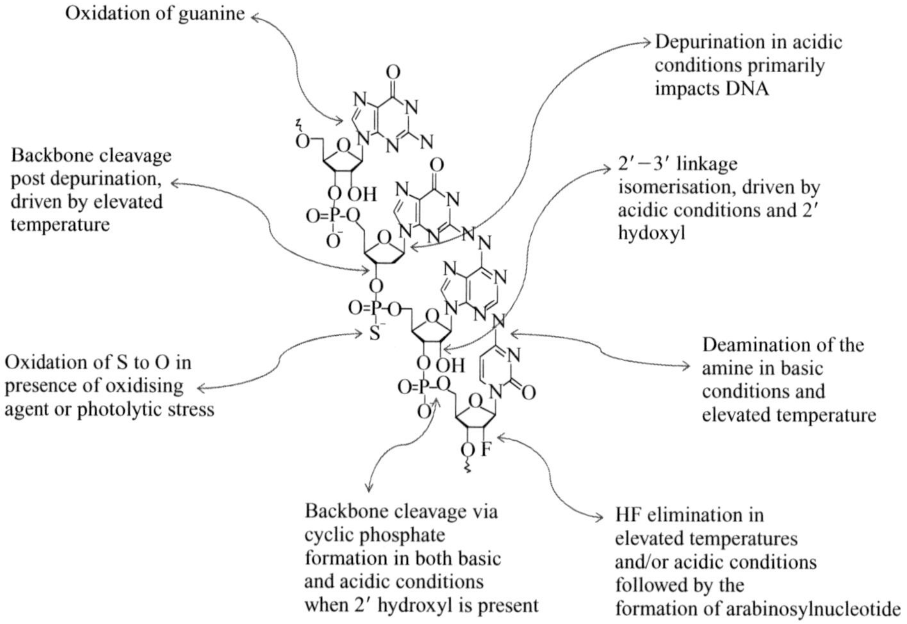

Figure 7.1 Several oligonucleotide functionalities which are labile to chemical degradation under certain environmental conditions.

for therapeutic activity, which is based on the selectivity and affinity of binding to a therapeutic target [3]. While likely not as critical as it is for aptamers, the secondary structure of double- and single-stranded-based oligonucleotide therapeutics can also be important for their activity [4, 5]. In this regard, consideration should be given to developing an understanding of the impact that changes in secondary structure on stability may have for each oligonucleotide.

The importance of secondary structure for non-aptamer-based oligonucleotide therapeutics can be seen in their sequence design. For antisense therapeutics, sequences which can result in intramolecular strand hybridising (folding up) are typically avoided at the design stage to achieve maximum therapeutic activity [6]. In addition, double-stranded oligonucleotides such as short interfering ribonucleic acid (siRNA) are designed to form a complementary duplex structure with an ideal binding affinity to achieve therapeutic activity [7–12]. While the design of sequence is critical in determining the resulting secondary structure of an oligonucleotide, it is not the only factor. The formation of oligonucleotide secondary structure is a thermodynamically driven event. The polyanionic backbone of oligonucleotides provides electrostatic repulsion between the phosphate or phosphorothioate groups. The presence of a counter ion such as sodium shields this repulsion, allowing for stronger interactions between the nucleobases. As such, salt concentration and temperature, as well as the type of ions present in solution can be critical to forming and maintaining the optimal secondary structure [11–13]. The solution pH is also an important factor as changes to the ionisation state of the bases will affect their ability to H-bond. In most cases, the formation of the secondary structure in oligonucleotides is considered to be a reversible process, whereby when the solution conditions such as temperature, salt content and pH are adjusted, the oligonucleotide secondary structure will change in accordance in a reversible manner. However, there are clearly manufacturing conditions and processes which are tightly controlled to ensure proper annealing of double-stranded oligonucleotides during the manufacturing process (such as salt concentration and rate of temperature change), indicating that if not properly followed then a non-optimal duplex product would be formed [11, 12]. Of particular interest is the warning found in several protocols that the process of lyophilisation, for a siRNA duplex solution, can lead to aggregated and/or large super structures [11, 12, 14]. In addition, one study with G-rich oligonucleotides indicated that the process of dialysis can result in the formation of aggregated structures [15].

The same factors which can alter oligonucleotide secondary structure during manufacturing can potentially alter it during storage. Factors such as pH, temperature and ionic strength can influence a change in secondary structure during storage. The critical piece to this, however, is the state of the secondary structure as it is delivered to the patient and whether it has any impact on biological activity and safety. In that regard, any sample preparation associated with the analysis of secondary structure of the final drug product should aim to mimic the concentration and solution conditions of the sample given to the patient. As an example, if microcalorimetry analysis is being used to investigate secondary structure of a drug product which simply contains oligonucleotide in water, then the analysis should be performed in water and not a buffered system. For aptamers, enzyme-linked immunosorbent assays are considered ideal for evaluating secondary structure due to the specificity of the secondary structure required to achieve binding to the substrate. However, for other single- and double-stranded oligos, thermal gradient ultra-violet (UV)-visible spectroscopy, microcalorimetry, nuclear magnetic resonance (NMR) spectroscopy and circular dichroism are techniques which may be able to indicate if changes to secondary structure have taken place. Chapters 5 and 6 provide valuable insights into the applicability of these techniques for monitoring secondary structure. A change in secondary structure could also be associated with the formation of aggregates. Aggregates are identified as a class of impurities/degradants and will be discussed in the next section in more detail.

7.3 Types of Degradation Products

7.3.1 Chemical Degradation

Oligonucleotides can be quite labile in certain environmental conditions, resulting in significant chemical degradation. It is recommended to investigate the formation of oligonucleotide degradation products early in the development process *via* a forced degradation study, as shown by Calvitt *et al.* [16]. Table 7.1 provides the type and structure of some of the known oligonucleotide degradation products as well as the environmental conditions that favour their formation. Modifications to the oligonucleotide backbone such as phosphorothioates in place of phosphates and 2′ O-methoxyethyl, 2′ O-methyl or 2′ F in place of ribose or deoxyribose sugars are commonly used to improve nuclease resistance and enhance systemic circulation. Many of the backbone modifications can impart their own unique degradation chemistry, whereas others clearly provide enhanced stability. The phosphorothioate modification can undergo desulfurisation back to a phosphate, and the 2′ F modification can form a 2,2′-anhydroribonucleotide under thermal stress or acidic conditions. However, the electron withdrawing nature of many of the 2′ modifications reduces the ability for acid or oxidative catalysed depurination compared to deoxyribose. As shown in Table 7.1, the presence of the 2′OH on the ribose construct can result in strand cleavage in alkaline conditions. Table 7.1 highlights that many environmental conditions, such as acidic, alkaline, thermal stress, oxidative stress and photolytic stress, can facilitate oligonucleotide degradation.

Given the infancy of oligonucleotides as therapeutics, there is a limited amount of literature describing the effects that environmental conditions, such as those described in the stress testing section of ICH Q1A(R2), can have on oligonucleotides. The majority of data available is related to degradation occurring during synthesis, which is typically not representative of environmental conditions that the drug substance or product would experience on storage or in use. Because of this, it is important for the sponsor to establish an understanding of the realistic degradation pathways that could be experienced for their particular oligonucleotide chemistry. The benefit to this is that once those pathways are well understood they should be applicable to all oligonucleotides with the same modification chemistry, regardless of sequence.

Table 7.1 List of several literature-identified degradants and the environmental drivers and structural considerations for the degradation pathway.

Degradant description	Structure	Environmental and structural considerations	References
Depurination	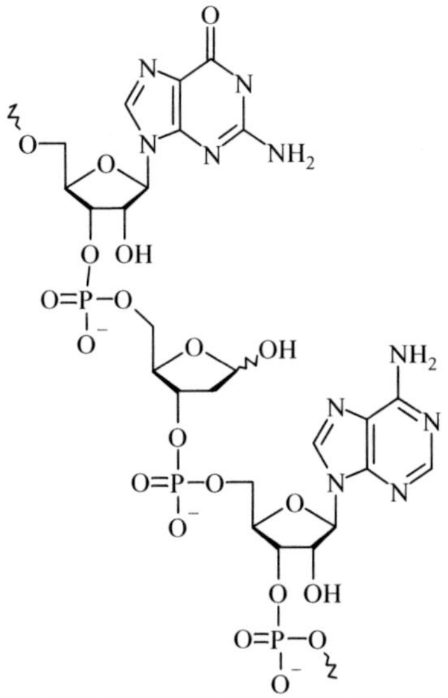	Driven by low pH, primarily impacts DNA, electronegativity of RNA and other 2′ modifications reduce this effect	[16–18]
Backbone cleavage post depurination	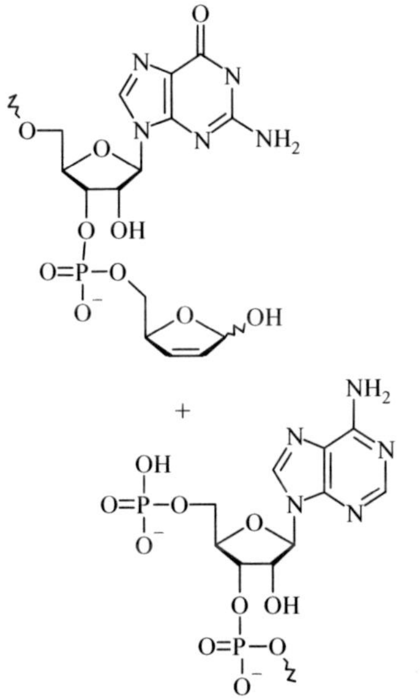	Driven by thermal stress	[19]

(*continued*)

Table 7.1 (*continued*)

Degradant description	Structure	Environmental and structural considerations	References
3′ to 2′ Isomerisation	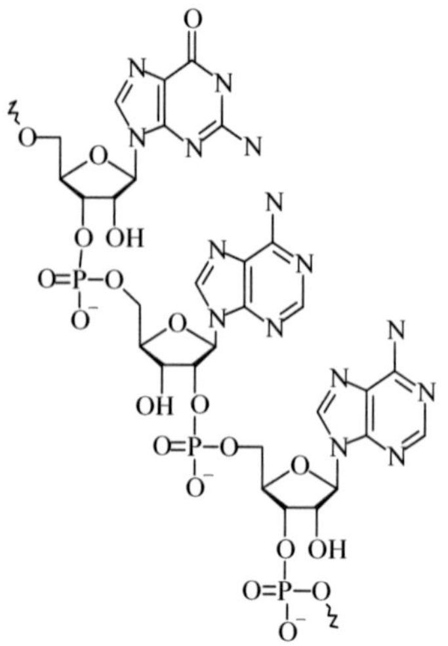	Driven by low pH, 2′ OH needed to facilitate reaction	[16, 20]
Backbone cleavage via 2′,3′ cyclic phosphate	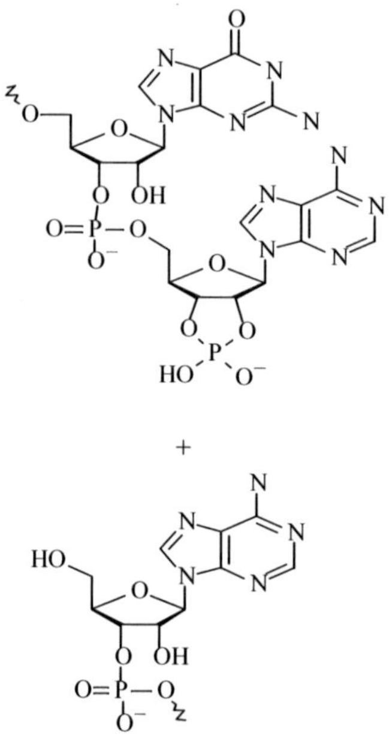	Driven by both low and high pH conditions, 2′ OH needed to facilitate reaction	[16, 20]

(*continued*)

Table 7.1 (*continued*)

Degradant description	Structure	Environmental and structural considerations	References
Deamination	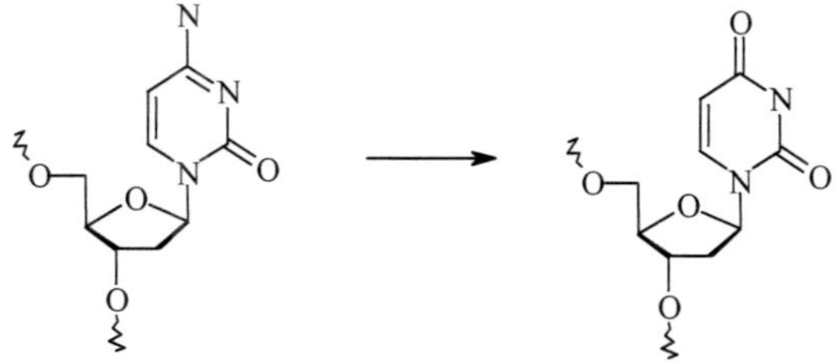	Driven by high pH and/or elevated temperature	[21, 22]
Loss of HF to form 2,2′-anhydronucleotides	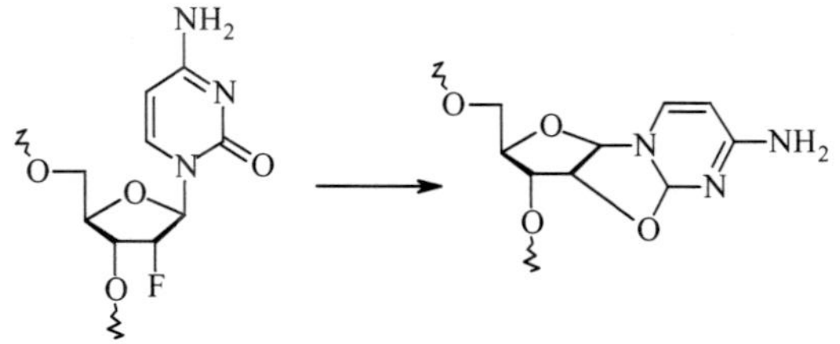	Driven by elevated temperature and/or low pH, reaction rate faster for cytodine than uridine	[16, 23]
Hydrolysis of 2,2′ anhydronucleotide to form arabinosylnucleotide	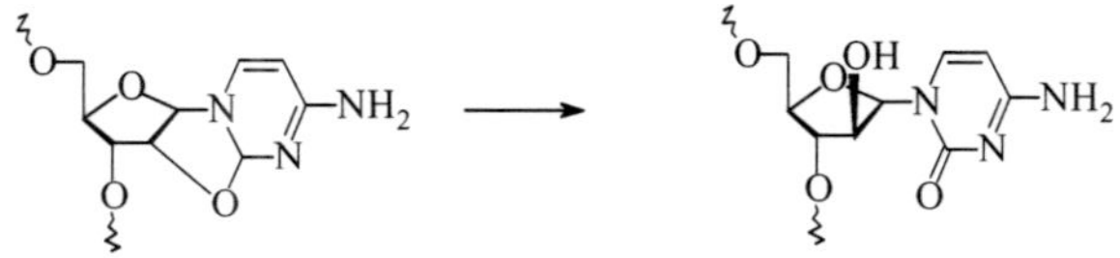	Driven by hydrolysis at neutral to high pH and elevated temperature	[16]

(*continued*)

Table 7.1 (*continued*)

Degradant description	Structure	Environmental and structural considerations	References
Guanine oxidation	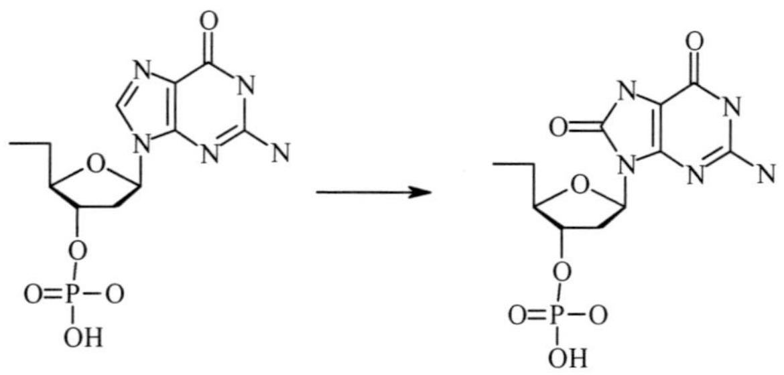	Driven by oxidative conditions, hydroxyl radicals	[16, 24]
Depurination of guanine via oxidation (ribonolactone)	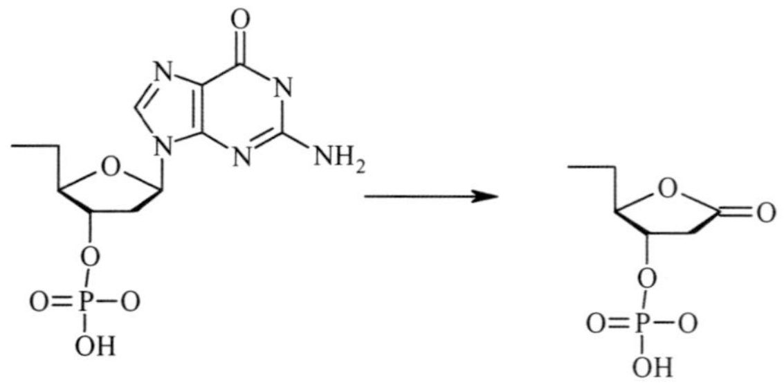	Driven by oxidative conditions, primarily impacts DNA, electronegativity of RNA and other 2′ modifications reduce this effect	[16]

(*continued*)

Table 7.1 (*continued*)

Degradant description	Structure	Environmental and structural considerations	References
Backbone cleavage post-formation of ribonolactone	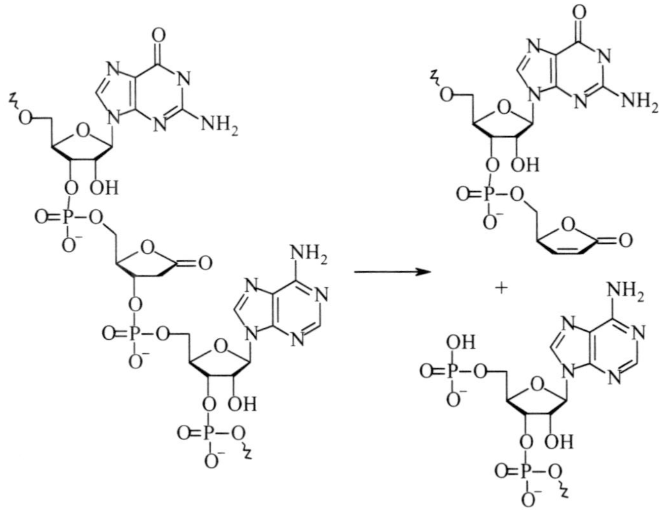	Driven by oxidative conditions, primarily impacts DNA, electronegativity of RNA and other 2′ modifications reduces this effect	[16]

(*continued*)

Table 7.1 (*continued*)

Degradant description	Structure	Environmental and structural considerations	References
Oxidation of phosphothioate	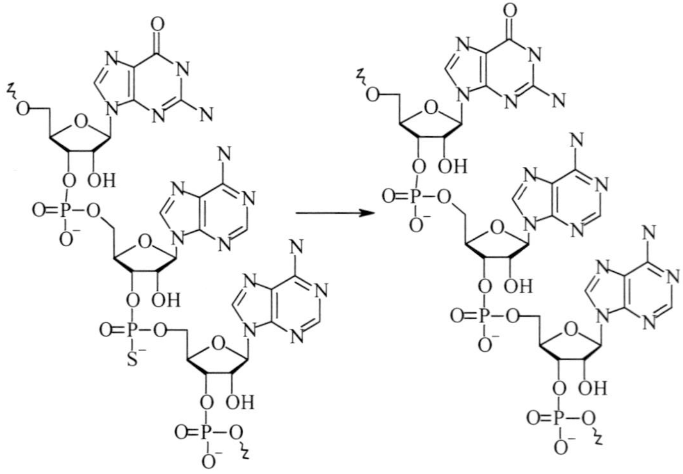	Presence of oxidising agent and/or photolytic stress	[19, 25]

7.3.2 Aggregate Degradants

In addition to oligonucleotide degradants which arise from the breaking and forming of covalent bonds, there is also the possibility that degradants can be formed through non-covalent interactions, resulting in the formation of aggregates or higher ordered oligonucleotide constructs such as dimers, trimers and tetramers. Aggregated oligonucleotide impurities may be of greater concern than other oligonucleotide impurities due to their observed link in eliciting an immunogenic response, and for this reason their presence in the drug substance or drug product should be carefully evaluated [26, 27]. To date, there is very little understanding around what drives the formation of aggregated impurities during manufacture or on storage. The present author has found that for double-stranded oligonucleotides, aggregates are more likely to form in the solid state than the solution state, and there is a correlation between the melting temperature (T_m) of the duplex and the degree of aggregate formation, with the lower melting temperature duplexes generating more aggregates [28]. This melting temperature dependency indicates that the sequence of the oligonucleotide is an important factor in the formation of aggregated impurities. It is well documented that single-strand oligonucleotide sequences with four guanidines (G-quartet) in a row have a greater propensity for forming aggregated species [29]. The ability of a single-strand oligonucleotide to hybridise with itself due to self complementation of sequence may be linked to the formation of aggregated impurities during manufacture or on storage. Analysts should use caution when developing methods to detect/quantify the presence of aggregated impurities, as sample preparation and method conditions can significantly alter the amount of aggregates present. As an example, some aggregated species may dissociate in pure water where they would otherwise remain intact in physiologically buffered conditions [30]. An ideal approach would be to use solution conditions representative of your final drug product as your sample solvent/diluent for analysis. Size exclusion chromatography (SEC) is typically used to evaluate for the presence of aggregated species in drug substance and drug product.

7.3.3 Oligonucleotide-Conjugate Degradants

Some oligonucleotide therapeutics exist as an oligonucleotide chemically conjugated to another molecule. The most prevalent type of chemically conjugated oligonucleotide is seen in the conjugating of polyethylene glycol (PEG) to aptamers, as is the case for Macugen, an anti-angiogenic PEGylated aptamer approved for age-related macular degenerative disease [31]. Other types of conjugates being explored in clinical development include small molecules [32], peptides [33] and proteins [34]. Consideration needs to be given to the degradation pathways of the conjugated molecule, as well as the chemical linker attaching the molecule to the oligonucleotide. As an example, many drug developers are exploring the use of di-sulfide bonds, which are susceptible to being reduced and broken under acidic conditions, to attach the conjugate to the oligonucleotide [35, 36]. In some cases it may be necessary also to evaluate the stability and degradation profile of the oligonucleotide and conjugated molecules independently if chromatographic specificity is not sufficient for the impurity/degradant analysis of the intact oligonucleotide-conjugate.

7.3.4 Excipient Degradants

While individual drug product excipients are typically not evaluated for chemical stability to the same extent that the drug substance is, some excipients may require the same level of scrutiny. More and more in oligonucleotide therapeutic development we are seeing the use of 'biologically active' excipients. That is, the excipient is required to elicit a biological response in the body, which is necessary for the oligonucleotide to be active. In most cases these excipients are being used to target specific cell or tissue types and/or facilitate cellular uptake of the oligonucleotide into the cytoplasm or nucleus [37]. The US Food and Drug Administration (FDA) *Guidance for Industry: Liposome Drug Products* is a valuable reference for the type of control likely to be required for such excipients [38]. In addition to the degradation of the excipients themselves, an understanding of how the excipients affect the degradation of the oligonucleotide should also be considered. As an example, Isis Pharmaceuticals found that a PEG-derived surfactant excipient being explored in a topical antisense formulation was responsible for desulfurisation of the phosphorothioate backbone [39].

With an increase in the development of complex formulations occurring throughout the oligonucleotide industry, control over excipient purity becomes a critical issue. The use of a wide variety of phospholipids, customised cationic lipids, cholesterol and PEGylated lipids has become commonplace in the area of oligonucleotide formulation development [40–42]. This increases the potential for reactive impurities from these excipients to be carried through into the final drug product. Oligonucleotide degradation products, which may be outside those covered in forced degradation testing of the API alone, can be generated as a result of these reactive impurities. Reactive impurities in batches of excipients such as PEG and polysorbate have been linked to the formation of new API impurities for several small molecule drug products [43, 44]. Forced degradation studies can be designed to focus solely on the interactions of a single excipient and the API to evaluate the potential for reactive impurities or excipient induced degradation.

Development of oligonucleotide therapeutics is in its infancy with regards to chemistry, manufacturing and controls. As such, there are no standard practices yet which sponsors can adopt for the control of impurities and degradation products. It is up to the sponsor to identify the risk associated with each likely degradant and build an appropriate control strategy around that risk. As an example, many degradants are also metabolites, possibly providing an early look into the safety profile of those impurities and a justification for setting appropriate reporting, identification and qualification thresholds [45, 46].

7.4 Considerations for the Analysis of Degradation Products

7.4.1 Quantifying and Reporting of Degradation Products

Quantifying levels of single-strand oligonucleotide impurities within drug substance and drug product is typically performed *via* reversed-phase, ion-pairing chromatography and/or anion exchange chromatography [47, 48]. Reversed-phase, ion-pairing, high-performance liquid chromatography (RP-IP-HPLC) has its advantages in that it is typically compatible with mass spectrometric detection, enabling structural characterisation and the ability to quantitate co-eluting impurities [49]. The use of ultra-performance liquid chromatography (UPLC) systems and columns has emerged as an attractive approach for impurity analysis of oligonucleotides owing to the significant improvement in chromatographic resolution over conventional HPLC. As an example, the use of a RP-IP-UPLC-based method enabled the baseline separation of several isomeric impurities (ca. 20 Da difference from parent) away from the parent peak and each other for a 21mer oligonucleotide (approximately 6.8 KDa) [50].

Both ICH-Q3A(R2), *Impurities in New Drug Substances* and ICH-Q3B(R2), *Impurities in New Drug Products*, highlight how degradation products should be reported, identified and qualified for small molecule drugs [51, 52]. While oligonucleotide therapeutics are explicitly excluded from these guidance documents, the concepts of reporting, identifying and qualifying degradation products are still applicable to the development of oligonucleotides. As an example, appropriate thresholds need to be in place with regards to reporting, identifying and qualifiying new degradation products which arise during storage stability studies, in particular registration stability batches. In most cases, any degradation product which is not also a process impurity would likely not be present in batches used for safety and clinical studies. This would trigger the thresholds necessary to report, identify and qualify this degradation product if it had not been previously qualified in other batches. It is important to keep in mind that oligonucleotide impurities and degradants exist as classes (i.e. $n - 1$, $n + 1$) and that all degradants within a class may or may not elute as a single peak. As such, it may be appropriate to establish thresholds for the entire class of impurity/degradant rather than each individual peak.

7.4.2 Method Specificity as Impacted by the Degradation Product Profile

Impurity analysis of oligonucleotide therapeutics is quite challenging owing to the magnitude of closely related process impurities which are not easily resolved from the main peak (chromatographically). As such, specificity is usually a concern for impurity analysis, and mass spectrometric characterisation of impurities is commonly needed to evaluate method specificity. The detection and quantification of degradation products in stability samples further adds to the analytical complexity. As mentioned in Section 7.3, the use of forced degraded samples early in development is valuable not just for understanding degradation pathways and products, but also for properly developing stability indicating methods. The use of stressed samples will provide the ability to fully interrogate the specificity of a method. It may be found that an impurity method can provide suitable specificity for release but that it may not adequately detect all the expected degradation products on stability. In such a case, a decision could be made early in development to utilise an additional orthogonal impurity method for stability studies, whereby the combination of the two methods provides suitable specificity.

7.4.3 Single-strand Degradant Analysis for Double-stranded and Conjugated Oligonucleotide Therapeutics

Depending on the type of oligonucleotide therapeutic, achieving suitable impurities analysis for a drug substance or drug product on stability can be significantly more challenging than at release. For release of double-stranded oligonucleotides, impurity testing is typically performed on the individual single strands prior to annealing [53]. This is primarily due to the fact that the chromatography is typically less complicated for a single strand compared to a double-stranded oligonucleotide *via* a denaturing method owing to the lack of co-eluting peaks from the 2 strands. Figure 7.2 (see colour insert) shows the overlay of chromatograms for individual sense and antisense siRNA strands on a denaturing RP-IP method showing the co-elution of impurities from both strands. However, depending on the siRNA chemistry and chromatography optimisation, it may be possible to develop denaturing RP-IP-HPLC methods capable of completely separating the sense strand impurities from the antisense strand impurities, as shown in Figure 7.3 (see colour insert). The significance of achieving the best possible separation can be seen in comparing resolution of the impurity peaks in Figures 7.2 and 7.3, and in particular the added peak complexity present for the stability sample in Figure 7.3. Regardless of the potential for co-elution of the impurities, a denaturing method to look at the single-strand impurity profile of a double-stranded oligonucleotide would be necessary to provide the best possible impurity/degradant analysis of stability samples. Because of this, the use of a denaturing method would be required at the initial stability time point, which in many cases coincides with release.

For oligonucleotides which utilise a conjugated moiety, such as a PEG-aptamer, purity and impurity analysis can be difficult owing to the lack of specificity with regards to detecting individual impurities. This is primarily due to the large size of the conjugate and in many cases its heterogeneity, which results in poor chromatographic resolution of impurities associated with small chemical differences on the oligonucleotide portion of the molecule. As a result, for conjugated oligonucleotides it may be advantageous to cleave off the conjugate prior to the analysis of stability samples to enable the detection of oligonucleotide related impurities. Another approach may be to place the unconjugated oligo upon stability in parallel with the conjugated oligo in order to achieve better control of the oligonucleotide degradant profile across the various storage conditions.

7.4.4 Use of Non-Denaturing Techniques for Stability Samples

Non-denaturing analytical techniques are used on stability for several reasons. One of the primary reasons is to monitor any changes in the formation of aggregated species. The potential for the formation of aggregates is a concern across all the different classes of oligonucleotide therapeutics. SEC is commonly used to detect and quantify the presence of aggregates as it is a size-based separation, provides good sensitivity and imparts less influence on the structural conformation of the analyte compared to other separation techniques (see Chapter 3 for more details).

Beyond the analysis of aggregates, non-denaturing chromatographic techniques are used on stability to measure any changes in the assay/purity of double-stranded oligonucleotides. One of the challenges for these types of methods is to demonstrate that they are actually stability indicating; that is, they can accurately detect a change in the assay/purity of the duplex content of the double-stranded oligonucleotide. For double-stranded oligonucleotides the majority of related impurities and degradants exist in the duplex state. Since SEC is a size-based separation technique, it only provides a determination of total duplex and is unable to detect changes in assay/purity for the target duplex of interest. As such, it has significant limitations when used to determine assay/purity of double-stranded oligonucleotides. The present author compared purity values obtained with three different methods for a lyophilised siRNA drug substance placed on stability at 25°C with 20% relative humidity (RH) headspace for 6 months [28]. The three methods were non-denaturing SEC, non-denaturing strong anion exchange (SAX) and denaturing RP-IP-UPLC. The non-denaturing SAX method was developed to provide as much resolution as possible for the target duplex of interest away from the other duplex variants/impurities. Figure 7.4 (see colour insert) shows the non-denaturing SAX chromatography for the siRNA at initial and after the six months on stability, highlighting the separation of the target duplex from the related duplex impurities. Figure 7.5 shows the plotted purity data for the three different methods over the 6-month period. The SEC method provides purity of the total duplex, the SAX method purity of the target duplex and the denaturing RP-IP-UPLC method combined purity of the sense and antisense strands. As can be seen, both the non-denaturing SAX and denaturing RP-IP-UPLC methods are able to quantify a significant change in purity over the 6-month period. The non-denaturing SAX method was the most sensitive stability-indicating method, detecting greater than a 20% nominal decrease in purity. In contrast, the SEC method was unable to detect any change in purity, clearly identifying that it was not a suitable stability-indicating method for the determination of assay/purity. In line with these findings, Murugaiah details the use of UV measurement corrected for water content and purity by SEC for assay determination of siRNA, whereby he notes that this approach is not a responsive indicator of stability owing to the lack of specificity offered by SEC [54].

Caution should be taken in attempting to use non-denaturing chromatographic techniques to determine if a loss in duplex secondary structure is occurring on stability. The difficulty arises in that there is a strong likelihood that the non-denaturing method being used to measure assay and purity of the duplex can bias the results. Non-denaturing techniques by nature attempt to stabilise the secondary structure of the analyte. SEC and SAX can use varying concentrations of salt in the mobile phase to stabilise the secondary structure of oligonucleotides. However, if the intent of the method is to be able to detect a decrease in the content of duplex secondary structure on stability, it may not be able to, as mobile phase conditions can enhance the duplex content 'on-column' prior to separation and detection. Because of this it is critical to understand method limitations with regards to their intended use. As discussed in Section 6.2, it may be appropriate to use additional techniques such as T_m analysis, microcalorimetry, circular dichromism or NMR, to identify if changes to secondary structure are taking place on stability.

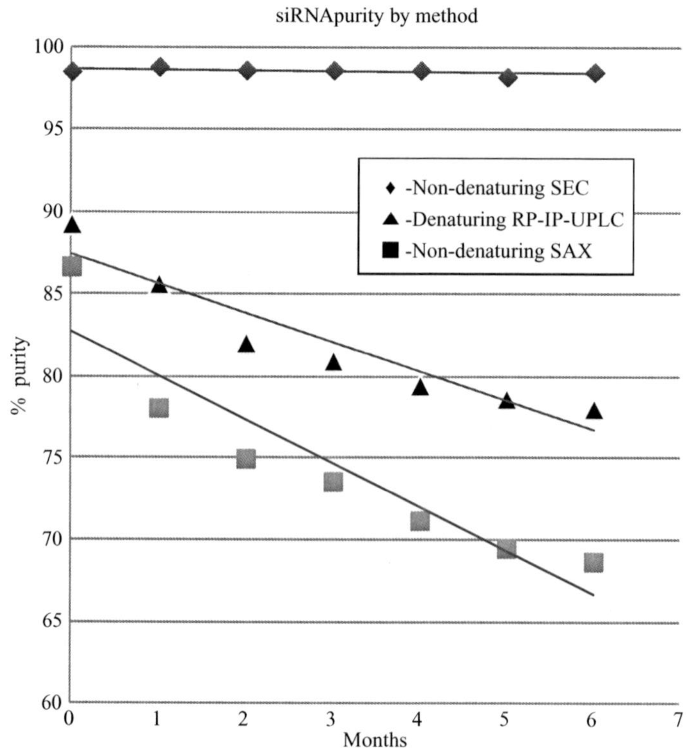

Figure 7.5 Comparison of purity values for a lyophilised siRNA API placed on stability for 6 months at 25°C with 20% RH headspace and analysed by non-denaturing SEC, non-denaturing SAX and denaturing RP-IP UPLC [28].

7.5 Storage of Oligonucleotides

Very few stability data exist in the public domain for oligonucleotide therapeutic drug substance and drug product. It is common for oligonucleotide API to be stored frozen as a lyophilised solid, and the expectation is that several years' stability is achievable for that condition. While not commonly employed, solution storage of the API can be a beneficial alternative, allowing for the removal of the costly and time-consuming manufacturing process of lyophilisation. In addition, control of the solution buffer may actually provide enhanced stability compared to the lyophilised solid which is typically in a desalted state. This can be particularly true for double-stranded oligonucleotides where the duplex conformation may be less susceptible to degradation than the dissociated single strands.

Thermo Scientific Dharmacon provided a product bulletin on the stability of their Accell siRNA reagents stored in both the solid and solution state. While this study only looked at the impact of product storage on *in-vitro* messenger RNA knockdown and does not account for changes in siRNA purity, it provides valuable insight into stability with regards to potency. They looked at both 1 mM and 10 μM solution concentrations in their siRNA buffer, RNase-free water and 1X PBS (phosphate bufferered saline with the same molarity as standard saline). The solid and solution samples were placed for up to nine months at −80°C, −20°C, 4°C, room temperature and 37°C. In general, no decline in knockdown activity was observed across all the tested conditions except 37°C, where a very slight decrease in knockdown can be seen trending over time for both the solid and solution state samples.

One study looked at the ability to predict the storage stability of plasmid DNA as a function of solution pH and temperature [55]. While the stability predictions for supercoiled plasmid DNA are not directly applicable to synthetic oligonucleotides, their approach of determining degradation rate constants for strand cleavage across a range of temperatures and using those data to predict solution stability at a given pH is applicable. Evans *et al.* [55] were able to demonstrate very good correlation between predicted and actual stability for a plasmid DNA in solution at three different pHs. For synthetic oligonucleotides there can be value in generating this type of predictive data based on the type of chemical modification chemistries being used.

With regards to storage and shelf life of drug product only two oligonucleotide products have been approved for marketed use, Vitravene and Macugen. Vitravene is a 6.6 mg/ml solution of the active Formivirsen sodium. Formivirsen is a 21mer phosphorothioated antisense oligonucleotide. The inactive ingredients of Vitravene include sodium bicarbonate, sodium chloride, sodium carbonate and water. The Vitravene solution for injection has an osmolality of 290 mOsm/kg and a pH of 8.7. The European Commission approved Vitravene back in 1999 with a shelf life of two years at 2–8°C, protected from light [56]. Macugen is a 3.47 mg/ml solution of the active Pegaptanib sodium. Pegaptanib is a 28mer aptamer with a 40 kDa branched PEG chain conjugated to the 5′ end. The oligonucleotide is not phosphorothioated, but does contain heavy use of the 2′-methoxy and 2′-fluoro modifications on the ribose moiety. Macugen is a solution for injection, and the inactive ingredients include sodium chloride, monobasic sodium phosphate monohydrate, dibasic sodium phosphate heptahydrate and water. The European Commission approved Macugen in 2006 with a shelf life of three years at 2–8°C (with a note not to freeze) [57].

As can be seen in the examples above, oligonucleotide therapeutics can be quite stable when stored at refrigerated conditions. As more products come to market it is likely to be possible to build correlations between the different types of oligonucleotide chemistries being used and their associated impact on storage stability.

7.6 siRNA Stability Case Study

This section will discuss some key findings from an investigational stability study performed at GlaxoSmithKline on three different siRNA compounds. The intent of the study was to evaluate the impact, if any, that the siRNA duplex binding affinity had on stability. As such, it provides examples for some of the issues already discussed in this chapter. The three siRNAs used in the study were composed of 21mer sense and antisense strands with 2-nucleotide overhangs on the 3′ end of each strand. All of the siRNA molecules incorporate 2′-methoxy and 2′-fluoro chemical modifications at selected sites, as well as inverted abasic end caps (deoxyribose and phosphate). The abasic endcaps are only used at the 3′ and 5′ ends of the sense strand and the O-methylation modifications are only used on the antisense strand. In addition, all three siRNAs incorporate three 5′ terminal RNA nucleotides. The three siRNAs have T_m values of 70°C, 78°C and 94°C in 1X PBS.

The stability of the three siRNAs was evaluated in both the solid and solution state. For the solid state storage, all three siRNAs were equilibrated to the same water content of 15% (w/w) and sealed with a 20% RH headspace for the duration of the study, unless the samples were to be exposed to the environment. In general, the storage conditions would be considered accelerated and were intended to elicit a change in purity to aid in understanding the dynamics of degradation. In that regard, a major focus of the study was to monitor the integrity of the duplex conformation upon exposure to the various solution storage conditions and correlate duplex integrity with chemical degradation. The duplex conformation of the three siRNAs was evaluated via both thermal gradient UV analysis and SEC over eight months at 25°C in water, 0.1X PBS and 1X PBS. The SEC analysis provides a quantitative value for the amount of duplex that has dissociated into the single-strand state (the SEC method was optimised to detect duplex dissociation occurring in the sample, 0.5X PBS mobile phase compared to 2X PBS normally). Whereas the UV analysis offers a qualitative look at the presence of duplex *via* plotting of the absorbance shift from the duplex to the single-strand state as a function of temperature, if no shift is present then it indicates the duplex is already predominantly dissociated in the sample solution.

The three different siRNAs with T_m values of 70°C (siRNA **A**), 78°C (siRNA **B**) and 94°C (siRNA **C**), all demonstrated no change in duplex integrity when stored in either 0.1X or 1X PBS for eight months at 25°C. However, significant differences in duplex integrity were observed when stored in neat water. The loss of duplex structure in neat water is the result of increased electrostatic repulsion between the phosphate backbones of the sense and antisense strands. In buffered solution, the cations present quench the repulsion between the backbones and stabilise the duplex structure. Figure 7.6 shows both the SEC chromatography and thermal gradient UV plots for all three siRNAs in water at the initial timepoint and after eight months of storage. siRNA **A**, with the weakest binding affinity, displayed a significant loss of duplex integrity at the initial timepoint. SEC revealed that 38% of the sample was detected as single strands as opposed to duplex and thermal gradient UV analysis demonstrated a shortened shift from the duplex to single-strand state, indicating that duplex dissociation was present. In contrast, when siRNA **A** is prepared in a buffered solution only 3% residual single strands are detected by way of SEC, indicating that the 38% observed in water is due to dissociation of the duplex conformation. After eight months of storage in water, siRNA **A** displayed very little data in support of any remaining duplex structure. The percentage of single strand as measured by SEC increased to 84% and no shift was present in the thermal gradient UV plot. siRNA **B** demonstrated a very slight loss in duplex integrity at the initial timepoint via SEC (9% single strands versus 3% when in buffered solution) but maintained a full shift via the UV plot. However, after eight months of storage a significant loss in duplex integrity had occurred, as demonstrated

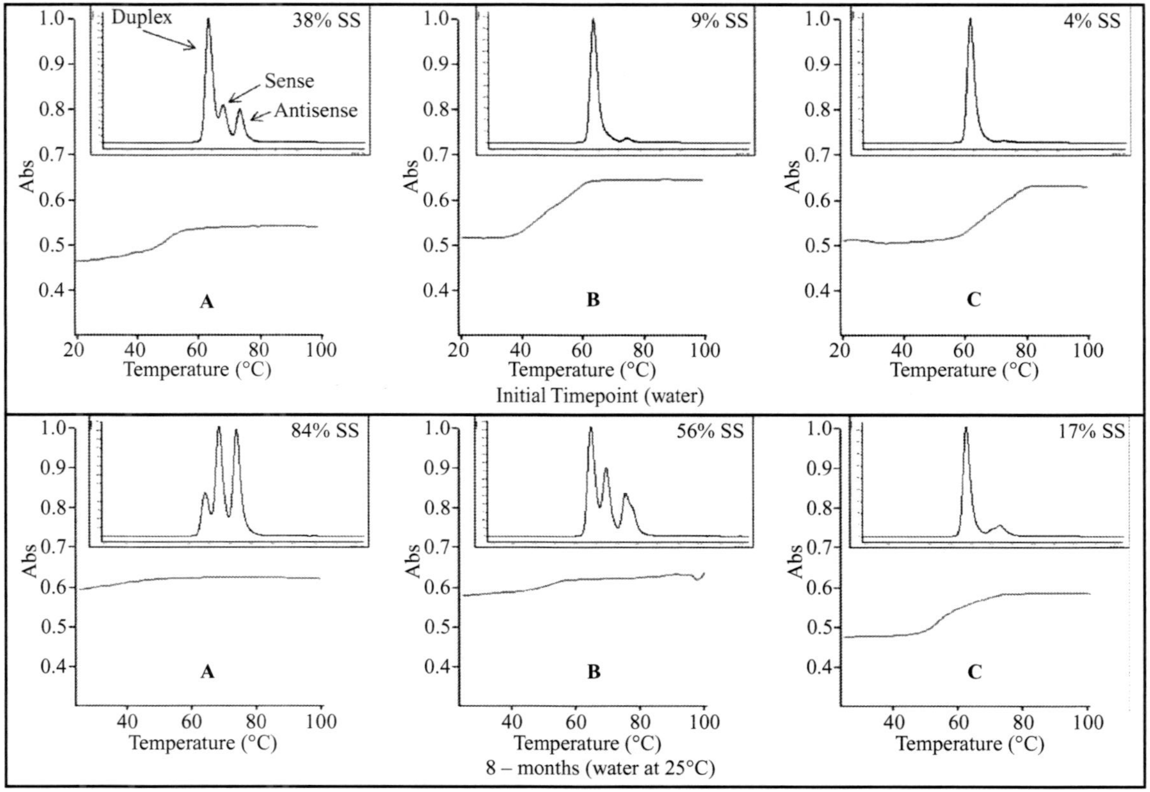

Figure 7.6 T_m plots of siRNAs **A**, **B** and **C** (representing low, medium and high binding affinities respectively) in neat water at initial (top row) and after 8 months' storage (bottom row) in water at 25°C. SECs of the same samples are inserted above each T_m plot showing the correlation between binding affinity and duplex integrity. SEC method conditions: Agilent 1100 HPLC system with UV detection at 260 nm, Waters BioSuite 125 5 μm HR column, 0.5X PBS mobile phase, 0.75 ml/min flow rate, 20°C column temperature, 10 μl injection volume (SS: single strand).

by 56% single strands and almost non-existent shift via the UV plot. siRNA **C**, with the highest binding affinity, demonstrated that its duplex conformation was well maintained at the initial timepoint and was only slightly impacted after eight months of storage compared to the other two siRNAs.

The before and after comparison of duplex conformation shown in Figure 7.6 is quite striking, but it does not address the dynamic nature of the transition. It is possible to imagine that a thermodynamically stable secondary structure is reached rather quickly in a solution state at 25°C (hours to days) and that eight months' storage is irrelevant. However, when the percent single-strand values from the SEC analysis are plotted over time, as shown in Figure 7.7, it is clear that achieving the lowest energy state of the siRNA secondary structure can be a very slow dynamic process. For all three siRNAs a gradual transition from a state of more to less duplex conformation can be seen over an 8-month period.

Chemical degradation of the three siRNAs was assessed via a denaturing RP-IP-UPLC method. The change in total impurities for each siRNA was monitored over time in the various storage conditions. Figure 7.8 shows the percent change in total impurities for each siRNA in either water, 0.1X PBS, or in the solid state at 25°C out to six months. No appreciable difference in degradation occurred between the 0.1X and 1X PBS solution conditions, so only the 0.1X data are plotted in Figure 7.8. As can be seen, the degradation profile across the different storage conditions looks the same for siRNAs **A** and **B**. That is, the most degradation occurred when stored in water. However, siRNA **C** was able to maintain chemical stability in water similar to

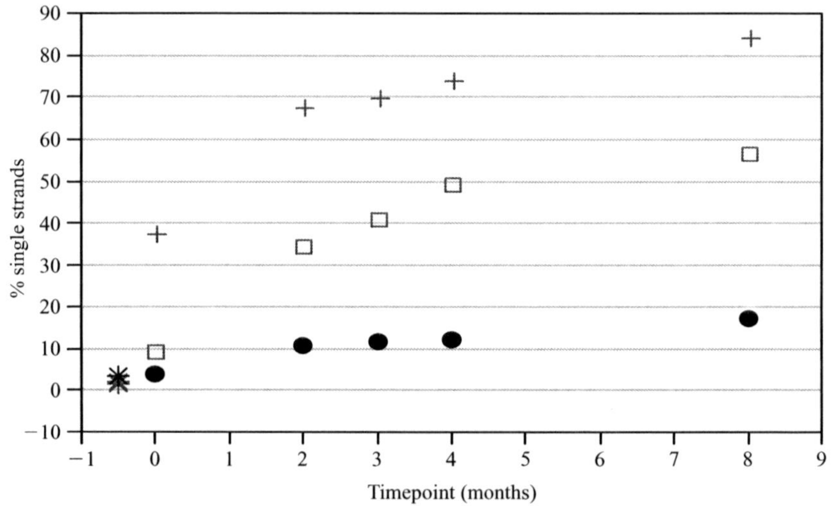

Figure 7.7 Percent single strand measured by SEC for siRNA: **A** (plus symbol (+) denotes data for low T_m), siRNA **B** (square symbol denotes data for middle T_m), and siRNA **C** (circle symbols denote data for high T_m) stored in water at 25°C over an 8-month period. The increase in single-strand content for all three samples over time demonstrates a continuous decrease in the amount of duplex conformation present in the samples. The star value preceding $t = 0$ is the percent single strands for the same siRNAs in 1X PBS at initial (all were around 3%).

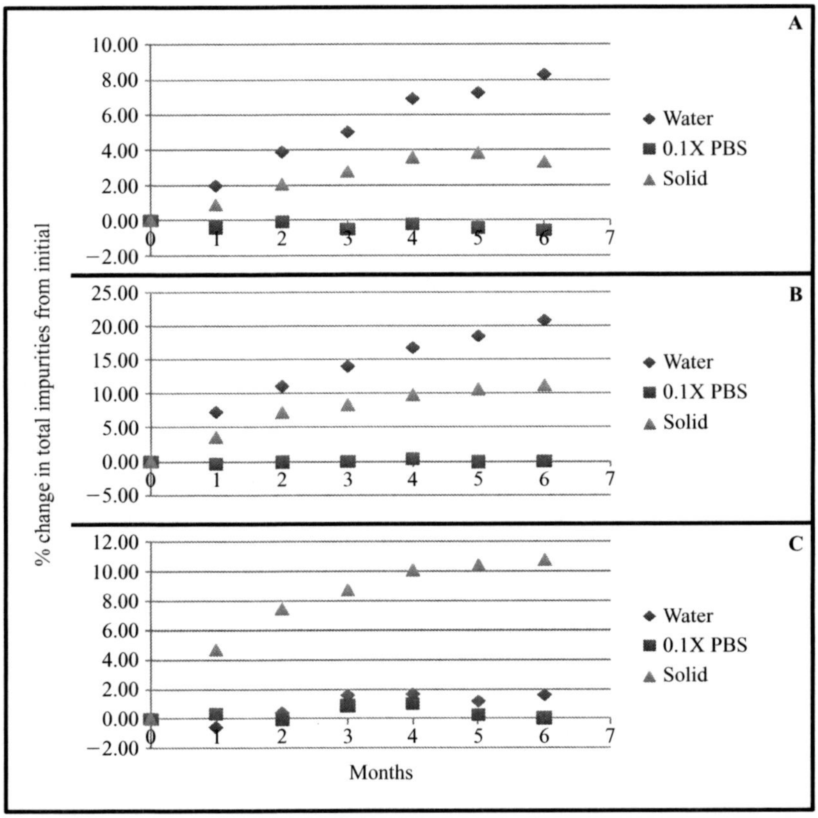

Figure 7.8 Percent change in total impurities measured by denaturing RP-IP chromatography for siRNA **A** (low T_m), siRNA **B** (middle T_m), and siRNA **C** (high T_m) stored either in water, 0.1X PBS, or as a solid at 25°C over a 6-month period. The stronger binding affinity of siRNA **C** is able to maintain the duplex conformation in water imparting protection from chemical degradation compared to siRNAs **A** and **B.**

that in 0.1X PBS. These findings correlate well with the differences in duplex stability demonstrated across the three siRNAs in Figures 7.6 and 7.7, whereby the duplex conformation for siRNAs **A** and **B** was significantly less stable in water than for siRNA **C**. Based on these results, one can infer that the ability to maintain the duplex conformation provides protection from chemical degradation. For all three siRNAs, degradation was significantly worse in the solid state than in a solution of 0.1X PBS, demonstrating that a stabilised duplex conformation in solution can provide more stability than storage of the lyophilised solid at the same temperature. This is even further emphasised by the fact that siRNA **C** is more stable in water than the solid state, the only siRNA able to effectively maintain the duplex conformation in water. These data suggest that the process of desalting and lyophilisation renders the duplex in more of a dissociated single-strand state regardless of the siRNA's binding affinity, making it more susceptible to chemical degradation via the water present in the solid state form. This finding is in line with previous reports whereby the process of DNA dehydration results in destacking of the bases and breaking of hydrogen bonds [58].

Figure 7.9 (see colour insert) shows the change in the chromatographic impurity profile upon storage of siRNA **B** for two months at 25°C in either 1X PBS, water, or the solid state. Interestingly, it can be observed in Figure 7.9 that, by stabilising the duplex in the 1X PBS solution, the formation of a main degradant group has been halted. The present author and co-workers have identified that several isomers of the anhydronucleotide impurity, resulting from the loss of HF (hydrogen fluoride gas), are present as both process impurities and degradation products (peaks observed in the 13–18 min and 34–38 min regions of the chromatogram). Once the duplex is stabilised in PBS, the reaction is sterically blocked from taking place. As a result, the anhydronucleotide impurities present at the initial time point disappear owing to a secondary degradation conversion to an arabinosyl nucleotide [16]. In addition, Figure 7.9 demonstrates that duplex stabilisation in the PBS solution reduces the chance for hydrolytic cleavage of the backbone, which can be seen only in the water sample associated with dissociation of the duplex (**C**, 44–46 min region), and not the 1X PBS sample (**B**).

An additional study compared the formation of aggregates as a function of siRNA binding affinities upon exposure of the lyophilised solid to high temperature and high humidity. Four siRNAs with T_m values of 63°C, 70°C, 78°C and 94°C in 1X PBS were exposed to 40°C/75% RH conditions for 2 weeks (water content increased to approximately 28% w/w for all). The solid samples were then prepared at a concentration of 0.5 mg/ml siRNA in 1X PBS and analysed by way of SEC for total percent aggregate impurities. Figure 7.10 shows the percent aggregates formed as a function of T_m for the four different siRNAs. A clear trend between a lower binding affinity and greater aggregate formation can be seen. While no additional data exist to explain this relationship, it may be that the lower binding affinity results in more non-specific intermolecular strand hybridisation, as an increase in both water content and temperature provides greater mobility for the single strands. As discussed previously in the context of chemical degradation, the siRNA conformation within the lyophilised

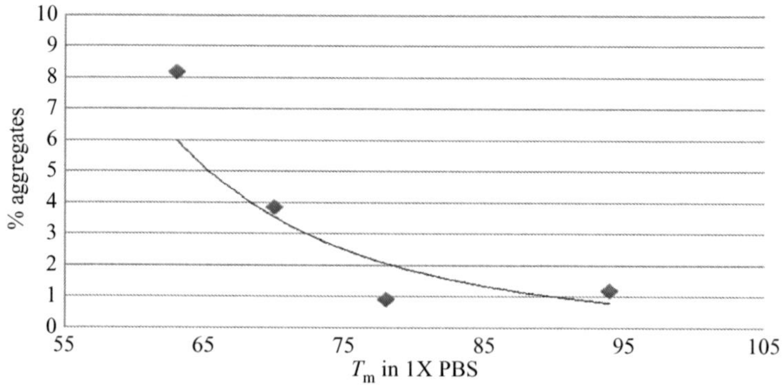

Figure 7.10 Percent aggregates measured by SEC plotted as a function of the siRNA T_m in 1X PBS. All 4 siRNA samples were stored for 2 weeks in the solid state at 40°C/75% RH.

solid may be in a more dissociated single-strand state rather than a duplex state. This supports a hypothesis that as the solid becomes hydrated, pockets of super-saturated single strands would become available to interact and, depending on the thermal energy and binding affinity of the siRNA, would either re-form a stable duplex conformation quickly or be susceptible to the formation of aggregated strands. The nucleobase sequence and resulting susceptibility to form non-specific intermolecular base pairing is likely to be a significant factor in the formation of aggregates, but was beyond the scope of this study to investigate.

This siRNA case study demonstrates that oligonucleotide stability can be quite complex if secondary structure is influencing the degradation. One of the primary challenges is to accurately detect changes in the secondary structure without the sample preparation and analysis procedures biasing the results. As an example, if the T_m measurements for the water samples in the case study were diluted in any type of salt solution versus water for analysis, the observed change in duplex integrity would not have been present. In addition, the SEC method was optimised with the lowest salt concentration mobile phase possible to prevent on-column annealing of dissociated single strands, providing a more representative picture of the actual duplex integrity in the sample solution. Optimising these two methods with the intent of being able to detect a change in duplex integrity enabled quantification of the changing duplex structure in the samples over an 8-month period. This enabled the correlation between changes in secondary structure as a function of siRNA binding affinity and chemical degradation to be made.

7.7 Summary

Vitravene and Macugen have demonstrated that oligonucleotide drug products can be stable for several years at refrigerated storage conditions. The class of oligonucleotide therapeutic, the modification chemistries used and the nucleotide sequence all impact the decisions made in establishing an appropriate stability control strategy. As discussed in this chapter, utilising forced degradation studies early on in development is critical to understanding degradation pathways and products associated with specific oligonucleotide modification chemistries and to develop stability indicating methods. Secondary structure of the oligonucleotide may be important to the stability of the API and drug product and should be evaluated. The case study given in this chapter highlighted the impact that secondary structure can have on stability, whereby the loss of duplex structure resulted in increased chemical degradation of the individual strands.

Acknowledgements

The author would like to thank Benjamin Shepperd and Claude Calvitt for their first rate contributions towards the analytical development of oligonucleotide therapeutics at GSK over the past several years.

References

1. Anon, ICH Q1A(R2), *Guidance for Industry: Stability Testing of New Drug Substances and Products*, Published in the FDA Federal Register, November 2003.
2. Anon, ICH Q5C, *Quality of Biotechnological Products: Stability Testing of Biotechnological/ Biological Products*, Published in the FDA Federal Register, July 1996.
3. Nimjee, S.M., Rusconi, C.P., Sullenger, B.A., Aptamers: An emerging class of therapeutics, *Ann. Rev. Med.*, 2005, 56, 555–583.
4. Vickers, T.A., Wyatt, J.R., Freier, S.M., Effects of RNA secondary structure on cellular antisense activity, *Nucleic Acids Res.*, 2000, 28, 1340–1347.
5. Wünsche, W., Sczakiel, G., The activity of siRNA in mammalian cells is related to the kinetics of siRNA-Target recognition *in vitro*: mechanistic implications, *J. Mol. Biol.,* 2005, 345, 203–209.
6. Matveeva, O.V., Mathews, D.H., Tsodikov, A.D., Shabalina, S.A., Gesteland, R.F., Atkins, J.F., Freier, S.M., Thermodynamic criteria for high hit rate antisense oligonucleotide design, *Nucleic Acids Res.*, 2003, 31, 4989–4994.
7. Meister, G., Tuschl, T., Mechanisms of gene silencing by double stranded RNA, *Nature*, 2004, 431, 343–349.
8. Katoh, T., Suzuki, T., Specific residues at every third position of siRNA shape its efficient RNAi activity, *Nucleic Acids Res.*, 2007, 35, 4, e27.
9. Petri, S., Dueck, A., Lehmann, G., Putz, N., Rüdel, S., Kremmer, E., Meister, G., Increased siRNA duplex stability correlates with reduced off-target and elevated on-target effects, *RNA*, 2011, 17, 737–749.
10. Lu, Z.J., Mathews, D.H., Efficient siRNA selection using hybridization thermodynamics, *Nucleic Acids Res.*, 2008, 36, 640–647.
11. Tuschl, T., Elbashir, S., Harborth, J., Weber, K., *The siRNA User Guide*, see http://www.rockefeller.edu/labheads/tuschl/sirna.html (accessed on 10/05/2011).
12. Elbashir, S.M., Harborth, J., Weber, K., Tuschl, T., Analysis of gene function in somatic mammalian cells using small interfering RNAs, *Methods*, 2002, 26, 199–213.
13. Bozza, M., Sheardy, R.D., Dilone, E., Scyinski, S., Galazka, M., Characterization of the secondary structure and stability of an RNA aptamer that binds vascular endothelial growth factor, *Biochemistry,* 2006, 45, 7639–7643.
14. *siRNA Synthesis and Purification Protocol*, see http://www.microsynth.ch/381.0.html (accessed 10/05/2011).
15. Spindler, L., Rigler, M., Drevenšek-Olenik, I., Hessari, N.M., Webba da Silva, M., Effect of base sequence on G-wire formation in solution, *J. Nucleic Acids*, 2010, doi: 10.4061/2010/431651.
16. Calvitt, C.J., Levin, D.S., Shepperd, B.T., Gruenloh, C.J., Chemistry at the 2′ position of constituent nucleotides controls degradation pathways of highly modified oligonucleotide molecules, *Oligonucleotides,* 2010, 20, 5, 239–251.
17. Kossel, A., Neumann, A., Ueber da Thymin, ein Spaltungsproduct der Nucleinsäure, *Ber. Drsch. Chem. Ges.*, 1893, 26, 2753.
18. Gut, I.G., Depurination of DNA and matrix-assisted laser desorption/ionization mass spectrometry, *Int. J. Mass Spectrom. Ion Proc.*, **1997,** 169–170, 313–322.
19. Capaldi, D., Podium presentation. Overview of impurities in synthetic oligonucleotides, *Drug Information Association: 3rd Oligonucleotide Based Therapeutics Conference*, Bethesda, MD, USA, March 2010.
20. Oivanen, M., Kuusela, S., Lomnberg, H., Kinetics and mechanisms for the cleavage and isomerization of the phosphodiester bonds of RNA by Bronsted acids and bases, *Chem. Rev.*, 1998, 98, 3, 961–990.
21. Lloyd, F., Mac Neela, J.P., Wolfenden, R., Transition state stabilization by deaminases: rates of nonenzymatic hydrolysis of adenosine and cytidine, *Bioorg. Chem.*, 1987, 15, 100–108.
22. Ehrlich, M., Norris, K.F., Wang, R.Y.-H., Kuo, L.-H., Gehrke, C.W., DNA cytosine methylation and heat-induced deamination, *Bios. Rep.*, 1986, 6, 387-393.

23. Mo, J., Podium presentation. Analytical challenges in characterizing oligonucleotides, *Drug Information Association: Oligonucleotide Based Therapeutics Conference*. Bethesda, MD, USA. April, 2007.
24. Evans, R.K., Xu, Z., Bohannon, K.E., Wang, B., Bruner, M.W., Volkin, D.B., Evaluation of degradation pathways for plasmid DNA in pharmaceutical formulations via accelerated stability studies, *J.Pharm. Sci.*, 2000, 89, 76–87.
25. Krotz, A.H., Mehta, R.C., Hardee, G.E., Peroxide-mediated desulferization of phosphorothioate oligonucleotides and its prevention, *J. Pharm. Sci.*, 2005, 94, 341–352.
26. Wu, C.C.N., Lee, J., Raz, E., Corr, M., Carson, D.A., Necessity of oligonucleotide aggregation for Toll-like receptor 9 activation, *J. Biol. Chem.*, 2004, 279, 33071–33078.
27. Suzuki, K., Doi, T., Imanishi, T., Kodama, T., Tanaka, T., Oligonucleotide aggregates bind to macrophage scavenger receptor, *Eur. J. Bioch.*, 1999, 260, 855–860.
28. Levin, D.S., Podium presentation. A comprehensive look at siRNA impurity formation as a function of melting temperature, and is water content a critical attribute for siRNA stability? *TIDES*, Boston, MA, USA, April 2010.
29. Williamson, J.R., Guanine quartets, *Curr. Opin. Struc. Biol.*, 1999, 3, 357–362.
30. Levin, D.S., Podium presentation. Challenges to validation of methods for double stranded oligonucleotides, *Drug Information Association: 3rd Oligonucleotide Based Therapeutics Conference*, Bethesda, MD, USA, March 2010.
31. *Macugen: EPAR – Scientific Discussion for Initial Marketing Authorisation of Medicine*, see http://www.ema.europa.eu/docs/en_GB/document_library/EPAR_-_Scientific_Discussion/human/000620/WC500026218.pdf (accessed 24/01/2012).
32. Uno, Y., Piao, W., Miyata, K., Nishina, K., Mizusawa, H., Yokota, T., High-density lipoprotein facilitates in vivo delivery of alpha-tocopherol-conjugated short-interfering RNA to the brain, *Hum. Gene Ther.*, 2011, 22, 711–719.
33. Venkatesan, N., Kim, B.H., Peptide conjugates of oligonucleotides: synthesis and applications, *Chem. Rev.*, 2006, 106, 3712–3761.
34. Rajur, S.B., Roth, C.M., Morgan, J.R., Yarmush, M.L., Covalent protein-oligonucleotide conjugates for efficient delivery of antisense molecules, *Bioconj. Chem.*, 1997, 8, 935–940.
35. Eguchi, A., Dowdy, S.F., siRNA delivery using peptide transduction domains, *Trends Pharmacol. Sci.*, 2009, 30, 341–345.
36. Bongartz, J.-P., Aubertain, A.-M., Milhaud, P.C., Lebleu, B., Improved biological activity of antisense oligonucleotides conjugated to a fusogenic peptide, *Nucleic Acids Res.*, 1994, 22, 22, 4681–4688.
37. De Fougerolles, A., Vornlocher, H.-P., Maraganore, J., Lieberman, J., Interfering with disease: a progress report on siRNA-based therapeutics, *Nature Reviews: Drug Disc.*, 2007, 6, 443–453.
38. US Food and Drug Administration (FDA), *FDA Guidance for Industry: Liposome Drug Products, Chemistry, Manufacturing, and Controls; Human Pharmacokinetics and Bioavailability; and Labeling Documentation*, see http://www.fda.gov/downloads/Drugs/Guidance ComplianceRegulatoryInformation/Guidances/UCM070570.pdf (accessed on 31/05/2011).
39. Krotz, A.H., Rahul, C.M., Hardee, G.E., Peroxide-mediated desulferization of phosphorothioate oligonucleotides and its prevention, *J. Pharm. Sci.*, 2005, 94, 341–352.
40. Singha, K., Namgung, R., Kim, W.J., Polymers in small-interfering RNA delivery, *Nucleic Acid Therapeutics* (formerly *Oligonucleotides*), 2011, 21, 133–147.
41. Zhi, D., Zhang, S., Wang, B., Zhao, Y., Yang, B., Yi, S., Transfection efficiency of cationic lipids with different hydrophobic domains in gene delivery, *Bioconj. Chem.*, 2010, 21, 563–577.
42. Wang, X.-L., Xu, R., Wu, X., Gillespie, D., Jensen, R., Lu, Z.-R., Targeted systemic delivery of a therapeutic siRNA with a multifunctional carrier controls tumor proliferation in mice, *Mol. Pharmac.*, 2009, 6, 738–746.
43. Wasylaschuk, W.R., Harmon, P.A., Wagner, G., Harman, A.B., Templeton, A.C., Xu, H., Reed, R.A., Evaluation of hydroperoxides in common pharmaceutical excipients, *J. Pharm. Sci.*, 2007, 96, 106–116.
44. Wu, Y., Levons, J., Narang, A.S., Raghavan, K., Rao, V.M., Reactive impurities in excipients:

profiling, identification and mitigation of drug-excipient incompatibility, *AAPS Pharm. Sci. Tech.*, 2011, 12, 1248–1263.

45. Wei, X., Dai, G., Zhongfa, L., Cheng, H., Zhiliang, X., Marucci, G., Chan, K.K., Metabolism of GTI-2040, a phosphorothioate oligonucleotide antisense, using ion-pair reversed phase high performance liquid chromatography (HPLC) coupled with electrospray ion-trap mass spectrometry, *AAPS J.*, 2006, 8, 4, article 84.
46. Gaus, H.J., Owens, S.R., Winniman, M., Cooper, S., Cummins, L.L., On-line HPLC electrospray mass spectrometry of phosphorothioate oligonucleotide metabolites, *Anal. Chem.*, 1997, 69, 313–319.
47. Thayer, J.R., Flook, K.J., Woodruff, A., Rao, S., Pohl, C.A., New monolith technology for automated anion-exchange purification of nucleic acids, *J. Chromatogr. B*, 2010, 878, 933–941.
48. Gilar, M., Fountain, K.J., Budman, Y., Neue, U.D., Yardley, K.R., Rainville, P.D., Russell II, R.J., Gebler, J.C., Ion-pair reversed-phase high-performance liquid chromatography analysis of oligonucleotides: Retention prediction, *J. Chromatogr. A*, 2002, 958, 167–182.
49. Capaldi, D.C., Scozzari, A.N., Manufacturing and analytical processes for 2′-O-(2-methoxyethyl)-modified oligonucleotides. In *Antisense Drug Technology*, 2nd edn, Crooke, S.T. (Ed.), 2006, CRC Press, Boca Raton, FL, USA.
50. Levin, D.S., Shepperd, B.T., Gruenloh, C.J., Combining ion pairing agents for enhanced analysis of oligonucleotide therapeutics by reversed phase-ion pairing ultra performance liquid chromatography (UPLC), *J. Chromatogr. B*, 2011, 879, 1587–1595.
51. Anon, ICH Q3A(R2), *Guidance for Industry: Impurities in New Drug Substances*, Published in the FDA Federal Register, June 2008.
52. Anon, ICH Q3B(R2), *Guidance for Industry: Impurities in New Drug Products*, Published in the FDA Federal Regiser, November 2003.
53. Capaldi, D., Ackley, K., Brooks, D., Carmody, J., Draper, K., Kambhampati, R., Kretschmer, M., Levin, D., McArdle, J., Noll, B., Raghavachari, R., Roymoulik, I., Sharma, B.P., Thuermer, R., Wincott, F., Quality aspects of oligonucleotide drug development: specifications for active pharmaceutical ingredients, *Drug Information Journal*, 2012, online 14 May, doi: 10.1177/0092861512445311.
54. Murugaiah, V., Stability indicating methods for oligonucleotide products. In *Handbook of Analysis of Oligonucleotides and Related Products*, 1st edn, Bonilla, J.V., Srivatsa, G.S. (Eds), 2011, CRC Press, Boca Raton, FL, USA.
55. Evans, R.K., Xu, Z., Bohannon, K.E., Wang, B., Bruner, M.W., Volin, D.B., Evaluation of degradation pathways for plasmid DNA in pharmaceutical formulations via accelerated stability studies, *J. Pharm. Sci.*, 1999, 89, 76–87.
56. *Vitravene Summary of Product Characteristics, Annex 1*, see http://ec.europa.eu/health/documents/community-register/1999/199907293331/anx_3331_en.pdf (accessed 24/01/2012).
57. *Macugen EMA Documents: Summary of Product Characteristics, Annex 1*, see http://www.ema.europa.eu/docs/en_GB/document_library/EPAR_-_Product_Information/human/000620/WC500026214.pdf (accessed 24/01/2012).
58. Bonnet, J., Colotte, M., Coudy, D., Couallier, V., Portier, J., Morin, B., Tuffet, S., Chain and conformation stability of solid-state DNA: implications for room temperature storage, *Nucleic Acids Res.*, 2010, 38, 1531–1546.

Index

Illustrations and figures are in *italics*. Tables are in **bold**. Roman numerals refer to the colour insert pages